Heavenly Numbers

HEAVENLY NUMBERS

Astronomy and Authority in Early Imperial China

CHRISTOPHER CULLEN

Needham Research Institute
and Darwin College, Cambridge
CRCAO, Paris
Sometime scholar of University College, Oxford,
and Research Fellow of Clare Hall, Cambridge

OXFORD
UNIVERSITY PRESS

OXFORD
UNIVERSITY PRESS

Great Clarendon Street, Oxford, OX2 6DP,
United Kingdom

Oxford University Press is a department of the University of Oxford.
It furthers the University's objective of excellence in research, scholarship,
and education by publishing worldwide. Oxford is a registered trade mark of
Oxford University Press in the UK and in certain other countries

First Edition published in 2017

Published in the United States of America by Oxford University Press
198 Madison Avenue, New York, NY 10016, United States of America

British Library Cataloguing in Publication Data
Data available

Library of Congress Control Number: 2017943729

ISBN 978–0–19–873311–9

吾生也有涯, 而知也无涯. 以有涯隨无涯, 殆已.

My life has a boundary, but knowledge has no boundaries. If we use the bounded to follow after the unbounded, there will be trouble.
Zhuang Zhou 莊周 (late fourth century BCE) in *Zhuang Zi* 莊子, Inner chapters 3, *Yang sheng zhu* 養生主 '*What matters in the nourishing of life*'

PREFACE

This book is a narrative history of astronomy and its practitioners in the region now known as China, from the late third century BCE, when the Qin 秦 (221–206 BCE) empire first united 'all under heaven' *tian xia* 天下, to the fall of Han 漢 (206 BCE–220 CE)—a period often referred to as 'early imperial China'. It centres mainly on the kind of astronomy that uses careful records of past observations of the movements of the sun, moon and planets as a basis for calculations aimed at predicting what those bodies will be seen to do in the future—what is commonly called 'mathematical astronomy'[1] in modern English.

The story told here outlines changes in ways of understanding the movements of the heavens and the heavenly bodies that took place during those four and a half centuries, and tells the stories of the institutions and individuals involved in those changes. The main narrative begins with an emperor who inaugurated a new astronomical system in the hope that this might make him immortal,[2] and ends with a commoner who constructed a method for predicting solar eclipses. In between, a great deal happened that is of interest to historians of astronomy in particular, and to historians of science and historians of China more generally. Throughout I have aimed to write in a way that is accessible to non-specialists, while still giving enough detail to satisfy those more familiar with the relevant research.

[1] I recognize the problems involved in using a modern term such as this to characterize an activity rooted in an ancient society whose structures—material, social and intellectual—were very different to those of the 21st century. I hope to bring out those differences sufficiently clearly to undo any harm caused by using such expressions from time to time, and in general I give preference to terms reflecting the language used by ancient writers whenever it is possible to do so without excessive artificiality.

[2] 'Immortal' here is to be taken in the literal and grammatical sense, rather than as merely 'renowned in history'.

In writing this book I have made use of the work of many scholars, from recent years and from previous centuries, as well as my own researches. Whenever I have been aware of adopting some particular idea from the writings of an individual scholar I have acknowledged it appropriately, but it is often hard to separate general influence from specific borrowing. I am grateful to all those who have worked on this field in the past, and am conscious of how much easier my task has been made by what they accomplished.[3]

Several scholars have responded to my requests for advice by giving generously of their time. By far the first amongst these is Catherine Jami, who has read and commented in critical but helpful detail on the entire book in draft. Any faults it may still have are no responsibility of hers—but without her it would certainly have had many more! Other expert scholars have read and commented helpfully on parts of the book. Here I must thank Hashimoto Keizō 橋本敬造, Marc Kalinowski, Lü Lingfeng 吕凌峰, Michael Nylan, Shi Yunli 石云里, John

[3] To give only a few examples, a classic study of pre-modern Chinese astronomical systems by a great Japanese scholar is Yabuuti Kiyosi 藪内清 (1969) *Chūgoku no tenmon rekihō 中国の天文暦法 (Chinese mathematical astronomy)*. Tokyo, Heibonsha. Perhaps the nearest Chinese equivalent is Chen Meidong 陈美东 (1995) *Gu li xin tan 古历新探 (New investigations of old astronomical systems)*. Shenyang, Liaoning educational press. A scholar of the younger generation has published two detailed discussions that cover different aspects of this area, Qu Anjing 曲安京 (2005) *Zhong guo li fa yu shu xue 中国历法与数学 (Chinese astronomical systems and mathematics)*. Beijing, Science Press and Qu Anjing 曲安京 (2008) *Zhong guo shu li tian wen xue 中国数理天文学 (Chinese mathematical astronomy)*. Beijing, Science Press. Turning to studies in western languages, a pioneering attack on some of the problems posed by ancient astronomical systems was made in Wolfram Eberhard and Rolf Mueller (1936) 'Contributions to The Astronomy of The Han Period III: Astronomy of The Later Han Period.' *Harvard Journal of Asiatic Studies* 1 (2): 194–241, reprinted with other work in Wolfram Eberhard (1970) *Sternkunde und Weltbild im alten China: gesammelte Aufsätze*. Taipei, Distributed by Chinese Materials and Research Aids Service Center. Joseph Needham seems to have made a conscious decision not to go into much detail in his discussion of astronomical systems, a topic whose importance he perhaps underestimated: see Joseph Needham and Wang Ling (1959) *Science and civilisation in China. vol 3: Mathematics and the sciences of the heavens and the earth*. Cambridge, Cambridge University Press, 390. The highly theoretical and deliberately non-historical treatment in Jean-Claude Martzloff (2009) *Le calendrier chinois: structure et calculs, 104 av. J.-C.-1644. Indétermination céleste et réforme permanente. La construction chinoise officielle du temps quotidien discret à partir d'un temps mathématique caché, linéaire et continu*, Paris, H. Champion, somewhat reduces its relevance to the present study. Nathan Sivin (2009) *Granting the Seasons: the Chinese astronomical reform of 1280, with a study of its many dimensions and a translation of its records.*, New York, Springer, deals with material a millennium later than the texts discussed in this book, but its clear introductory sections may still be profitably consulted. Other works by Sivin will also be referred to from time to time. A recent PhD thesis intersects with some of the issues discussed in this book, though in a rather different register: Daniel P. Morgan (2013). 'Knowing Heaven: Astronomy, The Calendar, And The Sagecraft Of Science In Early Imperial China.', University of Chicago, PhD; I have referred the reader to it as seemed useful.

Steele, Sun Xiaochun 孫小淳, and Tang Quan 唐泉.[4] In past years, Nathan Sivin and Geoffrey Lloyd have often helped me with advice on my research and writing on the topics discussed in this book. Once more, I assume all responsibility for remaining errors and omissions.[5] In the more distant past, I would like to express my particular gratitude to D.C. Lau, my kind, deeply learned and always helpful PhD supervisor at SOAS, for introducing me to the critical analysis of Chinese texts, and insisting that they should be translated into readable English. More remotely still, I hope the reader will join me in thanking Cai Yong 蔡邕 and Liu Hong 劉洪 for selecting and preserving the rich and fascinating material on which this book draws for much of its content, and Sima Biao 司馬彪 for appreciating its value and passing it down to us. Finally, I wish to thank Keith Mansfield, then Senior Commissioning Editor at Oxford University Press, for the kind welcome he gave to the proposal and outline for this book that I sent him in early 2013. He and his successor Daniel Taber showed exemplary patience and helpfulness in the years that followed, even allowing me to pause for a while to work on a different book for another publisher. I am most grateful for all they have done to bring this book to print.

[4] The order of names is alphabetical. John Steele is also the editor of the series in which Christopher Cullen (2017) *The Foundations of Celestial Reckoning: Three Ancient Chinese Astronomical Systems*, London, Routledge appeared. That book translates and explicates the main original sources on which the present work is based.

[5] Like all scholarly writing, this book had a horizon after which it was no longer possible to take account of subsequent research publications if the manuscript was ever to be placed in the publisher's hands. For most of the chapters, this point had already been reached by the beginning of 2016.

CONTENTS

INTRODUCTION

The narrative I construct in this book lays emphasis on technical practice in observation, instrumentation and calculation, and the steady accumulation of data over many years—but it centres on the activity of the individual human beings who observed the heavens, recorded what they saw, and made calculations to analyse and eventually make predictions about the motions of the celestial bodies. Some of these people had official posts that gave them responsibility for work of this kind; others held official rank without such responsibilities, but still played a major role in technical discussions about celestial phenomena. A few others held no official rank at all, but showed themselves well capable of talking and writing about the heavens at an expert level. It is these individuals, their observations, their calculations and the words they left to us that provide the narrative thread that runs through this work.

The last person whose work is discussed at length in this book is Liu Hong 劉洪 (c. 130–c. 210 CE); most of his activity that is of concern to us took place while he was in the second category sketched above. He was, as we shall see, the first person in his part of the world to construct a theory of the moon sufficiently detailed and accurate to make it possible to predict solar eclipses, not by any means infallibly, but with a useful level of confidence that an eclipse might be observed on the calculated date. One instance of such a prediction was claimed, some years later, to have had a striking effect on those who knew of it. It was, amongst other things, the accumulation of data by the generations of specialists who preceded him that ultimately made it possible for Liu Hong to make such an unprecedented attempt. But an account of the work of Liu Hong that presented his activities, and the work of his colleagues and predecessors, as motivated by the pure and simple pursuit of what today might be called scientific progress would be radically misleading.

Those subjects of the Han empire who, like Liu Hong, devoted themselves to careful observation, recording and calculation in relation to celestial phenomena did so in the context of a shared culture in which such phenomena were seen as a vehicle of meaning directed to the human world in general, but primarily to the ruler of that world—the emperor himself. The same person might be concerned with both aspects of the heavens at the same time: in that regard, the sky-watchers and calculators of the Han empire were no different from their contemporaries in the ancient Mediterranean world.[6] Thus, early on in our narrative, we shall discuss the activities of Gongsun Qing 公孫卿, who persuaded the emperor to launch the first major astronomical reform of the Han dynasty in 104 BCE. But the arguments he used to support his proposal had little or nothing to do with astronomical observation. Instead, Gongsun Qing presented the reform as a way for the emperor to emulate the mythical Yellow Emperor of remote antiquity by bringing himself into ritual accord with the cosmos—so that, like the Yellow Emperor, he might ascend to immortality. When around 10 CE Liu Xin 劉歆 used the basic system created in 104 BCE as the foundation for a greatly elaborated structure of astronomical calculation in the form of his Triple Concordance system *San tong li* 三統曆, he invested immense effort into deriving all the major constants underlying the movements of the heavenly bodies from the numerical cosmology associated with the ancient divinatory handbook, the *Book of Change, Yi jing* 易經.[7] His drive to show not only that the cosmos worked in a particular way, but also to demonstrate that there was a profound reason why it worked in that way parallels the work of Kepler in his *Mysterium Cosmographicum* of 1596. Liu Hong himself included in his Uranic Manifestation system detailed instructions for calculating which of the 64 'hexagram' divinatory symbols of the *Book of Change* would dominate the cosmic situation at any given time of the year. None of this would have seemed anomalous or irrelevant to their contemporaries.

Nor was it the case that making the motions of the celestial bodies predictable removed all ominous significance from those movements. In Liu Hong's

[6] Thus Ptolemy of Alexandria (c. 100 – c. 170 CE) left us two major works. One, nowadays called the *Almagest*, is the primary source for ancient western mathematical astronomy at its most complex, sophisticated and evidence-based. The other, the 'Fourfold book' *Tetrabiblos*, is an exposition and rationale of astrology. There are no signs that he regarded the latter work as less worthy of attention than the former. Astrology continued to be taken seriously in Europe up to early modern times; Galileo himself cast a horoscope for his future patron Cosimo de' Medici. See Owen Gingerich (2005) *The book nobody read: chasing the revolutions of Nicolaus Copernicus*. London, Arrow, 198–201.

[7] Liu Xin's system is discussed in chapter 4; it is translated in Cullen 2017, chapter 2.

own day, the renowned scholar Cai Yong 蔡邕 (who had worked with Liu Hong on the major editing project that gives this book some of its most important material) was under some circumstances prepared to argue against those who wanted to give what may be called scriptural considerations greater weight in astronomical matters than he felt was appropriate.[8] But although it is hard to imagine that Cai Yong was not aware of his associate's work towards predicting solar eclipses, he submitted a memorial to the emperor in response to an eclipse in 178 CE in which he said:

> 踐阼以來, 災眚屢見, 頻歲日蝕, 地動 [...] 修五事於聖躬, 致精慮於共御, 其救之也.
>
> Since Your Majesty ascended the throne, there has been a succession of natural calamities, with several years being marked by solar eclipses and earthquakes. [...] You should rectify matters concerning your sacred person, giving careful thought to self-restraint—that will save the situation. (*Hou Han shu*, zhi 18, 3370, commentary)

That which became predictable might not necessarily lose its ominous significance. Even when it had become the normal expectation that all solar eclipses would be predicted successfully, their ominous nature was never forgotten. Jesuit astronomers found this to be the case when in the mid-17th century they began to work for the last imperial dynasty, the Qing 清 (1644–1911). In 1692 the Kangxi 康熙 emperor (r.1661–1722 CE) reacted to a prediction of an eclipse on New Year's Day by suspending all official banquets for the festival—the cultural equivalent of cancelling Christmas festivities in a modern western country. He did this despite his deep personal study of the predictive methods of western astronomy, and his expressed doubts about the validity of astronomical prognostication.[9] The earliest similar example of a proposal to cancel New Year ceremonies because of a predicted solar eclipse occurred very shortly after the time of Liu Hong, in 212 CE.[10]

I have sketched here my reasons for insisting that the persistent and increasingly successful efforts of Han dynasty scholars to construct mathematical methods for astronomical prediction must be seen in their deep cultural context

[8] See chapter 7, section 7.2.3.

[9] See the account in Catherine Jami (2012) *The Emperor's new mathematics: Western learning and imperial authority in China during the Kangxi reign (1662–1722)*. Oxford, Oxford University Press, 222–9.

[10] See chapter 8, section 8.3.4. In that case, after discussion, it was decided to take the risk and go ahead—and in the event no solar eclipse was seen.

if we are not to misunderstand them or distort our view by looking at them through a modern lens. The proper field of comparison lies with pre-modern specialists in celestial observation and calculation in other cultures, rather than with the very different work of modern astronomers. If we set out to make such comparisons, what can the Chinese material contribute to constructing a world history of human understanding of the heavens?

The sources studied here give us access to an ancient tradition that appears—at least in its origins—to be effectively independent of those more familiar to historians writing in western languages. It is a tradition that has commonly been given little attention in the history of astronomy as recounted by such historians.[11] I hope that this book will prove to be useful and interesting to historians of science worldwide, and to general historians of China, as well as to those with more specialist interests in the history of astronomy, or the history of science in China. I have therefore written as far as possible with the needs of such broad and disparate readerships in mind.

The Chinese imperial state was an institution that habitually conducted its business in writing rather than orally, and that maintained archives of all significant documents thus generated. As we shall see, it was also an institution that placed a high value on reflective historical writing that made use of those archives. The combination of that emphasis on documentation and the importance of astronomy for the state means that our records of astronomical activity in early imperial China are rich, detailed and relatively continuous. This has enabled me to follow the principle of 'show, don't tell': instead of telling the reader what people thought on the topic of this book two thousand years ago, I am often able to let the ancient actors' own words speak for them.[12] My role is to introduce, and to interpret, so far as interpretation is both desirable and possible. Readers who want direct access to the main sources used in this book may consult another book that appeared not long before this one—*The Foundations of Celestial Reckoning: Three Ancient Chinese Astronomical Systems* (Cullen, 2017; London, Routledge). There the reader will find my principal sources translated in their entirety, with detailed introductions and explanatory

[11] Even Joseph Needham took the deliberate decision to pay only brief attention to mathematical astronomy in China: Needham and Wang Ling (1959), 390–408. This may perhaps to some extent excuse the neglect of subsequent authors.

[12] Alexandre Koyré wrote, 'As far as possible, I have allowed authors … to speak for themselves in these studies. … For a history of scientific thought … nothing can take the place of the original sources and texts. They alone enable us to catch the spiritual and intellectual atmosphere of the period under study; they alone enable us to appreciate the motives and incentives which guided and impelled the authors of them … ' Alexandre Koyré (1973, first published 1961 in French) *The astronomical revolution: Copernicus, Kepler, Borelli*. London, Methuen, 10–11.

commentary.[13] As part of my interpretive effort, I have from time to time cited parallels and divergences between the practices of ancient astronomers in the Mediterranean world and those at the other end of Eurasia.[14]

What can the historian gain from the study of such material? In the first place, our sources often put us in the fortunate position of being able to read what appear to be close to the actual words written by a number of named individuals who were the principal actors in the story. We are mostly not obliged to piece together fragmentary quotations from lost works, or evaluate references by later authors to what someone is said to have said. I do not make this distinction to dismiss or diminish the latter mode of writing history; many skilled authors have shown how much can be achieved by working in that way.[15] It is simply that the case is different for early imperial China; there we are able to read documents from government records—edited no doubt, and subject to the bias of those who selected them, but ultimately written by the actors themselves. This is not just the case for those persons who held official posts, as did most of the principal people who play a role in this book, but also for a number of private individuals whose views were sought by the imperial court, or who decided to put their views forward of their own accord. When such individuals made submissions setting out their views and proposals, these too were copied into official records for later reference. In many cases we also know in what institutions people who wrote on astronomy worked, what posts they held, and even (in some cases) the salaries they earned. For much of the literature of ancient astronomy from elsewhere in the world, such information is wholly lacking.[16]

So much, then, for the nature of our sources. But what do those sources tell us about what ancient Chinese astronomers knew, how that knowledge was validated, and how it was actualized in practice? Paradoxically, the interest of the early imperial Chinese sources for a world history of astronomy does not lie in

[13] Where I cite ancient Chinese texts in the present book, I generally give two references, one to the original source in a standard modern edition, and one to the page in *Foundations of Celestial Reckoning* where that source is translated, and in some cases explained at some length. Readers who compare translations in this book with those in *Foundations* may occasionally detect small differences; sometimes these are due to my need to make a short free-standing extract in this book comprehensible in itself, and sometimes they are attempts at clarification that have suggested themselves on revisiting translations made some time ago.

[14] The sources I cite were mostly written in Greek, and some in Latin. Given a larger measure of time, space and talent, perhaps I might have been able to attempt a wider comparative canvas—one that took in Mesopotamia, ancient India, and perhaps the Arabic world and the pre-Columbian Americas. But I have limited myself to material of which I have at least an informed outsider's level of contextual grasp, combined with some ability to access the sources in their original language.

[15] See for instance the detailed, analytical and carefully sceptical account of early Greek astronomy given in Book IV of Otto Eduard Neugebauer (1975) *A history of ancient mathematical astronomy*. Berlin; New York, Springer-Verlag, 571–776.

the extent to which the work of Chinese astronomers was in advance of what was happening elsewhere in the same period. On the contrary, one of the most interesting features of the Chinese documentary record is that it enables us to follow developments in the observation and analysis of celestial phenomena from a considerably less advanced stage than does the record from the ancient western world – a fact that was first pointed out by Henri Maspero in the 1930s.[17]

Two comparative examples may serve to illustrate this.

In the ancient Greek-speaking world, the evidence relating to the early history of the concept of the ecliptic as the apparent path of the sun round a circle inclined to the celestial equator is difficult to trace. By the time we have any account of the mathematical implications of solar, lunar and planetary motion round (or close to) an inclined ecliptic, the discussion is already so detailed and advanced as to make it clear that much of the earlier history is simply missing. Thus, according to fragmentary quotations from a *History of Astronomy* written by Eudemos in the fourth century BCE as part of the research programme of Aristotle's Lyceum, a certain Oenopides who lived about a century before Eudemos was the first to introduce the notion of the ecliptic as the sun's inclined path.[18] But that is all we are told. By the time Plato wrote the *Timaeus*, perhaps around 360 BCE, the notion of an equator and ecliptic inclined to one another was sufficiently well accepted for it to be worked into the 'likely story' of the creation of the cosmos that Plato puts into the mouth of Socrates.[19] We have no sign of any discussion surrounding the adoption of the concept.[20]

[16] A partial exception to this is the case of the 'scribes of Enūma Anu Enlil' of ancient Mesopotamia: see chapter 7, section 7.2. But the nature of our information about their lives and work, and the register in which the documentary evidence of their activities is written, are both very different to what is found in the documents studied in this book.

[17] See Maspero, Henri (1950, but written c. 1932-1939). 'L'astronomie dans la Chine ancienne: histoire des instruments et des découvertes' *Mélanges posthumes, volume III Études historique.* P. Demiéville. Paris, Musée Guimet: 13-34, page 15.

[18] See Hermann Diels (1903) *Die Fragmente der Vorsokratiker: Griechisch und Deutsch.* Berlin, 239, fragment 29:7. The reference in question is found in the writing of Theon of Smyrna, c. 100 CE, who is therefore reporting what Eudemos wrote over four centuries earlier. But see also István M. Bodnár (2007). 'Oenopides of Chius: A survey of the modern literature with a collection of the ancient testimonia (preprint 227)', Berlin, who points out how difficult it is to be sure what, if anything, Oenopides did or did not discover.

[19] *Timaeus; Critias; Cleitophon; Menexenus; Epistles, with an English translation by R.G. Bury.* (1989). Plato (428/427 or 424/423–348/347 BCE), Cambridge, Mass, Harvard University Press, 36C, 70–1.

[20] Neugebauer (1975), 593 refers to another possibility, based on a passing reference in *Natural history (vol 1) [Loeb Classical Library]*. (1938). Pliny (Gaius Plinius Secundus 23–79 CE), H. Rackham tr. and T. E. Page ed., London, II.vi.31, page 188–9, which associates the first reference to the ecliptic with Anaximander and Cleostratus in the sixth century BCE. Neugebauer does not mention Pliny's reference to Anaximander, but agrees that the ecliptic was a well-known concept by c. 300 BCE.

In China, however, the evidence is more revealing. We have detailed records of a controversy that took place from 92 to 103 CE, when official astronomers were the target of efforts to persuade them to move from the idea of a sun that, in modern terms, moved at a steady rate in right ascension to a sun that moved at a steady rate in longitude—i.e. along the ecliptic rather than along the equator (chapter 6, section 6.3.2). This eventually resulted in the addition of an ecliptic ring to the armillary sphere used by official astronomical observers. Since we have quite full documentation of these persuasive efforts, we can not only identify the person who advocated the use of the ecliptic (Jia Kui 賈逵), but also read the arguments he used to win his case. Moreover, we can also see how Chinese astronomers tried to cope with the mathematical implications of this change in terms of (in effect) the interconversion of ecliptic and equatorial coordinates, at a time when they did not have anything like Ptolemy's spherical trigonometry at their disposal.

Another example is the question of the complex motion of the moon—including the inclination of its orbit to the ecliptic, the steady motion of the nodes at which that orbit crosses the ecliptic, and the moon's varying speed along its orbit. In the Greek-speaking world, the first extensive writing we have on the topic consists of the already highly developed and detailed theory provided by Ptolemy of Alexandria in the second century CE, containing a sophisticated analysis of the relations of that theory to observation, and telling us exactly how it may be applied to the calculation of solar eclipses.[21] Ptolemy's writing is an impressive monument of ancient science by any standard. But anybody reading it will inevitably want to know who were the predecessors who provided the foundations on which this monument was erected. Here we face the problem that the *Almagest* was so successful in summing up the highest achievements of mathematical astronomy in its day that it seems to have greatly reduced the chances that anybody would subsequently bother to copy and preserve the work of those predecessors.[22] The difference may perhaps be summed up by saying that from the world of Ptolemy we have a number of books that were successful enough to survive—but from early imperial China we have an archive.

[21] In *Almagest* IV to VI: G. J. Toomer (1998) *Ptolemy's Almagest*. Princeton, N.J., Princeton University Press, 173–320.

[22] See Toomer (1998), 1–2. As James Evans has reminded me (private communication), modern discoveries of fragmentary documents from Egypt, dated not far from the time of the Almagest and written in Greek on papyrus, have given us tantalizing glimpses of ways of doing astronomical calculations that owed more to the algorithm-based traditions of Mesopotomia than they did to the geometrical techniques recorded by Ptolemy: see Jones, Alexander (1999), *Astronomical papyri from Oxyrhynchus*: (P. Oxy. 4133–4300a) / edited with translations and commentaries by Alexander Jones. Philadelphia, Philadelphia: American Philosophical Society, 1999.

Thus, we know from what Ptolemy tells us that he is drawing on the work of Hipparchus four centuries earlier, and that Hipparchus himself was in turn drawing on 'observations made by the Chaldeans'—but the actual writings of Hipparchus on this topic are lost to us.[23] In China, however, the fact that a later document added to the archives was thought to take astronomical theory and practice to a higher level did not imply that older documents should be thrown away. Because of this, we are still able to read what seems to be a record of the first systematic observations designed to measure variations in lunar speed, with an initial analysis of the variation found, followed up a century later by the first complete theory of lunar motion dealing with both speed and latitude variation, and what may be a record of the first time this theory was successfully applied to the prediction of a solar eclipse: chapter 6, section 6.3.5, and chapter 8, section 8.3.3. Elementary as were these developments compared with what Ptolemy gives us, they do add an important new chapter to the history of astronomy in a world context.

Different readers will no doubt want to use this book for different purposes; there is more than one way to read it, and no one right way to read it. To facilitate reader choice, the following paragraphs give some guidance on the contents of this book, and possible ways of approaching it.

Chapter 1 explains the importance of the calendar in the historical self-image of the Chinese imperial state, and outlines the basic structures of calculation that underpinned it. It looks too at the way in which the calendar acted as a means of social control over imperial subjects. *Chapter 2* looks at the ancient Chinese documents that we may today call 'calendars', outlines their structure and contents, and explains the ways that officials and the population as a whole related to them in their daily lives.[24]

[23] *Almagest* IV.2: Toomer (1998), 175. Elsewhere Toomer suggests that it may be possible to identify a surviving cuneiform clay tablet that contains data identical to that used by Hipparchus: Toomer (1998), 224 n14. The only surviving work attributed to Hipparchus is a *Commentary on the Phenomena of Aratus and Eudoxus*, for which see Hipparchus (1894) *Ipparchou tōn Aratou kai Eudoxou phainomenōn exēgēseōs biblia tria = Hipparchi in Arati et Eudoxi phaenomena commentariorum libri tres/ad codicum fidem recensuit, germanica interpretatione et commentariis instruxit Carolus Manitius*. Lipsiae, Lipsiae: in aedibus B.G. Teubneri, 1894; see Neugebauer (1975), 274–343 for an account of what can be known of Hipparchus' work on astronomy.

[24] Earlier treatments of some of the material in these two chapters were published in Christopher Cullen (1996) *Astronomy and mathematics in ancient China: the Zhou bi suan jing*. Cambridge; New York, Cambridge University Press, 1–27; my views on the day-books, excavated calendars and their role developed in response to an invitation from Marc Kalinowski in 2011 to join a book project, whose results will be published as Donald Harper and Marc Kalinowski, Eds. (2017, forthcoming). *Books of Fate and Popular Culture in Early China. The Daybook Manuscripts of the Warring States,*

After the ground has been prepared by the preceding chapters, *Chapter 3* begins the main narrative by telling the often surprising story of the way in which an emperor's search for personal immortality led in 104 BCE to the first great astronomical reform in Chinese history, with wide-ranging consequences that included the humiliation of the leading astronomical official of the imperial government. As we trace the aftermath of this affair, we see for the first time a distinctive feature of the way that astronomical controversy was dealt with in early imperial China, through the creation of an expert group charged with carrying out a careful programme of observational testing of the merits of the claims advanced by the contending sides.[25] *Chapter 4* contains full details of the structure and theoretical underpinning of China's first complete extant system of astronomical calculation, dating from around 10 CE, that was constructed on the basis of the 104 BCE reform, as well as discussing the ways in which this system played its part in supporting the *coup d'ětat* through which a new (but short-lived) dynasty seized power.[26]

Chapter 5 begins from the experience of the first individual directly concerned with astronomy in imperial China who speaks to us in his own voice about his personal concerns and views—and tells us about his arguments about astronomy with a friend and colleague who thought differently. This leads us in two directions—firstly towards an attempt by another individual (myself) to repeat the kind of observations that the person in question says that he made, using the apparatus he describes, and secondly towards the question of what the size and shape of heaven and earth may be, and how the form of the cosmos may

Qin, and Han. Leiden, Brill. I first discussed the question of what I initially called a 'bottom-up' history of Chinese astronomy making use of this material in Christopher Cullen (2011c), 11 October 2011, 'Daily life and cosmic time: excavated calendrical documents and their significance', The Fitzwilliam Museum conference on Life and Afterlife of Han China, Cambridge, in which I also discussed the importance of the edict of 5 CE studied in Charles Sanft (2011) 'Edict of Monthly Ordinances for the Four Seasons in Fifty Articles from 5 C.E.: Introduction to the Wall Inscription Discovered at Xuanquanzhi, with Annotated Translation.', *Early China* (32 (dated 2008–9, actually published 2011)): 125–208.

[25] I first studied the main topic of this chapter at length in Christopher Cullen (1993) 'Motivations for Scientific Change in Ancient China: Emperor Wu and the Grand Inception Astronomical Reforms of 104 BC.' *Journal for the History of Astronomy* 24 (3): 185–203. The core of the material relating to Zhang Shouwang first appeared in Cullen (1996), 30–1.

[26] This chapter draws on and develops the detailed discussion of Liu Xin's system in Christopher Cullen (2004) 'The birthday of the Old Man of Jiang County and other puzzles: work in progress on Liu Xin's *Canon of the Ages*.' *Asia Major* xiv (2): 27–70. A fully commented translation of this system is given in Cullen (2017), Chapter 2. My views on the *Wu xing zhan* and associated questions were first set out in Christopher Cullen (2011a) 'Wu xing zhan 五星占 "Prognostics of the Five Planets"', *SCIAMVS* 12: 193–249, and Christopher Cullen (2011b) 'Understanding the Planets in Ancient China: Prediction and Divination in the Wu xing zhan', *Early science and medicine*, 16: 218–51.

relate to the instruments used to observe it. Like the preceding chapter, this one centres on events early in the first century CE.[27]

In *Chapters 6 and 7* we move from the private to the official: a long series of archived documents enables us to see how experts argued for their point of view, often through the preparation of written documents, but sometimes, it appears, in public confrontations in the presence of large numbers of their fellow officials. Once more, as in the later part of chapter 3, we see recourse to the procedure of subjecting new proposals to prolonged tests of their predictive powers. Here, however, the emperor plays no real role in the discussion, except to lend his name to edicts that were no doubt drafted for him by astronomical experts.[28] We bring the story of this book to an end when *Chapter 8* recounts the work of one man—Liu Hong—leading up to what may have been the first successful East Asian solar eclipse prediction based on careful calculations of the moon's apparent motion relative to the sun in two dimensions.[29] The chapter concludes with evidence that it was not long after the time of Liu Hong that astronomers began the practice of announcing in advance when they thought solar eclipses were likely to occur. The book ends with a brief epilogue, looking ahead through the following centuries to the persistence of remnants of the ancient Chinese astronomical tradition up to the present day.

For historians of astronomy wondering how to read this book, I would recommend that they should follow the advice of the King of Hearts in *Alice in Wonderland*: 'Begin at the beginning ... and go on till you come to the end: then stop'. Other readers, such as historians of China, may choose to do differently. I recognize that while some will want to follow the technical practices of the Chinese astronomer as closely as possible, others will find such details unnecessary for their purposes. In order to avoid both repelling the generalist and

[27] Earlier versions of the material in this chapter will be found in Christopher Cullen (1981) 'Some further points on the shih.', *Early China*, 6: 31–46, and Cullen (1996), 35–65; my experiments with gnomons and waterclocks were reported in Christopher Cullen (1982a), September 1982, 'Early Chinese measurements of right ascension before the armillary sphere', First International Conference on the History of Chinese Science, Louvain, Belgium.

[28] These two chapters draw on my work in Christopher Cullen (2000) 'Seeing the Appearances: Ecliptic and Equator in the Eastern Han.' *Zi ran ke xue shi yan jiu* 自然科學史研究 *(Studies in the History of Natural Sciences)* xix (4): 352–82, Christopher Cullen (2007a) 'Huo Rong's observation programme of AD 102 and the *Han li* solar table.', *Journal for the History of Astronomy*, 38 (1): 75–98, and Christopher Cullen (2007b) 'Actors, networks and "disturbing spectacles" in institutional science: 2nd century Chinese debates on astronomy', *Antiquorum Philosophia*, 1: 237–68.

[29] The survey of Liu Hong's lunar theory enlarges on Christopher Cullen (2002) 'The first complete Chinese theory of the moon: the innovations of Liu Hong c. AD 200', *Journal for the History of Astronomy*, 33: 1–24.

disappointing the specialist, I have frequently given the main points of a technical discussion in the main text, while putting the detailed calculations in a box that can be ignored by those not interested.

This introduction has used such terms as 'astronomy' and 'mathematical astronomy' to describe its topic, as indeed does the title of this book. This usage is continued in much of the first chapter, and occasionally elsewhere. I think there is no impropriety in this choice, since it helps the reader to get an immediate general idea of what this book is about, and what to expect from reading it—as well making it possible for what I write to be usefully tagged and linked with similar texts by non-human agents such as internet search engines. I am, however, amongst those who feel that the uncritical use of such terms can have a misleading effect on the reader, since they may elide or conceal the large differences of purpose and nature between ancient activities that apply calculations to the movements of the natural lights in the day and night sky, and the modern science of astronomy. I have put forward elsewhere similar arguments for caution in talking about 'ancient Chinese mathematics' rather than using the main term for their activity used by the people whose work is most usually studied under that heading—*suan* 算, a word whose range of coverage in its historical context differs significantly from that of the modern English word 'mathematics'.[30] So what are we to call the topic discussed in this book? This problem is not a new one. In the preface to his *Tabulae Rudolphinae* 'Rudolphine Tables' of 1627, Kepler drew a distinction that, he pointed out, had not been made in the ancient western world:

> Duas habet Astrorum scientia partes: prior est de Motibus, posterior de Effectibus Siderum in natura sublunari. Utramque veteres communi vocabulo Astrologia soliti sunt appellare.
>
> The study of the stars has two parts, of which the former is concerned with [their] movements, and the latter with the effects of the celestial bodies on the sublunar world. The ancients were accustomed to call both of these by the same term, '*astrologia*'. (Kepler, Johannes and Brahe, Tycho 1627: 1, Praefatio)

Kepler, however, states that he prefers to call the first of these parts '*astronomia*' and to reserve '*astrologia*' for the second. However, if Kepler had been working in the Chinese tradition, his problem would not have arisen, since the first

[30] In Christopher Cullen (2009), 'People and numbers in early imperial China', in *Oxford Handbook of the History of Mathematics*, Oxford, Oxford University Press: 591–618.

extant systematic Chinese classification of books on the topic, from close to the beginning of the common era, draws a clear distinction between two genres of writing about the heavens, *tian wen* 天文 and *li* 曆, which, while not exactly paralleling his *astronomia/astrologia* distinction, does much the same job. The first of these expressions may be rendered as 'heavenly writings/patterns/signs', and may be applied to texts concerned with interpreting the significance of celestial phenomena that were not thought to be predictable, such as meteor showers or comets. The second may be rendered as '[astronomical] systems' and is primarily concerned with performing calculations that predict the motions of the celestial bodies (see chapter 1, section 1.5 for a more detailed discussion). The topic of this book is the story of developments that fall under the second of these headings, and for the most part I shall refer to the relevant topic as *li*, '[astronomical] systems', or when the context is clear, simply 'systems'.

Other terminological matters are less significant, though still worthy of note. Firstly, in the body of this book I have called the political entities that governed major parts of the East Asian landmass by the names that they used for themselves during the period considered in this book—Qin 秦 (221–206 BCE), followed by Han 漢 (206 BCE–220 CE). Calling both of them 'China' without qualification begs a number of important historical, cultural and political questions. On more technical matters, I have sought non-anachronistic translations of terms relating to *li* that avoid implying assumptions absent from the original texts. Thus, despite the temptation to say that *sui* 歲 means 'tropical year', I have instead coined the term 'solar cycle', since in the period studied in this book no distinction was drawn between the tropical year and the sidereal year (see chapter 1, section 1.3). This also allows me to escape potential confusion with calendrical years, which in China consisted of whole numbers of lunar months as well as whole numbers of days. Sometimes I have had to resort to translating the same word by two different English words depending on the context, as when *yue* 月 can mean both 'month' in the sense of the calendrical unit consisting of a whole number of days (either 29 or 30), and in other contexts the non-integer quantity (about 29 ½ days) representing the length of the moon's cycle of phases from one luni-solar conjunction to the next, which I have called a 'lunation'.[31] All such decisions are discussed and justified in detail as they first become relevant.

[31] Given a free choice, I might have preferred simply to render *yue* for both periods by its root meaning 'moon'. But the use of that term in exoticizing phrases such as 'many moons ago' rules that out.

The texts used in this book are mostly well preserved, and are available in good modern editions, which make use of the cumulative critical scholarship of the last two millennia. I have generally adopted the emendations made by modern editors without comment, while showing such emendations as are marked in the Chinese text, with round () brackets for a deletion, and square brackets [] for an insertion or substitution. On some occasions I have proposed emendations of my own, and in such cases I have noted the reasons for my decision in detail.

On frequent occasions, I refer to astronomical phenomena, such as eclipses, lunar phases or the rising and setting of planets, that would have been visible at a given place and time. Modern astronomical software available for personal computers enables one to obtain data on such events that is acknowledged to be accurate to well within the precision of ancient naked-eye observation and time-keeping. I currently use Starry Night Pro™ for this purpose, mostly because of the ease of user input to display ancient Chinese asterisms instead of those from the western tradition. In some cases, where higher precision and flexible tabular output are more useful, I have made use of the 'Horizons' ephemeris software made available online by NASA.[32] A computer screen is, however, no substitute for reality. I urge any reader who has not done so recently to spend a little while looking at the night sky—without a telescope, naturally—taking note of the changes visible as one night progresses, and comparing the view at the same time of night over the course of a few weeks. Some time spent this way will often make it considerably easier to appreciate what the ancient sky-watchers were doing and thinking.

Finally, as I said in the preface, this book is a narrative history. A narrative is, ultimately, a story that somebody chooses to construct and to tell. Stories about history have meanings, explicit or implicit: they are more than an attempt at arranging factual statements about the past in chronological order. Sometimes the meaning of the story comes from an overarching 'grand narrative' of which the author believes his writing forms a part. In such cases, we are at least clear about what the author's intentions are, even if we may be out of sympathy with them. In other cases, the author may only be able to say what his story means—if it means anything at all—when he steps back and reads what he has written.

Some of the materials from which I have constructed this story are, I think, uncontroversial matters of fact. Somebody lived and died; something was observed and measured, or calculated, and even though the language and

[32] See http://ssd.jpl.nasa.gov/horizons.cgi.

technical concepts used in our sources require translation, we can be sure what things and activities are being described. But our sources are triply selective: in the first place they are selective on the basis of what people in antiquity chose to write about, and how they chose to write.[33] After that, we have the selectivity of the survival or disappearance of texts over the last two thousand years. And finally, a modern historian looks at the surviving evidence, decides what is of interest to him, and chooses what to write about, and how to write about it.

Further, it is next to impossible to describe what somebody did, without having at least a provisional answer to the question 'why did he do that?'. Thus, why did an emperor not renowned for his interest in technical matters invest so much of his authority into the astronomical reform of 104 BCE? What was Liu Hong's intention, near the end of the second century CE, in making an elaborate mathematical model of the behaviour of the moon? Neither of these people has left us answers to such questions, and we cannot answer them ourselves without an imaginative effort to place ourselves in their situation. But imagination is not an objective process, especially if it has to start from a very imperfect knowledge of what that situation actually was.

So there can be no pretence that a book like this is simply an account of the past 'as it really was'.[34] It is, at best, a construction that hopes to succeed in part in being a reconstruction. At the end of the book he based on a course of lectures on the 'Exact Sciences in Antiquity' delivered at Cornell University in 1949, Otto Neugebauer wrote some sentences that have been frequently cited, but which bear repeating. Recalling an illustration of a captured unicorn taken from a renaissance tapestry in the Metropolitan Museum, New York, he wrote:

> At the end we see the miraculous animal captured, gracefully resigned to his fate, standing in an enclosure surrounded by a neat little fence. This picture may serve as a simile for what we have attempted here. We have artfully erected from small bits of evidence the fence inside which we hope to have enclosed what may appear as a possible, living creature. Reality, however, may be vastly

[33] An example of that selectivity is the fact that Cai Yong and Liu Hong give us virtually nothing on the subject of Zhang Heng, whom later writers have chosen to regard as a figure of major importance in the field with which this book deals: see section 6.5. Were they ignoring him deliberately for some reason that had nothing to do with his astronomical talents? Or was he perhaps simply not so important in his own day as his later admirers have imagined? We cannot tell. The best we can do is to make this selectivity part of the story we tell.

[34] To get some sense of the ambition behind this project (famously set out by Leopold von Ranke, 1795–1886), one need only ask oneself whether any of us would feel able to describe the present 'as it really is', even given that our access to information about that present is infinitely greater than our access to information about the past.

> different from the product of our imagination; perhaps it is vain to hope for anything more than a picture which is pleasing to the constructive mind when we try to restore the past. (Neugebauer (1969), 177)

One thing that Neugebauer does not mention is that the unicorn in the picture he cites appears to be bleeding from wounds inflicted by the hunters who captured it. Perhaps after all the unicorn is not all that 'resigned to his fate'? I hope the reader of this book will not think I have done too much damage to the past that I have tried to capture in this book.

CHAPTER 1

The astronomical empire

This chapter begins by sketching the historical and cultural foundations of the role played by astronomy in the self-presentation of the early Chinese imperial state, principally through its claim to have the right to structure the time of its subjects by issuing a luni-solar calendar. This claim was presented as fulfilling a need on the part of the population; we shall discuss how far this claim was an accurate reflection of the power relations involved. There follows a preliminary account of the main components of such a calendar, and of how they formed an integrated whole. Finally there is an introductory review of the main written sources on *li* 曆 '[astronomical] systems' on which much of this book will be based.

1.1 Building astronomy into the foundations

In 1888 the Austrian astronomer and sinologist Franz Kühnert remarked ironically that one reason that 'civilized' Europeans considered the Chinese to be barbarians was no doubt because they treated astronomers—'a completely useless little bunch' (*ein höchst unnützes Völkchen*)—with honour, and even gave them high official rank.[1]

No doubt Kühnert's statement puzzled many of his readers. But it had a solid basis in fact. From the time when the Qin 秦 dynasty (221–206 BCE) brought most of the East Asian landmass into a unified empire, astronomers and astronomy

[1] Franz Kühnert (1888) 'Das Kalendarwesen bei den Chinesen.' *Österreichische Monatsschrift für den Orient* (8): 111–116, 116. Joseph Needham used an English translation of the closing passage of Kühnert's article on the title page of Needham and Wang Ling (1959), from which it has frequently been quoted.

were recognized as indispensable to the work of the imperial government. And this state of things lasted for most of the next two millennia, well into the last imperial dynasty, the Qing 清 (1644–1911 CE). The Jesuit missionary Dominique Parrenin complained in 1730 of what he considered to be the un-enterprising and conservative spirit of the staff of the imperial observatory in Beijing, which he interpreted as being in part due to a lack of competition amongst experts. But he still characterized China as a country that considered astronomy important enough to make it conceivable that it might 'go to war over an almanac'.[2] In the preceding century the observatory had been re-equipped with the best instruments that money could buy and that imported Jesuit expertise could construct, as was appropriate to its status as the last in a long line of state observatories. The remains of a state observatory from the first century CE can be still be seen south of the site of the Eastern Han dynasty capital, Luoyang 洛陽, while that of the Qing dynasty, with its 17th century instruments in place, is still visited by many tourists in Beijing. See Figures 1.1 and 1.2.

Figure 1.1 The rammed-earth foundations of the state observatory (*Ling tai* 靈臺 'Numinous terrace') of the Eastern Han dynasty at Luoyang (first to third centuries CE), as seen in 2007. Photograph supplied by Professor Lü Lingfeng and reproduced with his permission. The remaining mound measures about 40 m by 30 m, and is about 8 m high. Its original stepped form bore three levels of buildings, with an open central area at the top.

[2] Isabelle Vissière and Jean-Louis Vissière, (1979) *Lettres édifiantes et curieuses de Chine: 1702–1776*. Paris, Garnier-Flammarion, 362.

Figure 1.2 The state observatory on a section of the city wall of Beijing, as renovated and re-equipped in 1674; from (Verbiest, Ferdinand 1674). Despite some changes in arrangement of the instruments, this illustration bears a close resemblance to what can still be seen by 21st-century visitors to the site.

China was a distinctively astronomical empire, and this book tells the story of how the astronomical expertise that served that empire—what has been called the 'Han paradigm'[3]—was created. As we shall see, that paradigm was shaped in the context of intense competition, even conflict, between experts—circumstances very different from those of which Parrenin complained. All this took place during what has been called the 'early imperial period' when the empire conquered by the Qin was consolidated and developed by the Han 漢 dynasty (206 BCE–220 CE), and enduring patterns of political and intellectual

[3] I owe this expression to a conversation with Professor Sun Xiaochun 孙小淳 in late 2015.

culture were established.[4] Both Qin and Han governed their empires through a corps of non-hereditary civil officials, who were ultimately responsible to the emperor. These officials had to have a number of practical administrative skills, but in view of the wide responsibilities they carried, their moral and political training was also of vital concern to successive emperors and their advisors. Important elements in that training served to underline the role played by astronomy in the imperial state.

Thus, in 136 BCE the emperor is said to have set up five 'professorships' *bo shi* 博士 (literally 'scholars of comprehensive [learning]'), one in each of the 'Five Classics', *wu jing* 五經. These works came to be seen as embodying the core doctrine of the group of scholars known as *Ru* 儒, who are commonly (and somewhat misleadingly) called 'Confucians' by later writers in both the west and China.[5] The texts of the Five Classics were to become central to the training that formed the ideology of imperial administrators, and they retained their importance for most of the time that the empire continued to exist. As the dynasty progressed, arrangements were made for an increasing number of students to study in what amounted to a state college, *Tai xue* 太學, designed to produce well-educated recruits for the administration.[6] One of the Five Classics to be studied by Han dynasty students contains documents said to date from

[4] For a comprehensive account of this period, see Denis Crispin Twitchett, Michael Loewe and John King Fairbank (1986) *The Cambridge history of China. Vol.1, The Ch'in and Han Empires, 221 B.C.–A.D. 220*. Cambridge, Cambridge University Press.

[5] See *Han shu* 6, 159 and 19a, 726; see also A. C. Graham (1989) *Disputers of the Tao: philosophical argument in ancient China*. La Salle, Ill., Open Court, 31–33 and Twitchett, Loewe and Fairbank (1986), 754–755. On the *Han shu*, see section 1.5. Confucius himself lived from c. 551 to 479 BCE. For reasons of respect, use of his personal name, Kong Qiu 孔丘, was avoided by his disciples. One of the alternative titles by which he was known in later ages was Kong Fu Zi 孔夫子 'Master Kong', later Latinized by western scholars as 'Confucius'. Although the importance later given to the Five Classics is indisputable, Michael Nylan points out that they came to have that importance through a complex process of advocacy and cultural change that began in Western Han and continued thereafter: see Michael Nylan (2009). 'Classics without canonization, learning and authority in Qin (221–210 BC) and Han (206 BC—AD 220)' in *Early Chinese religion*, Leiden; Boston, Brill. **1**: 721–76. The retrospective view of scholars in later imperial times may have ascribed more significance to the events of 136 BCE than they were perceived to have had when they took place.

[6] See *Han shu* 6, 159 and 171–172; Twitchett, Loewe and Fairbank (1986), 756–7 and 769. It must not be forgotten, however, that at this period the principal method of recruitment to the civil service was still by recommendation, often in response to imperial requests that appropriate candidates should be brought forward. Officials above a certain rank also had the right to recommend family members for recruitment. In both cases the person recommending was held responsible for the performance of their favoured candidate. While regular checks on both qualifications and performance were built into the system under the Han, we are still far from the large-scale open examinations that by the Song 宋 dynasty (960–1279 CE) had become the dominant path into the civil service.

the time of the sage rulers of high antiquity. Very significantly, it begins with the story of an emperor who began his reign by gathering round him a group of specialists in astronomical calculation and observation, and setting them to work.

This text is the *Shang shu* 尚書 ('*Honoured Documents*'), known in later centuries as the *Shu jing* 書經 ('*Document Classic*' or more commonly '*Book of Documents*'). The part of the *Book of Documents* discussed here may date from the fourth or fifth centuries BCE, although it describes legendary events that later traditional historiography would place around 2300 BCE.[7] Figure 1.3 comes

Figure 1.3 The emperor Yao commissions his astronomers (Sun Jianai 孫家鼐 et al. (edited) and Zhan Xiullin 詹秀林 et al. (illustrations) 1905: 1, 8a).

See Michael Loewe (2006) *The government of the Qin and Han Empires, 221 BCE–220 CE*. Indianapolis and Cambridge, Hackett 71–76. As for the *Tai xue* itself, the very large numbers of 'students' recorded later in the Han dynasty does raise the possibility that only a proportion of them can have been engaged in supervised study, whatever its original purpose may have been: Nylan (2009), 747–8.

[7] For an account of the modern view of the formation of this text and the dates of its various components, see M.A.N. Loewe (1993) *Early Chinese texts: a bibliographical guide*. Cambridge, 376–89.

from an illustrated edition of the *Book of Documents* published under imperial patronage in the last few years of the empire, and designed to encourage use of the text in schools. The emperor Yao 堯 is seated at the left of the scene, and is addressing the members of the Xi 羲 and He 和 clans, each represented by a father and his two sons.

The fathers are labelled as *Xi shi* 羲氏 'Mr Xi' at upper centre, and opposite him stands *He shi* 和氏 'Mr He'. The sons are labelled too, as *Xi zhong* 羲仲 'middle brother Xi', *Xi shu* 羲叔 'younger brother Xi', and similarly for the He brothers. First the emperor addressed the heads of the two families:

> 乃命羲和: 欽若昊天, 厤象日月星辰, 敬授民[8]時.
>
> Thereupon he ordered Xi and He: 'Accord reverently with august Heaven, sequence and delineate the sun, the moon and the stellar markers, and thus respectfully bestow the seasons on the people.' (*Shang shu* 2, 9a–10b in (Ruan Yuan 阮元 (1764–1849) 1973 reprint of original of 1815: vol. 1, 21–1 & 21–2))

Once the fathers have been given their instructions, the sons are sent off to the four quarters of the world, where, the emperor tells them, they may determine the times of the solstices and equinoxes by observing certain stars. (We may note in passing the implication that Yao rules over the whole of the known world, which was also at least the theoretical pretension of the sovereigns of the early empire and their successors.) To one of the Xi brothers, sent to dwell in the east, the emperor says:

> 日中星鳥. 以殷仲春.
>
> [When] the day is of middle [length], the star is *Niao* ('The Bird'). By that you may fix mid-spring. (*ibid.*)

To the other, who is to dwell in the south, he says:

> 日永星火. 以正仲夏.
>
> [When] the day is prolonged, the star is *Huo* ('The Fire'). By that you may check mid-summer. (*ibid.*)

To the He brother sent to the west, he says:

> 宵中星虛. 以殷仲秋.
>
> [When] the night is of middle [length], the star is *Xu* ('The Barrens'). By that you may fix mid-autumn. (*ibid.*)

[8] I read *min* 民 rather than *ren* 人, though both would in this context be rendered as 'people'. As the *Shi san jing zhu shu* collation note states, the former word is found in earlier texts.

To the other, sent to the north, he says:

> 日短星昴. 以正仲冬.
>
> [When] the day is short, the star is *Mao* ('The Mane'). By that you may check mid-winter. (*ibid.*)

Exactly how these stars are to be observed is not stated. However, from the late third century BCE onwards we begin to find lists of stars which will, we are told, be 'centred' *zhong* 中, i.e. crossing the meridian due south of the observer, at dusk and at dawn at different times of the year. Interpreting the emperor's words in this way, we may check whether the stars named would have shown such phenomena around the time when the ideas in the *Yao dian* might have taken shape, say around the middle of the first millennium BCE.[9] The first 'star', Bird, is not a very good test, since it is usually taken to refer to a large grouping of stars known as the 'Red Bird' *zhu niao* 朱鳥, extending between the modern asterisms Corvus and Gemini.[10] The stars in question were, however, more or less equally spread on either side of the meridian an hour after sunset at the spring equinox in 500 BCE.[11] The Fire star is a better test, since it has always been identified as the conspicuous red star Antares, α Scorpii. This was close to the meridian one hour after sunset at the summer solstice of 500 BCE. 'Barrens' is the name of a small asterism of which the principal star is β Aquarii, and that was also close to the meridian one hour after sunset at the autumn equinox of 500 BCE. 'Mane' refers to the Pleiades cluster, which is close to the meridian about 2 ½ hours after winter solstice sunset in 500 BCE. The instructions given in the *Yao dian* would therefore not have seemed obviously anomalous at the time that the text was probably written down—and indeed would still have been similar to what would have been seen during the Western Han. We shall meet both Barrens and Mane later in this book as two of the 28 'lodges' into which the circumference

[9] Here I invert the traditional approach to these data, which has been to use modern calculations to determine the epoch at which they would have been closest to observation. For examples of the results, see Needham and Wang Ling (1959), 17 and 245–6. All such attempts are in my view rendered unreliable because we know nothing for certain about the observation procedures that the writers of this text had in mind, and quite small changes in such variables as the time of observation can shift a 'best fit' date estimate by many centuries. See the further discussion of 'centred stars' in chapter 5, section 5.3.

[10] On the early history of this grouping, see for instance David W. Pankenier (2013) *Astrology and cosmology in early China: conforming earth to heaven*. New York, Cambridge University Press, 75–8 and elsewhere.

[11] For the present purpose, I define sunset as seen from the Western Han capital of Chang'an長安, modern Xi'an西安, whose latitude is close to that of other important ancient centres.

of the heavens was thought of as being divided during the period with which we are principally concerned.[12]

We may pause here to consider one of the words used by the emperor, since it will occur frequently in this book: the word translated earlier in this section as 'sequence', written 厤. This word, pronounced *li* in modern standard Chinese (*putonghua*普通話) is more commonly written as 曆 or 歷. Anciently pronounced **rêk*,[13] it has a group of related senses, such as 'count, calculate', 'sequence', and often refers more particularly to the sequence of data that is called a calendar in English. It is significant that when this passage is rendered by one Western Han source, the phrase *li xiang* 厤象 translated earlier in this section as 'sequence and delineate [the sun, the moon and the stellar markers]' is replaced by the much more transparent expression *shu fa* 數法 'calculate and take as a pattern'.[14] *Li* can also refer to the entire system of constants and algorithms that generates a calendar, and can thus be rendered as 'calendrical system', or (following Nathan Sivin) 'astronomical system'.[15] This last rendering takes account of the fact that a *li* may not simply enable us to calculate when months and years will begin and end, but may also enable us to calculate the times when eclipses might be expected, and even the motions of the planets. I shall often just use the translation 'system' when the context makes the reference clear, or else simply transliterate as *li*.

Finally, the emperor turns back to the fathers:

咨汝羲暨和! 朞三百有六旬有六日. 以閏月定四時, 成歲.

[12] The 28 lodges *er shi ba xiu* 二十八宿 were associated with a band of asterisms of unequal widths that marks out a circuit of the sky in the general region of the celestial equator, and served as a reference system for the movements of the sun, moon and planets as they moved from west to east relative to the stars. For a discussion of the lodge system, see chapter 5, sections 5.2.1 and 5.3.

[13] Axel Schuessler and Bernhard Karlgren (2009) *Minimal Old Chinese and later Han Chinese: a companion to Grammata serica recensa*. Honolulu, University of Hawai'i Press, 132. The asterisk is conventionally used in historical phonology to indicate a hypothetical reconstructed form for which there is no direct attestation.

[14] See *Shi ji* 1, 16. On the *Shi ji*, see section 1.5. The author of this text was, amongst other things, the principal imperial official responsible for fulfilling the duties allocated by Yao to the Xi and He families.

[15] See the careful discussion in Sivin (2009), 38–40. Sivin distinguishes four senses in which the word *li* may be used: mathematical astronomy in the sense of 'the art of computing the times or locations of certain future or past phenomena in the sky', the systems of calculation that make such forecasts possible ('astronomical systems'), a treatise embodying such a system, and finally 'the products of computational treatises, namely ephemerides published as almanacs.' While I agree that 'astronomical system' is often a useful rendering, I am less insistent than Sivin on not using the word 'calendar' in this context, perhaps because of the rather different situations in the periods studied in this book and Sivin's. For further discussion, compare Martzloff (2009).

> O you Xi and He! The period is of three hundreds, and six tens, and six days.[16] Use intercalary months[17] to fix the four seasons correctly, and to complete the year. (*ibid.*)

The language in which the *Book of Documents* is written would have been seen as strikingly archaic in early imperial times; the impression of venerable antiquity given to a Han reader might have been similar to that made on a 20th-century English speaker reading the opening chapter of the Book of Genesis in the 17th-century King James translation of the Bible. But that was no doubt an essential part of the impact of the text as an ancient scripture that was to guide conduct in a much later age. And the message was clear enough—it was the responsibility of the emperor to ensure that his subjects had access to a well-run luni-solar calendar, and he might fulfil this responsibility by choosing appropriately talented people from amongst his subjects to handle the matter on his behalf. When Yao's chosen successor Shun 舜 took the throne, his very first act is described as follows:

> 正月上日受終于文祖; 在璿璣玉衡以齊七政
>
> On the first day of the first month he accepted the abdication [of Yao] in the [shrine of] the Accomplished Ancestor; he attended to the *xuan ji yu heng* in order to regularize the seven governors. (*Shang shu* 2, 4a, 35–2)

As we shall see later in this book,[18] there have been various interpretations of what 'attending to the *xuan ji yu heng* in order to regularize the seven governors' might have meant to whoever wrote this part of the *Book of Documents*. But the documentary record shows clearly that in Han times the interpretation of this passage was always given in astronomical terms—the *xuan ji yu heng* being taken to be the names of stars in Western Han, and being glossed as names of the parts of an armillary sphere in Eastern Han.

Those who studied the *Book of Documents* would certainly absorb the clear lesson that the first two emperors recorded in that text regarded astronomical

[16] The commentators take it that 'the period' *qi* 期 of 366 days given here is intended as an approximation to the length of the solar cycle (the interval between winter solstices in Chinese practice), which is close to 365 ¼ days. See note 40. The *Shi ji* (1, 17) renders this passage with *sui* 歲 in place of *qi* 期; as we shall see, *sui* is the normal term for the solar cycle in astronomical writing: see section 1.3.

[17] The reference to the need to insert intercalary months (rather than days) in order to keep the calendar working properly suggests strongly that these are lunar months—as we shall see this was in fact universal in pre-modern Chinese practice.

[18] See chapter 6, section 6.3.3.

matters as having great importance to their rule. There is plenty of evidence that the efficient running of an astronomical system, *li*, was already seen as an important topic in the pre-Qin age of feudal states from which the *Book of Documents* originated. Confucius himself was said to have been consulted on problems caused by mismanagement on the part of those responsible for running the system. The story in question, set in 484 BCE and taken from the *Zuo zhuan* 左傳 chronicle,[19] illustrated what could go wrong if such officials failed to match up to the demands of their job:

> 冬. 十二月. 螽. 季孫問諸仲尼. 仲尼曰: 丘聞之, 火伏而後蟄者畢. 今火猶西流. 司厤過也.
>
> Winter. The 12th month. [There were] grasshoppers. Jisun asked Confucius [about this unseasonable event]. Confucius said 'I have heard that [insects] are all in hibernation after the Fire [Antares] has set [i.e. it has ceased to be visible after sunset]. Now the Fire is still sinking in the west. This is a mistake by those in charge of *li*.[20] (*Zuo zhuan*, Duke Ai, 12th year, 59, 5a in (Ruan Yuan 阮元 (1764–1849) 1973 reprint of original of 1815: vol. 6, 1027–1))

The implication here is that the failure by 'those in charge of *li*' to insert an intercalary month at the right time has caused the lunar calendar to get in advance of the seasons, a fact that Confucius verifies by pointing to an obvious

[19] On the *Zuo zhuan*, see Loewe (1993), 67–76. The *Zuo zhuan* is an annual chronicle recounting events in the complex and often violent interplay between the so-called 'feudal states' of the 8th to 5th centuries BCE. It is often (and somewhat misleadingly) viewed as a commentary on one of the Five Classics, the *Chun qiu* 春秋 'Spring and Autumn [annals]', a much more terse chronicle of roughly the same period originating in Confucius' home state of Lu 魯 (in modern Shandong 山東 province), and said to have been edited by him. In the context of *li* we may note that the *Chun qiu* contains 37 dated records of solar eclipses, most of which have been shown by modern calculations to be records of real astronomical events. These were frequently used by later *li* specialists as long-term tests of the accuracy of the predictions of their systems. Both chronicles are translated and discussed in James Legge (1872a) *The Chinese Classics, Volume V: The Ch'un Ts'ew, with the Tso Chuen. Part 1: Dukes Yin, Hwan, Chwang, Min, He, Wan, Suen and Ch'ing, and the Prolegomena.* Hong Kong and London and James Legge (1872b) *The Chinese Classics, Volume V: The Ch'un Ts'ew, with the Tso Chuen. Part 2: Dukes Seang, Ch'aou, Ting, and Gae, with Tso's Appendix; and the Indexes.* Hong Kong and London.

[20] The 12th (lunar) month referred to in the heading of this entry is not the 12th month according to the month-count used in imperial times (the 'Xia [dynasty]' count, whose first month nowadays marks 'Chinese New Year': see section 4.4.5.3), but refers to the so-called Zhou [dynasty] count that begins two months earlier. This 12th 'Zhou' month would normally begin in late October or early November, by which time the sun would then be passing close to Antares in its annual cycle of eastwards motion round the heavens, so that this star would have ceased to be visible in the west after sunset, as it had been a month earlier. If Antares were still visible in this month, it would have been clear that an intercalation was needed to allow the seasons to catch up with the calendar.

astronomical event, in addition to the unseasonable presence of grasshoppers in a month when one would have expected them to have gone into hibernation.

Mencius (Meng Zi 孟子, early 4th to early 3rd centuries BCE), the thinker regarded by the *Ru* scholars as second only to Confucius, is recorded as making a claim that shows that in his time it was taken for granted that long-term predictions about celestial phenomena were possible:

> 天之高也, 星辰之遠也, 苟求其故, 千歲之日至可坐而致也.
>
> Although the heavens are high and the stars are far away, if you seek out the antecedents,[21] you can sit down and predict the solstices for a thousand years.
> (*Meng Zi* 8b, 3b in (Ruan Yuan 阮元 (1764–1849) 1973 reprint of original of 1815: vol. 8, 152–1))

Clearly such a claim presupposes a body of expertise in celestial calculation that was both widespread and well established, and it was on this foundation that the experts of the early imperial age built in their turn.

1.2 Delivering the seasons—how 'respectfully'?

We shall soon turn to the question of exactly how those in charge of the astronomical system were expected to do their jobs, including what was meant by 'completing the year', and how intercalary months were to be managed. But let us look first at a larger question—how was the canonized scripture of the *Book of Documents* actualized in terms of the way that the imperial government presented itself to its subjects, and how it sought to manage them? And what was meant in practice by giving the seasons to the people 'respectfully'?

An excellent example of what it might mean when an imperial government claimed to be acting in accordance with the pattern laid down by Yao and

[21] The suggestion is that if you know when past solstices have occurred, you will be able to work out when future ones are to be expected. There is a similar use of *gu* 故 in the sense of 'antecedent' or 'precedent', i.e. the past events that lie behind present and future events, in the *Xici* 繫辭 '*Appended explanations*' treatise (probably dating from around the third century BCE) now attached to the *Yi jing* 易經 '*Book of Change*': 仰以觀於天文, 俯以察於地理, 是故知幽明之故. '[The sage] looks up to see the signs in the heavens, and down to examine the principles of earth, and thus he knows the antecedents of darkness and light.' *Yi jing* 7, 9a in *Shi san jing zhu shu* 十三經註疏 *(The thirteen classics with commentaries and subcommentaries)*. (1973 reprint of original of 1815). Ruan Yuan 阮元 (1764–1849), Taipei, Yiwen Press, vol. 1, 147–1. On the significance for the construction of an astronomical system of this ancient divinatory manual and the cosmological material that became associated with it, see chapter 4, section 4.4.

Shun can be found in the ruins of what seems to have been a general purpose administrative post at Xuanquan 懸泉, near Dunhuang on the northwest frontier of the Han empire, an establishment which appears to have been active from the second century BCE to the first century CE. From the remains of one of the collapsed walls of the post, we can see that it once bore a lengthy document painted onto its plaster in a part of the building where it would have been clearly and conspicuously visible to those passing through the post. This was the text of an imperial edict dated to 5 CE, the *Si shi yue ling* 四時月令 'Monthly ordinances for the four seasons'. It was a conspicuous and clearly lettered piece of writing, 2.2 m wide by 48 cm high, bearing at its end a label in bold characters 使者和(中)[仲]所督察: 詔書四時月令五十條 'Inspected by the messenger, Younger Brother He:[22] Document of the Edict of Monthly Ordinances for the Four Seasons in Fifty Articles'. Figure 1.4 is a reconstruction of the original appearance of the edict on the wall by a modern editor and translator of the document.[23]

What was the purpose of this edict, and what kind of information did it set out to convey? The edict opens with a clear statement of the thinking on which it is based:

> 太皇太后詔曰: 往者陰陽不調, 風雨不時, 降惰農自安, 不勤作勞, 是以數被菑害, 惻然傷之. 惟聖明王, 靡不躬天之曆數, 信執厥中, 欽順陰陽, 敬授民時, 勸耕種, 以豐年穀, 蓋重百姓之命也. 故建羲和, 立四子 . . . 時以成歲, 致憙 . . . 其宜, □[24] 歲分行所部各郡.
>
> 詔條
>
> 元始五年五月甲子朔丁丑, 和仲普使下部郡太守, 承書從事下當用者, 如詔書, 書到言. 從事史況.

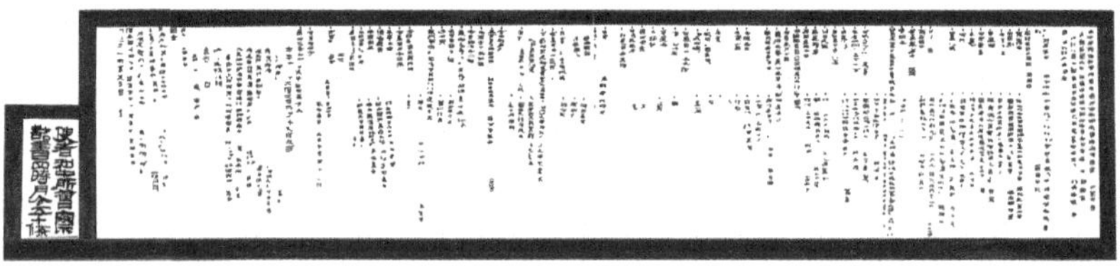

Figure 1.4 The *Si shi yue ling* 四時月令 edict of 5 CE; reconstruction of original layout supplied by Charles Sanft, and reproduced with his permission.

[22] Sanft translates as 'Master of Autumn'—since this was the son charged with checking the autumn equinox.

[23] Sanft (2011), 177. I am grateful for the author's permission to reproduce this figure.

[24] Sanft follows the suggestion that the obliterated or missing character here should be *mei* 每.

> The Grand Empress Dowager[25] proclaims: Recently Yin and Yang have not been in harmony. Wind and rain have not come at the proper times, and the lazy farmers have been at ease, not striving at their work. Because of this, we have repeatedly suffered disaster. I worry sadly about this. I think of the sagacious emperors and perspicacious kings, who all personally followed heaven's sequence (*li*) of reckonings, and sincerely held to its core. They reverently accorded with Yin and Yang, 'respectfully bestowed the seasons on the people', and encouraged agriculture in order to make the harvest plentiful—because they gave importance to the fate of the common people. So I have established [the office of] Xi He and set up the Four Sons[26] . . . time in order to properly 'complete the year' and bring happiness . . . what is fitting, and shall travel separately to each commandery in their jurisdiction [every] year.
>
> Instruction: In the fifth year of the Yuanshi 'Origin Initiation' reign period [5 CE][27], in the fifth month, which had day *jiazi*.1 [May 27] as its first day, on the day *dingchou*.14 [June 9],[28] the Younger Brother He, [whose given name was] Pu, had this distributed to his subordinate commandery grand administrators. When you receive this document, carry out its tasks and distribute it to those who should use it, as specified in the document of the edict. When this document arrives, report. Attendant Clerk, Kuang.[29]

The tradition of the *Book of Documents*—which is quoted directly at two points—is clearly operating here as much more than a pious scriptural reference. Failure

[25] This was Wang Zhengjun 王政君 (71 BCE–13 CE), Empress Yuan 元. Although the edict was issued in her name, it almost certainly originated with her nephew, the powerful courtier and later emperor of a short-lived new dynasty, Wang Mang 王莽 (c. 45 BCE–23 CE): see Sanft (2011), 145. Wang Mang's rise to power is outlined at the start of chapter 4. As noted there, he was to echo some of the words of the Empress Dowager's document in his accession proclamation.

[26] Sanft renders this as 'Four Masters', playing on the fact that *zi* 子 can be a honorific reference to a respected person as well as meaning 'son'—in this case clearly referring back to the four sons of Xi and He.

[27] The Yuanshi 元始 'Origin Initiation' reign period mentioned here ran from 1–5 CE. On reign periods, see chapter 2, section 2.1.2.

[28] The terms *jiazi* and *dingchou* are the names of two days in the cycle of 60 days that ran continuously from at least as early as the last few centuries of the first millennium BCE. The appended numbers show the position of the day in the sequence: for more on these sexagenary day names, see chapter 2, section 2.1.3. All ancient western style dates in this book are given in the Julian calendar, normally used by modern historians as a common reference system for the time before the Gregorian reform of 1582 CE. This calendar is based on the system introduced at Rome by the reforms of Julius Caesar in 45 BC, and modified by Augustus 40 years later; see James Evans (1998) *The history and practice of ancient astronomy*. New York; Oxford, Oxford University Press, 163–6. For convenience of dating, scholars project back the use of the Julian system into periods long before the time of Julius Caesar: this is the so-called 'proleptic' Julian calendar.

[29] Sanft (2011), 178–9: slightly modified at some points The dots '. . .' indicate missing sections of text.

to 'complete the year' and otherwise fulfil the ruler's duty to order the people's time and work by 'delivering the season' is the root cause of natural disasters and social disorder. In this case, the authorities have even borrowed the names of Xi, He and their sons as names for the officials charged with managing the astronomical system.[30]

But the ambitions of the imperial government did not end with the issuance of a calendar. The edict goes on to prescribe a very fine-grained system of management which is to be imposed upon the population (the 'lazy' farmers) in virtue of the emperor's hegemony over the time of his subjects, based on his authority to issue the calendar. Many of the instructions to be given are intended to control interactions between the populace and the natural environment. In some cases the people are to be told what not to do, and in others what they must do:

> 毋彈射蜚鳥, 及張羅, 為它巧, 以捕取之.
>
> Do not shoot birds with pellets, spread nets, or use other techniques to capture them (third month of spring). (Sanft, Charles 2011: 182)
>
> 乃勸種麥, 毋或失時. 失時, 行罪毋疑.
>
> Then exhort them to plant wheat. Do not permit them to miss the proper season; if they miss the proper season, carry out the punishment without a doubt (second month of autumn). (Sanft, Charles 2011: 184)

It is clear from this wording that the edict is not so much a direct address to the mass of the population, most of whom would not have been able to read it, but an instruction to the officials at the Xuanquan administrative post telling them what orders to give to the people at different times of the year. Perhaps the text of the edict was placed on public display as a form of pre-emptive justification for the orders that were to be given?

Other articles of the edict might be said to be about general personnel management such as:

> 毋聚大眾; 謂聚民繕治也.
>
> Do not assemble large groups. [Note in text:] This refers to assembling common people to do repairs (first month of spring). (Sanft, Charles 2011: 180)

[30] As we shall see in section 4.1, the great expert in astronomical systems Liu Xin 劉歆 held the office of Xi He: *Han shu* 12,359. The use of this particular title for those charged with administering such matters was confined to the time when Wang Mang was influential, *Han shu* 12,351, but the responsibilities it carried continued to be seen as essential.

Personnel management does not seem to have taken much notice of what today we might see as essential boundaries of privacy and personal autonomy:

> 日夜分, 靁乃發聲, 始電 [. . .] 先靁三日, 奮鐸以令兆民曰: 靁勿懷妊 [. . .]不戒其容止者, 生子不備, 必有災.
>
> When day and night are equal, then thunder will make its sound and start the lightning. [. . .]. Three days before the thunder, ring the bell to command the people, saying: When there is thunder, [do not] become pregnant. [. . .] Those who do not control their comportment will give birth to imperfect children and are sure to suffer calamity (second month of spring). (Sanft, Charles 2011: 180–1)

In other words, the officials were to 'command' the people not to have sex. It appears that the 'respect' that the officials were to manifest in 'delivering the seasons' was not directed towards the emperor's subjects.

In one instance the instruction relates to the performance of ritual acts:

> 告有司大難旁磔, 出土牛, 以送寒氣. 謂天下皆以 . . . 歲終氣畢以送之. 皆盡其日.
>
> Command the responsible officials to carry out the great exorcism and the directional sacrifice and make the clay ox to send off the cold pneuma.[31] This refers to the entire realm using . . . the year is concluded and the pneuma finished, in order to send them off. All last the entire day (last month of winter). (Sanft, Charles 2011: 186)

Here, however, the recipients of the order are not the 'lazy farmers' whom we have seen being micro-managed into carrying out their allotted tasks, but a group of officials, who must perform a ritual to ensure the harmony of the human world with the cosmos at the close of the year. Paradoxically, while the farmers may have had their doubts about the practical utility of many of the orders issued to them, it is likely that they shared the conviction of the Empress Dowager and her officials that rituals such as the one described were vital to the well-being of the empire. A ritual wrongly conducted, or conducted at the wrong time, might be worse than useless. As one excavated text dating from around 300 BCE puts it:

> 民人弗知歲, 則無攸祭

[31] This translation by Sanft is one commonly used for the term *qi* 氣, a word that I normally prefer to transliterate without looking for an English equivalent. Here *qi* refers to something that combines both the vapours that predominate at this season, and the cosmic energies expressed through them.

If the people do not know the year, they will have no means of conducting sacrifices.[32]

This was no mere matter of propriety: for example, in 175 CE two officials sparked a full-scale debate involving the highest levels of government when they claimed that because of failures to calculate the calendar properly:

故妖民叛寇益州, 盜賊相續為害 [. . .] 受虛欺重誅

Therefore evil folk are rebelling and thieving in Yizhou, and robbers and bandits make endless trouble. [. . .] [Those responsible for the calendar] should receive heavy punishment for empty deceptions.' (*Hou Han shu, zhi* 2, 3,037; Cullen 2017, 404)[33]

Such insistence on the proper timing of rituals and other activities according to a well-run calendar makes a striking contrast with the situation in ancient Athens, where the failure of the city authorities to ensure that religious festivals took place at the right time of the lunar month was so notorious in the fourth century BCE that the city's most famous comic playwright, Aristophanes, made a joke of it in his comedy *The Clouds*, where the leader of the chorus says:

When we were ready to set forth on our trip here, the Moon happened to run into us and told us first to say hello to the Athenians and their allies, but then she expressed her annoyance at the awful way she has been treated . . . you don't keep track of your dates correctly, but scramble them topsy-turvy, so that the gods scold her, she says, every time they're misled about a dinner and go home having missed the festival that was specified in the calendar.[34]

The problem was that the calendar for festivals was promulgated by the city officials, known as 'archons', who may have begun their planning on the basis of the actual phases of the moon, but then displaced the beginnings and ends of months, and thus the dates of sacrifices ('dinners' from the point of view of the gods), to suit their administrative convenience, leading to the complaints voiced by the moon on behalf of her fellow gods and goddesses.[35]

[32] Marc Kalinowski (2004) 'Fonctionalité calendaire dans les cosmogonies anciennes de la Chine.' *Etudes chinoises* 23: 87–122, 115; see also Donald John Harper (1999). 'Warring States Natural Philosophy and Occult Thought' in *The Cambridge history of ancient China: from the origins of civilization to 221 B.C.*, Cambridge, UK; New York, Cambridge University Press: 813–84, 845–7.

[33] For full discussion of the controversy that followed, see chapter 7, section 7.2.3.

[34] *The Clouds.* (1998). Aristophanes (c. 446–c. 386 BCE) and Jeffrey Henderson (tr. and ed.), Cambridge, Mass.; London, Harvard University Press, 605–15.

[35] I follow here the account given by Alan Edouard Samuel (1972) *Greek and Roman chronology; calendars and years in classical antiquity.* München, Beck, 57–8. Robert Hannah (2005) *Greek and Roman calendars: constructions of time in the classical world.* London, Duckworth, 47–52 does not

1.3 'Completing the year': running a luni-solar calendar in the second century BCE

But let us return to the technical tasks that the emperor allocates to the Xi and He families. Just how are intercalary months to be placed so as to 'complete the year'? And what exactly is the 'year' (*sui* 歲) that is to be completed? We can find that information in a book that was completed very close to the time that the *Book of Documents* was canonized by the Han dynasty. This is the *Huai nan zi* 淮南子 'The Master of Huainan', a compendium of knowledge prepared under the patronage of the Han prince Liu An 劉安 (c. 179–122 BCE) and offered to the throne in 139 BCE—only three years before the 'professorships' in the Five Classics were set up. We are thus in a position to understand how the work of the Xi and He might have been understood by a reader in the early imperial age.[36]

First, what about the *sui*? We are told:

> 日行一度, 以周於天. 日冬至峻狼之山, 日移一度, 凡行百八十二度八分度之五, 而夏至牛首之山. 反覆三百六十五度四分度之一而成一歲

> The sun moves one *du* [in a day], and thus makes its circuit of the heavens. At the winter solstice the sun is at High Wolf Mountain, then shifts one *du* in a day, and after moving 182 *du* and ⅝ *du* in all, at summer solstice it is at Ox Head Mountain.[37] It returns through 365 *du* and ¼ *du* to make a whole *sui*. (*Huai nan hong lie ji jie* 3, 94–5)[38]

A *du* 度 is a measure of displacement of a heavenly body against the background of the stars, equal in this period to the amount of the sun's daily motion,

agree with Samuel's sharp distinction between a 'regulatory' luni-solar calendar and the archon's calendar. But the situation that produced Aristophanes' joke certainly had no parallel in imperial China.

[36] A full commented translation of chapter 3 of this book, to which we shall refer here, is given in John S. Major, with an appendix by Christopher Cullen (1993) *Heaven and earth in early Han thought: chapters three, four and five of the Huainanzi* Albany, State University of New York Press. The translations given here are my own.

[37] These mountains are unknown to geography. The commentary of Gao Xiu 高誘 (fl. 210 CE) says that they are the mountains of the extreme south (High Wolf) and extreme north (Ox Head). Presumably the reference is to the extremes of the sun's north-south displacement, and the idea was that the sun was overhead at noon as seen from these positions at winter and summer solstice respectively.

[38] All citations from *Huai nan zi* are from the punctuated and commented edition *Huai nan hong lie ji jie* 淮南鴻烈集解 *(Collected commentaries on the great work of [the prince of] Huai nan).* (1989). Liu An 劉安 (c. 179–122 BCE), Beijing, Zhong Hua press.

then taken as being constant throughout the year.[39] As we shall see in chapter 5 (section 5.2.1), the 28 lodges *xiu* 宿 into which the circumference of the heavens is divided are themselves subdivided into *du*. A period of 365 ¼ days (365.25 days), twice 182 and ⅝ days, for the interval between winter solstices is not a bad approximation to the value that modern astronomy would give, which is 365.243 days. In modern terms, this period is called the 'tropical year', to distinguish it from the slightly longer 'sidereal year', the time taken for the sun to return to the same apparent position relative to the stars.[40] But in the early imperial period, there was as yet no systematic recognition of the difference between the two periods, and so I shall simply use the term 'solar cycle' to refer to the *sui* as the conflation of the two—the interval after which the sun supposedly both returned to winter solstice and returned to the same position relative to the stars. The value of the solar cycle given here is found in many other early imperial sources. It obviously differs from the value of 366 days for 'the period' given by Yao to his astronomers, which is not attested in any other ancient source. Shortly after the end of the Han dynasty the scholar Wang Su 王肅 (195–256 CE) suggested that Yao's value represented an inclusive count of the 365 ¼ day cycle in whole days, which seems the simplest interpretation.[41]

Sui is to be distinguished from another word normally translated as 'year', which is *nian* 年. In the sources studied in this book, *nian* refers to a calendrical

[39] Liu Zhuo 劉焯 (544–610 CE) seems to have been the first person in East Asia to deal systematically with the inequality of solar motion throughout the year. See Chen Meidong 陈美东 (1995), 311 ff. and Sivin (2009), 297–8. But Zhang Zixin 張子信 (active around 560 CE) had already noticed that the speed of the sun (presumably along the ecliptic) changed in the course of a year: see *Sui shu* 20, 561. He said that it was 'slow after the spring equinox, and fast after the autumn equinox'—which was roughly correct, given the dates in his own day (early June and December) when the earth was at maximum and minimum distances from the sun (aphelion and perihelion) and hence the sun's apparent motion was slowest and fastest. As the *Sui shu* notes, Zhang spent 30 years on an island where he had taken refuge from the disorders of the time, and concentrated on systematic observations of the sun, moon and planets with an armillary sphere. On the inequality of the seasons in the ancient west, see for instance the account of Ptolemy in the second century CE, describing the work of Hipparchus, second century BCE, *Almagest* III.4, Toomer (1998), 153–4.

[40] In the ancient west, the cycle of the seasons was normally counted from the spring equinox rather than winter solstice. For a more precise discussion of what exactly the term 'tropical year' has meant in the past, see Jean Meeus and Denis Savoie (1992) 'The history of the tropical year.' *Journal of the British Astronomical Association* 102 (1): 40–2. The sidereal year is 365.256 days long; this is about twenty minutes longer than the tropical year because of the phenomenon of precession, which causes the seasonal positions of the sun relative to the stars to shift steadily westwards in a cycle not far from 26,000 years in length (on which see chapter 5, section 5.1). As a result, the solstices and equinoxes 'come to meet' the sun in its annual eastwards circuit, and thus shorten the time it takes for the sun to return to them.

[41] Wang's view is quoted in a seventh century commentary on the Book of Documents, *Shang shu*, 2, 18a in *Shi san jing zhu shu*, vol. 1, 25–2.

unit, consisting of a whole number of lunar months (usually 12), counted from the beginning of a month customarily selected as the start of the administrative year, and never to the solar cycle. It is in this sense that it was used in the text of the edict discussed in section 1.2, where the date was given as *yuan shi wu nian* 元始五年 'the fifth *nian* of the Yuanshi [reign period]'. For precision, we may on occasion translate *nian* as 'civil year'.

As for months, a month in any calendar used in imperial China was always a lunar month, whose first day, designated by the term *shuo* 朔 'beginning' should ideally contain the instant of new moon, when the sun and moon are so closely aligned from the point of view of a terrestrial observer that the moon cannot be seen. A clear statement of how long the cycle of lunar phases (the lunation) should last is found in *Huai nan zi*:

> 二十九日九百四十分日之四百九十九而為月.
>
> 29 days and $^{499}/_{940}$ days make a *yue*. (*Huai nan hong lie ji jie* 3, 105)

Since a calendar month must contain a whole number of days, it is clear that *yue* here is being used in its sense of 'lunation', rather than 'month'. *Yue* also means simply 'moon'; unfortunately the use of this term in English as a time measure ('many moons ago') has associations of orientalist exoticism that render it inappropriate as a translation in this context.

A modern value for the long-term average length of the lunation is 29.53059 days (to seven significant figures); this value is commonly known as the 'mean synodic month'.[42] Since 29 $^{499}/_{940}$ = 29.53085 to the same precision, the two values agree to within 0.0003 day, just under half a minute.

The *Huai nan zi* passage cited earlier continues in a way that makes it clear how intercalations are to be used to 'complete the year':

> 而以十二月為歲，歲有餘十日九百四十分日之八百二十七. 故十九歲而七閏.
>
> . . . so if you use 12 *yue* for the *sui*, the *sui* will have a remainder of 10 days and $^{827}/_{940}$ days. For that reason, there are 7 intercalations, *run* 閏, in 19 *sui*. (*Huai nan hong lie ji jie* 3, 105)

[42] The interval between new moons is in fact not constant but can vary from 29.18 to about 29.93 days. In the period we shall be discussing, no attempt was made to vary calendrical month-lengths to reflect this: the month followed the 'mean lunation' rather than the 'true lunation'. This would rarely lead to any obviously visible problem, since the precise moment of conjunction is only obvious when a solar eclipse occurs. See, however, the work of Liu Hong, discussed in chapter 8, section 8.2.1 for the first successful efforts to calculate the moment of true rather than mean conjunction.

We have already seen that *sui* in such contexts refers to the precise length of the solar cycle between successive instants of winter solstice, which is not a whole number of days. This passage is pointing to a problem in making a year by taking 12 lunar months, which is that this period will be just under 11 days short of the solar cycle, the interval at which the seasons repeat. More precisely:

$$[365\ \tfrac{1}{4} - 12 \times (29\ \tfrac{499}{940})]\ \text{days} = 10\ \tfrac{827}{940}\ \text{days}$$

as stated in the text. As a result, after three years the deficit will have built up to over a month, and it will be necessary to pause for a month (an 'intercalary' month, *run yue* 閏月) before starting the month-count of the next year, in order to get back in step with the seasons. Now if we make seven intercalations in 19 years, the total number of months is $19 \times 12 + 7 = 235$. Comparing the total lengths of the cycles of the sun and moon completed, we see that they are identical:

$$19\ \text{solar cycles}\ \textit{sui} = 19 \times 365\ \tfrac{1}{4}\ \text{days} = 6{,}939\ \tfrac{3}{4}\ \text{days}$$
$$235\ \text{lunations}\ \textit{yue} = 235 \times (29\ \tfrac{499}{940})\ \text{days} = 6{,}939\ \tfrac{707}{940}\ \text{days} = 6{,}939\ \tfrac{3}{4}\ \text{days}$$

This 19-year intercalation cycle is commonly known in later sources as a *zhang* 章 'Rule', although that term is not used in *Huai nan zi*.[43] It is clear then, that we have a trio of interlocking values:

Solar cycle: 365 ¼ days
Intercalation cycle: 19 solar cycles, equal to 235 lunations
Lunation: 29 $\tfrac{499}{940}$ day

An astronomical system that used this set of data was commonly known as a *si fen li* 四分曆 'four parts system' (often rendered as 'quarter remainder system') from the fact that the fractional part of the solar cycle was ¼ day, for which the Chinese term is *si fen ri zhi yi* 四分日之一 'one of four parts of a day'. Such systems only differed from one another in specifying a different initial starting point (what was later called a *li yuan* 曆元 'system origin') for calculations of solar and lunar cycles, so that one system would predict moments of winter

[43] The *Huai nan zi* gives us no indication of how the required seven intercalary months are to be distributed through the 19 years of the cycle. As we shall see later, this is one of the points that has to be specified by any full description of a *li*.

solstice or luni-solar conjunction at the same intervals as any other quarter remainder system, but displaced from the other system's predictions by some constant amount. By the beginning of the Common Era, it was thought that a number of different 'quarter remainder' systems had been used in the pre-imperial age, and details were given of each of them. It is, however, unclear how far these systems were simply speculative reconstructions rather than systems that were actually put to use.[44]

In *Huai nan zi*, there is a description of the conditions from which calculations might start, but since it is not stated in what year those conditions are supposed to have applied, we are not given sufficient information to carry out practical calculations to define the calendar in any given year thereafter:

> 天一元始, 正月建寅, 日月俱入營室五度. 天一以始建七十六歲, 日月復以正月入營室五度無餘分, 名曰一紀. 凡二十紀, 一千五百二十歲大終, 日月星辰復始甲寅元.
>
> At the Origin Initiation of the Celestial Unity,[45] in the standard month, for which the [Dipper] establishment is yin.3,[46] the sun and moon both enter the 5th *du* of the lodge House.[47] Once Celestial Unity has marked 76 years, the sun and moon go back to entering the fifth *du* of the lodge House in the first month, with no remaining fractions. This is called an Era [period]. After twenty Eras, making 1,520 years, there is a Grand Conclusion, and the sun, moon and constellations once more begin a *jiayin*[48] Origin. (*Huai nan hong lie ji jie* 3, 95).

[44] See chapter 3, section 3.4.1. Some authors render *li yuan* as 'epoch'; I prefer a more literal translation. Typically, a system origin specified a moment when all calendrical variables of interest were at their starting points—thus the instant chosen would simultaneously be midnight, winter solstice and the moment of luni-solar conjunction beginning the first day of the first month as counted by astronomers. That day would also normally be the first of the sixty-day cycle that was foundation of the dating system: see chapter 2, section 2.1.3.

[45] Elsewhere in the same chapter, we are told that Celestial Unity is the same as *Tai Yin*太陰 'Grand Yin', elsewhere called *Tai sui* 太歲 'Grand Year' which marks the passage of the 12-year 'Jupiter cycle'; on this see Cullen 2017, 108ff. The cycle was based on the fact that Jupiter completes a circuit of the heavens in close to 12 years. Unlike Jupiter, *Tai sui* moves from east to west, thus passing in order through the divisions of the heaven corresponding to the 12 'heavenly branches', the sequence of cyclical signs running from *zi* 子 to *hai* 亥. For this reason, western scholars have sometimes called it 'Counter-Jupiter': see Needham and Wang Ling (1959), 402–3.

[46] See chapter 2, footnote 35 on these terms. This is the Xia first month.

[47] On the lodge system, see note 46, and chapter 5, sections 5.2.1 and 5.3. At the time of the compilation of *Huai nan zi*, the star α Pegasi marked the start of the lodge House, which extended a further 16 *du* eastwards up to the beginning of the next lodge, Wall, which began with γ Pegasi. A list of the names of the lodges and their extents is given in *Huai nan hong lie ji jie* 3, 123; see Table 5.1.

[48] This term is the 51st of a cycle of 60 terms used for dating purposes: see chapter 2, section 2.1.3 for further details.

The interest of this passage is that it shows us for the first time that use of the 'quarter remainder' pattern implies other cycles in addition to the 19-year Rule cycle used for intercalations. Let us look at the cycles in turn.

As seen in section 1.3 the Rule cycle *zhang* 章 implies that:

19 solar cycles *sui* = 235 mean lunations *yue* = 6,939 ¾ days:

If (for instance) a mean luni-solar conjunction and winter solstice coincide at midnight at the start of the first month of the first year of a Rule, they will be predicted to coincide again on the first day of the next Rule. But since the Rule is not a whole number of days, this coincidence will not occur until ¾ day after the midnight beginning that day. Since a whole number of solar cycles has elapsed, this conjunction will, however, take place at the same position amongst the stars as before—in this case 'the 5th *du* of the lodge House'.

Next we have the Era cycle *ji* 紀: 4 Rules = 76 solar cycles = 940 mean lunations = 27,759 days.

As we have seen, each Rule cycle reproduces the starting conditions of the preceding Rule, but ¾ day later. After four Rules, this shift has built up to three whole days, and so starting conditions will recur at the original time of day. Unlike the Rule cycle, the Era cycle is a whole number of days, civil months and civil years. The equivalence above therefore refers with equal precision to civil years and civil months, and to solar cycles and lunations, as do the equivalences that follow it.

Finally we have what we may call the Origin cycle *yuan* 元: 20 Era cycles = 1,520 years = 18,800 months = 555,180 days.

Since 555,180 = 9,253 × 60, an Origin cycle is a multiple of 60 days. As mentioned earlier (see also section 2.1.3) a sixty-day cycle was an important part of the ancient Chinese dating system, so this ensures that the Origin begins on the same day of the cycle as when the system started.

All the systems discussed in this book use such cyclical repetitions to simplify their calculations by casting out completed cycles. The three cycles used here—19, 76 and 1,520 years—were also used in a system of the quarter remainder type adopted in 85 CE, although they were there called by different names.[49]

[49] See chapter 6, section 6.2.2. The 19-year period remained a 'Rule' *zhang* 章 as we have called it here, but the 76-year period became an 'Obscuration' *bu* 蔀, 1,520 years became an 'Era' *ji* 紀, and three Eras made up a new period, the 'Origin' *yuan* 元 of 4,560 years designed to include a whole number of 60-year cycles. Since in later practice it was customary to designate years as well as days with a sexagenary cyclical name, this ensured complete repetition of system origin conditions, including the year name.

As we shall see, between 104 BCE and 85 CE a different system was adopted, with different basic constants and consequently different cycle lengths.

1.4 The origin of 'quarter remainder' values

In the *Huai nan zi* we are given a precisely consistent trio of statements of solar cycle length, lunation length and of the equivalence of simple multiples of these quantities in an intercalation cycle. What might be the origin and basis of these data? Faced with such a question, we may start by reviewing what we know of the situation in other parts of the ancient world. Let us begin with the question of the intercalation cycle.

The question of intercalation did not arise for the ancient Egyptian calendar, which had a fixed year length of 365 days. Although it drifted slowly ahead of the seasons, its unchanging length made it the standard time reference for astronomers from ancient times until the last few centuries.[50] Herodotus (c. 484–c. 425 BCE) said that he found this system to be 'a juster one than that of the Greeks', while more recently Neugebauer called it 'the only intelligent calendar which ever existed in human history'.[51] But where luni-solar calendars were used, the problem of intercalation had to be confronted.

For the Near East, John Britton concluded on the basis of excavated cuneiform texts that a regular 19-year intercalation cycle for lunar months was in place from at least as early as 484 BCE. It is, however, possible to trace back some kind of rule-governed intercalation method (though not a 19-year cycle) to the Assyrian astronomical compendium MUL.APIN in the late second millennium BCE.[52] On the length of the solar cycle, a considerable number of estimates are known. The first one that is reasonably close to a modern tropical year value appears to date from the beginning of the sixth century BCE. This value is

[50] Thus in the 16th century Copernicus found it normal to use the Egyptian calendar for finding the number of days between ancient observations and those made in his own day—see Evans (1998), 175.

[51] Herodotus (c. 484–c. 425 BCE) (1920) *Histories, Vol 1: Books I and II [Loeb Classical Library 117]*. New York, II, 4, p. 279; Otto Neugebauer (1969) *The exact sciences in antiquity*. New York; London, Dover Publications; Constable, 81.

[52] J.P. Britton (2007). 'Calendars, Intercalations and Year-Lengths in Mesopotamian Astronomy' in *Calendars and Years: Astronomy and Time in the Ancient Near East*, Oxford, Oxbow Books: 115–132, 119–122; Hermann Hunger and David Edwin Pingree (1989) *MUL.APIN: an astronomical compendium in cuneiform*. Horn, Austria, F. Berger. The capitalized form MUL.APIN indicates the presence of the Sumerian 'logograms' for 'star' and 'plough' in the otherwise syllabic cuneiform script.

expressible in sexagenary notation as 365; 10 days, or 365 $^{10}/_{60}$ in fractional form. An ephemeris written around 560 BCE gives 365;16 days, which is only $^{1}/_{60}$ day longer than 365 ¼ days.[53]

In the Greek-speaking world, we have a full review of the question by Ptolemy of Alexandria c.150 CE in *Almagest* III.1.[54] Much of what Ptolemy writes on this topic is a resumé of work by Hipparchus (c. 190–c.120 BCE), who, he tells us, concluded that the solar cycle did not vary in length, and was $^{1}/_{300}$ day less than 365 ¼ days. Hipparchus, Ptolemy states, derived this result from observations of two summer solstices, one in 280 BCE (observed by Aristarchus) and one observed by himself in 135 BCE. Since these events were 145 years apart, one only had to divide the number of days between these events by the years elapsed to get the desired value. Ptolemy claims to have confirmed this value by observations he made himself. He also reports that Hipparchus mentions lengths of 365 + ¼ + $^{1}/_{76}$ days adopted before his time by Meton c. 432 BCE, and of 365 ¼ days by Kallipos (c. 329 BCE).[55] As for intercalations, a 19-year cycle is said to have been introduced in Athens, probably for astronomical rather than practical calendrical purposes, around 432 BCE. In most modern accounts, the introduction of this cycle is attributed to Meton, following Diodorus Siculus (fl. 65–30 BCE):

> When Apseudes was archon in Athens . . . Meton, the son of Pausanias, who had won fame for his study of the stars, revealed to the public his 19-year cycle, as it is called, the beginning of which he fixed on the 13th day of the Athenian month of Scirophorion. In this number of years the stars accomplish their return to the same place in the heavens and conclude, as it were, the circuit of what may be called a Great Year; consequently it is called by some the Year of Meton. . . . even down to our own day, the larger number of the Greeks use the 19-year cycle and are not cheated of the truth. (Diodorus Siculus (fl. 65–30 BCE) and R. M. Geer (tr.) 1,947: 12, 36, pp. 447–449).[56]

[53] Britton (2007), 126.

[54] *Almagest* III.1 in Toomer (1998), 131–141; see also Evans (1998), 207–209.

[55] See Toomer (1998), 139, also 12. Centuries later, a person of unknown date writing under the name of Thabit Ibn Qurra (836–901 CE), known as 'Pseudo-Thabit' used observations of solstices by Meton and Ptolemy in combination with a summer solstice observation made in 832 to obtain a value of the tropical year. Because of the greater interval between the observations the value obtained was correspondingly more accurate, at close to 365.2426, very close to the modern value: O. Neugebauer (1962) 'Thâbit ben Qurra "On the Solar Year" and "On the Motion of the Eighth Sphere".' *Proceedings of the American Philosophical Society* 106 (3): 264–299.

[56] On Meton, see G. J. Toomer (1974a). 'Meton' in *Dictionary of Scientific Biography*, New York, Scribner's: 337–339.

This 19-year cycle was, however, not quite the same as that specified in *Huai nan zi*, for we are told by Geminus (c. 70 BCE) that the Athenian cycle (which he ascribes to 'the astronomers around Euctemon', without mentioning Meton) consisted of 235 months, made up of 110 months of 29 days, and 124 of 30 days, so that the total length was 6,940 whole days.[57] Thus the value for the solar cycle came out to be:

$$6{,}940/19 \text{ days} = 365 + 5/19 \text{ days} = 365 + 1/4 + 1/76 \text{ days}$$

as stated by Hipparchus, with a implied mean lunation length of:

$$6{,}940/235 \text{ days} = 29\ 500/940 \text{ days.}$$

Geminus next informs us of the 76-year cycle introduced by Kallipos in 330 BCE, which in effect made four 19-year cycles last a total of 27,759 days, one day less than 4 × 6,940 days, implying a solar cycle of 27,759/76 = 365 ¼ days. This 'Kallipic cycle' was therefore identical to the Era cycle of *Huai nan zi*.

Returning to *Huai nan zi*, there is no indication of how its data on the solar and lunar cycle might be justified, or where or by whom they were first put forward, and I know of no earlier Chinese historical source that makes any claims similar to those found in Greek sources. Analysis of early calendrical data in sources such as the *Chun qiu* annals (see note 53) does seem to suggest that the basic scheme of seven intercalations in 19 years embodied in the Rule cycle was used consistently from some time in the early sixth century BCE.[58] As for the length of the solar cycle, an obvious way of finding a value for that would be to look at the interval between two identical events in the cycle separated by a number of years. If one looks at two early winter solstice dates in the *Zuo zhuan*, with dates equivalent to 25 December 656 BCE and 25 December 523 BCE, the interval between the dates given is 48,578 days, implying a mean interval between solstices of:

[57] *Geminos's introduction to the Phenomena: a translation and study of a Hellenistic survey of astronomy.* (2006). Geminos (c. 70 BCE), tr. Evans and Berggren, Princeton, N.J.; Oxford, Princeton University Press, VIII, 50–57, pp. 183–184. Before his account of the 19-year cycle, Geminus refers to earlier patterns of intercalation said to have been used, such as the eight-year *octaeteris*. Geminus is, however, writing long after the events he describes, and it is not clear how far his writing is a useful guide to actual calendrical practice, rather than a retrospective attempt to impose system where there was none. See Hannah (2005), 34–41 and 55–58.

[58] Yabuuti Kiyosi 藪内清 (1969), 278–280.

$^{48,578}/_{133}$ days = 365.248 days to six significant figures, very close to 365 ¼ days.[59]

Were these solstices actually observed on the dates given, or were one or both of them simply calculated using a 365 ¼ day length for the solar cycle? That question naturally suggests that we should examine the means used to observe the principal moments in the solar cycle, such as the summer and winter solstices, and the spring and autumn equinoxes.

In the ancient Greek-speaking world, both solstice and equinox observations are recorded. The oldest datum given in Ptolemy's discussion of year-length in *Almagest* III.1 refers to an observation of the summer solstice by 'the school of Meton and Euktemon' at a date equivalent to 27 June 432 BCE 'at dawn'. (Toomer, G. J., 1998: 138). Although he does not state explicitly that Meton's solstice was determined by observing the shortest noon shadow of a gnomon, his references elsewhere to gnomon shadows at the solstices make it clear that this was the instrument he expected to have been used (Toomer, G. J., 1998: 82–88). However, as Ptolemy remarks at the beginning of the passage just cited, gnomon shadows are not easy to measure, and the end of the long winter solstice shadow is particularly indistinct. Taken with the fact that gnomon shadows change most slowly near their maximum (winter solstice) and minimum (summer solstice) lengths, it will be clear that simply looking for a longest or shortest noon shadow will not be likely to give a very reliable determination of the date of a solstice. Nakayama Shigeru has estimated that a 1 cm uncertainty in the winter solstice shadow of a gnomon 2 m in height could lead to a 4 to 5 day uncertainty in determining the date of winter solstice—with an even worse performance at summer solstice.[60] Further, since a solstice is an instant in time (when the sun's latitude is maximum or minimum), it will not in general fall at the noon when the shadow reaches an extreme. The fact that Meton is recorded as finding the time of the solstice as being at dawn suggests that he was using a technique involving interpolation between noon shadow observations over a number of days.[61]

[59] The two dates are given in the original as Duke Xi year 5, month 1, sexagenary day *xinhai*.48, and Duke Zhao year 20, month 1, sexagenary day *xinhai*.48. See *Zuo zhuan* 12, 18a and 49, 2a in the edition of *Shi san jing zhu shu*, vol. 6, 205–2 and 852–2. The use of the 60-day cycle makes it a simple matter to identify the date of the solstice in the Julian calendar.

[60] Shigeru Nakayama (1969) *A history of Japanese astronomy. Chinese background and Western impact.* Harvard University Press, Cambridge Massachusetts, pp. 242–3.

[61] See Nakayama (1969), 123–4 for this technique.

Modern astronomical software makes it an easy matter to determine to good accuracy the instants of events in the solar cycle at any time in the historic past. Meton's summer solstice in fact occurred around 11:00 (Athens local time) on 28 June, so his determination was about 30 hours too early. The two solstices from the *Zuo zhuan* cited earlier occurred at close to 21:00 (local time in Shandong) on 27 December 656 BCE rather than on 25 December, and close to 05:00 on 27 December 523 BCE rather than on 25 December. Since no time of day is given in the original record, we can only say that the predictions given were in error by about the same order of magnitude as in Meton's determination. Both errors are within Nakayama's theoretical estimate.

Although Ptolemy is familiar with gnomon shadow determinations of solstices, it is clear that he regards the determination of equinoxes as superior in accuracy. He states that both he and his predecessor Hipparchus observed equinoxes by using a ring mounted in the plane of the celestial equator, and lists several of the resulting dates and times.[62] If the ring is perfectly flat and accurately aligned, and the equinox occurs during the hours of daylight, then at the instant the sun passes the equinox the illumination will change from one side of the ring to the other, with an intervening short period when both sides are equally illuminated. But as he notes, a distorted or misaligned ring can easily lead to errors of more than ¼ day. Two particular observations that he claims to have been made 'accurately' by Hipparchus come out quite close to modern predictions: the autumn equinox of 147 BCE, and the spring equinox of 146 BCE.[63] The first is said to have been observed at midnight 26/27 September (and thus must be the result of interpolation), but actually fell 6 hours earlier at Alexandria local time, and the second is said to have been observed at dawn (thus close to 06:00) on 24 March, whereas in fact it fell closer to 15:00.

In early imperial China, however, solstices are the only instants in the solar cycle said to have been found as a result of observation, and the simple gnomon is the instrument used for that purpose. The *Huai nan zi* includes gnomon shadows in its description of the main events at the solstices:

> 日冬至, 井水盛, 盆水溢. 羊脫毛, 麋角解, 鵲始巢. 八尺之修, 日中而景丈三尺. 日夏至而流黃澤, 石精出, 蟬始鳴, 半夏生. 蟁蝱不食駒犢, 鷙鳥不搏黃口. 八尺之景, 脩徑尺五寸. 景脩則陰氣勝, 景短則陽氣勝.

[62] *Almagest* III.1 in Toomer (1998), 132–134.

[63] *Almagest* III.1 in Toomer (1998), 137–138. Note that Toomer gives the two years using the astronomical convention (which has a year zero, equivalent to 1 BCE), as −146 and −145.

> When the sun is at the winter solstice, wells are full of water and basins overflow. Goats shed their hair, deer antlers fall, and magpies begin to nest. [Using a pole of] 8 *chi* length,[64] when the sun culminates [at noon] the shadow will be one *zhang* [= 10 *chi*] and 3 *chi*. When the sun is at the summer solstice, flowing yellow [essence] enriches [the soil], and the essence of stones comes forth, cicadas begin to call, and the 'half-summer' [herb] grows. Flying insects do not bite foals and calves, and birds of prey do not seize nestlings. The shadow of an 8 *chi* [pole] is 1 *chi* 5 *cun* [= 1.5 *chi*] in length. When the shadow is long, Yin *qi* predominates; when the shadow is short, Yang qi predominates. (*Huai nan zi hong lie ji jie*, 3, 98).

The values given here correspond to a latitude of about 34.5°, which is close to that of several ancient Chinese centres in the Yellow River basin. Sima Qian, whose job it was to know about such things, stresses that the gnomon is the decisive means for finding a solstice:

> 冬至短極, 縣土炭, 炭動, 鹿解角, 蘭根出, 泉水躍, 略以知日至, 要決晷景.

> At winter solstice [the daylight] is at its shortest. If you balance earth and charcoal, the charcoal will move;[65] deer antlers fall; orchid roots appear; water rises in the springs—and by [such phenomena] you can tell [the date of] the solstice roughly. But the most important indication is the shadow of the gnomon. (*Shi ji* 27, 1342)

The gnomon, of various lengths and with various refinements, was to remain the key instrument for solstice determination in China for centuries to come.

1.5 Records of astronomical systems under the early empire

In the rest of this book, we shall look at the ways that experts in astronomical systems did their work in early imperial China, how that work related to the structures of imperial power that they helped to support, and how in turn that power shaped the directions taken by their activity.[66] Our task will be facilitated by the fact that the people responsible for managing these systems were very

[64] The *chi* (sometimes rendered as 'foot') of Han times was about 23 cm, 0.8 British imperial feet. It was divided into ten *cun* 寸 'inches' (literally 'thumbs'). So the total length of the gnomon was about 184 cm, comparable with the height of a person.

[65] The significance of this phrase is obscure; it may be that charcoal was thought to be denser at wet seasons compared with dry times of the year, whereas earth changed less.

[66] This brief introductory survey of the main sources used in this book, which is intended purely as background to the following chapters, may be supplemented by the treatment of the topic given at greater length in Morgan (2013), chapter 1.

largely state officials—although, as we shall see, they were not the only ones with expertise in this area. But there is another factor that works in our favour, which is the long Chinese tradition of systematic record keeping and historical writing. Since many historical works were written under imperial patronage, those who wrote them often had access to government archives, from which they were able to copy and condense large amounts of material from reports and memorials on a wide range of topics—amongst which was the topic of *li*. Despite the obvious risks that what we read has been conditioned by the nature of the sources and subsequent editorial bias, this is still an immense advantage to the modern historian who seeks to get as close as possible to the views of the main actors in the events he or she is trying to understand. And as we shall see, even non-official actors with appropriate expertise are not entirely absent from the records that officials compiled.

The first of the major historical sources whose coverage includes part of the early empire was completed about 90 BCE by Sima Qian 司馬遷 (c.145 or 135–86 BCE).[67] Sima Qian was in a uniquely privileged position to write history. It was his father, Sima Tan 司馬談 (c. 165–110 BCE) who had conceived the project of a universal history, dealing with the known world from high antiquity to recent times, and passed it on to his son together with the office of *Tai shi* 太史 'Grand Clerk'[68] at the Han imperial court in the capital, Chang'an 長安 (modern Xi'an 西安). Since the *Tai shi* was, amongst other things, responsible for the custody of state archives, this post gave him unparalleled access to the written records of the past, and Sima Qian's writing therefore reveals to us the view of history formed by a critical, richly informed and reflective mind early on in the period we shall be studying. The book that Sima Qian wrote—the *Shi ji* 史記 'Records of the Clerk/Historian' or 'Historical Records'—was not widely circulated at first, but eventually became the model for a series of histories of

[67] Like most important figures of the Qin and Han period up to around the beginning of the Common Era, Sima Qian has a useful biography in Michael Loewe (2000) *A biographical dictionary of the Qin, former Han and Xin periods (221 BC–AD 24)*. Leiden, Brill. For people who lived after that time, biographies will be found in Rafe de Crespigny (2007) *A biographical dictionary of Later Han to the Three Kingdoms (23–220 AD)*. Leiden, Brill.

[68] My renderings of Chinese official titles are eclectic. Sources on which I have drawn include: Charles O. Hucker (1985) *A dictionary of official titles in imperial China*. Stanford, Stanford University Press; Twitchett, Loewe and Fairbank (1986); Loewe (2000); and De Crespigny (2007). I have preferred 'Grand Clerk' to Hucker's 'Grand Scribe' or 'Grand Astrologer', partly because, like *shi*, 'clerk' in old-fashioned English usage can mean anything from a lowly pen-pusher to important officials such as the Chief Clerks in the British Foreign Office and the early US Department of State, and partly because it reminds me of the evocative usage of 'star-clerk' by Joseph Needham.

successive dynasties (the so-called *zheng shi* 正史 'standard histories') compiled over the next two millennia. After over a thousand years of scribal transmission, the *Shi ji* was first printed around 1035 CE. Fortunately we still have a text of this work that, while it may contain some interpolations, seems to be close to what Sima Qian originally wrote. In a modern collated and punctuated edition, it fills ten paperback volumes in a concise classical Chinese idiom that takes up considerably less space than an English translation.[69] The core of the work is the *ben ji* 本紀 'Basic Annals', twelve chapters that follow the reigns of successive rulers of *tian xia* 天下 'all that is under heaven'. Other divisions of the work include *shi jia* 世家 'Hereditary Houses', recounting the history of great feudal clans, *lie zhuan* 列傳 'Ordered Accounts', which mainly give biographies of individuals, and eight *shu* 書 'Treatises' giving accounts of important topics such as water conservancy and ritual, but also of astronomical and calendrical matters, and the interpretation of heavenly phenomena. There are also *biao* 表 'Tables' which help the reader by (for instance) summarizing the histories of different pre-imperial states in parallel columns.

The *Shi ji* is the first of a series of historical works that have ensured that our knowledge of the early Chinese empire and its successors is much systematic and continuous than our knowledge of most other ancient cultures, such as those of the Mediterranean area. We shall frequently refer to the next two of the standard histories that broadly follow the model of the *Shi ji*, although they differ from it in restricting their coverage to a single historical period. The first of these is the *Han shu* 漢書 literally 'Writings on the Han', whose coverage is limited to the first two centuries of the Han dynasty, usually called the 'Western Han' since its capital was in the west of China at Chang'an (modern Xi'an). The *Han shu* was edited by Ban Gu 班固 (32–92 CE), and was probably completed in its present form c.110 CE. Its coverage of the first century or so of the dynasty overlaps that of Sima Qian, and the relationship between the two texts remains a topic of debate amongst scholars.[70] For the rest of the dynasty, often called the 'Eastern Han' since its capital had moved eastwards to Luoyang 洛陽, our main source will be the *Hou Han shu* 後漢書 'Writings on the Later Han' or possibly 'Later Writings on the Han'. Fan Ye 范曄 (398–445 CE) wrote the main text of

[69] There is a complete translation of the *Shi ji* into French: tr. Chavannes et al. Sima Qian 司馬遷 (c. 145–86 BCE) (2015) *Les mémoires historiques de Se-ma Ts'ien (translation of Shi ji 史記 'Records of the Historian', completed c. 90 BCE).* Paris, You Feng. English language translations of parts of the book by Burton Watson have been widely reprinted, and William H. Nienhauser has already brought out parts of what is planned as a complete translation. See William H. Nienhauser (1994–) *The grand scribe's records.* Bloomington, Indiana University Press.

[70] See for instance Loewe (1993), 130.

this work, but the monographs on subjects such as the heavens and the calendar were taken from work by Sima Biao 司馬彪 (c. 240–c. 306 CE).[71] In addition to the standard histories, we shall also refer in this book to a number of other works which like the standard histories were transmitted at first by scribes and later printed, and to texts excavated by archaeologists in recent decades.

If we look at the series of histories that begin with the *Shi ji*, it will soon become clear that the people who composed these works saw writing about the heavens as falling into two broad categories, often distinguished as *tian wen* 天文 'Heavenly writings/patterns/signs', and *li fa* 曆法 'Methods for astronomical systems'. To understand this division, we may read the words of the father and son team responsible for a great bibliographical project to catalogue the imperial library that was launched in 26 BCE, a date that is centrally located in the period we shall be studying.[72] That catalogue is preserved in abbreviated form as the monograph on literature in the *Han shu*. After giving the titles and number of chapters of twenty-two books under the category *tian wen*, the catalogue explains:

> 天文者, 序二十八宿, 步五星日月, 以紀吉凶之象, 聖王所以參政也.
>
> In *tian wen*, one sets in sequence the 28 lodges, and follows the movement of the five stars [*sc*. 'planets'] and the sun and moon, so as to set in order the phenomena [that show] good or evil fortune, [all of which] the sage ruler uses to calibrate his government. (*Han shu* 30, 1765)

The titles of the books listed include: *Chang Cong on vapours [associated with] the sun, moon, and stars; On clouds and rain—by miscellaneous writers associated with the [deity] Grand Unity; Verifications of prognostications from shooting stars made under the Han*; and *Divination from the direct and retrograde motion of the five stars [by those who dwell in (?)] the midst of the sea*. Clearly the emphasis is on the observation and interpretation of the ominous significance of celestial portents—as we might have expected from the literal meaning of *tian wen* given earlier, with its strong suggestion that events in the heavens convey information.[73] Most of the events discussed are, in modern terms, unpredictable. But,

[71] See B. J. Mansvelt Beck (1990) *The treatises of later Han: their author, sources, contents, and place in Chinese historiography*. Leiden, Brill.

[72] On this project, see chapter 4, section 4.1. The two men who worked on it, Liu Xiang 劉向 and Liu Xin 劉歆, both played important roles in the events discussed in this book.

[73] Compare the title chosen for Francesca Rochberg (2004) *The heavenly writing: divination, horoscopy, and astronomy in Mesopotamian culture*. Cambridge, Cambridge University Press, which refers to a rendering of a Babylonian term. See Rochberg (2004), 64. Sima Qian's monograph on *tian wen*, to which he gives the title *Tian guan* 天官 'The celestial offices' has been translated in Pankenier (2013), 458–511.

as we shall see, some (such as planetary motion, direct and retrograde) were seen as in principle predictable—their divinatory significance largely lay in their departures from predicted behaviour.

We are then given a list of 18 more books, this time classified as *li pu* 歷譜. Of this topic, we are told:

> 歷譜者, 序四時之位, 正分至之節, 會日月五星之辰, 以考寒暑殺生之實. 故聖王必正歷數, 以定三統服色之制, 又以探知五星日月之會. 凶阨之患, 吉隆之喜, 其術皆出焉.
>
> In *li* [and] *pu*, one sets in sequence the placing of the four seasons, corrects the divisions of equinoxes and solstices, and brings together the mark-points for the [positions of] the sun, moon and five stars, in order to examine the reality [behind the timings] of cold and heat, killing and giving life. So the sage ruler must set right the reckoning of *li*, in order to fix the ordinances of the three concordances [here meaning heaven, earth and man] and the colours of ritual dress, and must also investigate and comprehend the conjunctions of the five stars and the sun and moon. The sorrows caused by baleful difficulties, or the happiness of the bestowal of good fortune—the procedures [that produce them] all come from this. (*Han shu* 30, 1767)

The fact that *li* and *pu* are names for two separate things is clear from the book titles that precede this explanation, in which there is no instance of a title containing the phrase *li pu*. Instead we have two titles containing *pu* in which it clearly bears its usual sense of 'listing, tabulation'—in this case, a chronological listing: *Listings of the generations of emperors, kings and feudal lords*; and *Listings of the [regnal] years of emperors and kings since antiquity*. As for *li*, 'astronomical systems' we have titles such as *The system of the Yellow Emperor according to five schools, The systems of Zhuan Xu* and *The systems of the Xia Shang and Zhou [dynasties] and of [the state of] Lu*. From one title, *Reckoning methods for systems and musical harmonics* it is clear that the matters covered by *li* involve calculation, a point reinforced by the last two titles in the list, which are *Xu Shang's calculation procedures* and *Du Zhong's calculation procedures*. The topic of astronomical systems is also associated with quantitative observations of the heavens, as we see from books in the *li pu* section entitled *Geng Chang on the degrees of the moon's motion* (on which see chapter 5, section 5.8) and *Writings on the solar shadow*. From all this, it seems obvious that the techniques falling under *li* 'systems' involve performing calculations that predict the movements of the celestial bodies, and hence the sequence of seasons, days and months, and gathering quantitative data against which calculation may be checked. And although we cannot nowadays check this interpretation against the contents of the books listed in the *Han shu*, since none of them has survived, there are many other extant documents from early imperial times dealing with *li fa* 曆法 'methods for astronomical systems' that enable us to be sure that this interpretation is correct.

In this book, our interest centres on writings that deal with *li fa*, although we shall also find ourselves at times making use of texts that would fall into the other category, *tian wen*. Thus, for instance, records of solar eclipses, which also provided vital reference points and test cases for *li fa*, might be found in treatises on *tian wen*,[74] and as already indicated the motions of the planets have a place in both categories. As is clear from the descriptions above, both categories of writing might be concerned with predictions of 'good or evil fortune'. But now let us leave generalities and look at what the application of *li fa* meant in practice, and how it related to the lives of people in general.

[74] The *Hou Han shu* places solar eclipses in part of its long collection of portents of all kinds gathered under the heading of *Wu xing* 五行 'The Five Phases', rather than in its *tian wen* chapters.

CHAPTER 2

Li in everyday life: dates and calendars

In the preceding chapter we noted the importance of the role of specialists in astronomical systems, *li* 曆, as part of the machinery of the Chinese imperial state, and sketched the kind of mathematical and astronomical structures that they used in their work, as well as reviewing the historical sources through which modern historians can study that work. In this chapter we shall examine more closely the structures and forms of *li* as they were presented to the subjects of the empire, and at the ways in which the mass of the population made use of the data that *li* put before them.

2.1 Looking at a date

First, let us consider the question of dates and dating. The management of an astronomical system involved predicting the dates on which certain astronomical events would occur. In order to understand the way such systems worked, we therefore need to understand the way that dates were defined. As we have seen, both Sima Qian and his father Sima Tan held imperial office that gave them responsibility for the calendar and for recording and interpreting omens of all kinds, especially celestial omens. For each omen a date was an essential part of the record. Here is one example of a particularly important type of dated omen that may well have been entered into the record by Sima Tan, since it fell in the period when he was in office, although the text in which we find it today forms part of the *Han shu*:

元鼎五年四月丁丑晦, 日有食之, 在東井二十三度.

> Yuanding ['Epochal Tripod'] fifth year, fourth month, [day] *dingchou, hui*. The sun was eclipsed [literally: 'there was something that consumed the sun'], in *du* 23 of [the lodge] Eastern Well. (*Han shu* 27b(2), 1503)

This date contains a number of terms—such as 'Epochal Tripod', *dingchou* and *hui*—that demand explanation. Let us consider all the elements of the record in turn. Doing so will give us access to the concerns and skills of early imperial celestial calculators.

2.1.1 Lunar phases

Taking the last element first, we see that the day on which the eclipse falls is designated by the term *hui* 晦. As the monograph on astronomical systems in the *Hou Han shu* explains, this word is one of a series of terms designating the stages of a lunar month during which the moon catches up and overtakes the sun as both move from west to east against the background of the stars:

> 日月相推; 日舒月速, 當其同[所],[1] 謂之合朔. 舒先速後, 近一遠三, 謂之弦. 相與為衡, 分天之中, 謂之望. 以速及舒, 光盡體伏, 謂之晦.

> "The sun and moon give place to one another";[2] the sun is slow and the moon is rapid. When they are in the same position, that is called 'joining at *shuo*'. When the slow one is ahead and the fast one is behind, and they are one [quarter] close and three [quarters] distant, that is called *xian* ['bowstring'].[3] When they are opposite one another, so that they divide heaven in the middle, that is called *wang* ['gazing'].[4] When the fast [moon] is [just] catching up with the slow [sun], so that its brightness is exhausted and its shape invisible, that is called *hui* ['darkening']. (*Hou Han shu, zhi* 3b, 3055)

[1] Here, as in most cases, I reproduce the text modifications proposed by modern editors in the version cited. Square brackets [] indicate an insertion into the text, whereas round brackets () indicate a character that should be deleted. Editorial modifications may be based on a range of evidence, from early quotations that show differences from the received version, to attempts to make sense of a passage that has apparently been garbled in transmission. Where necessary, I discuss such modifications in my text; otherwise the reader may consult the editions cited.

[2] This is a quotation from the cosmological treatise incorporated in the *Book of Change*: see *Yi jing* 8, 9a in *Shi san jing zhu shu*, vol. 1, 169–1. See chapter 4, section 4.3 and 4.4 for further discussion of the significance of this text in connection with *li*.

[3] This is the 'last quarter', when the moon (fast) is three quarters of a revolution past the sun, and still has one quarter to go before it passes the sun (slow) again. 'Bowstring' clearly refers to the appearance of the half-moon, which resembles a strung bow. The same expression is also used for the first quarter.

[4] This is full moon, when the moon is half a circuit of the heavens from the sun, and is hence facing it in 'opposition'.

In a calendar of this kind, naming a day amounts to making an astronomical statement. The two terms commonly used when defining days of the month in the early imperial age are the first day of the month, *shuo* 朔,[5] which should contain the moment of conjunction of sun and moon, when the moon is not visible, and the last day of the month, *hui*, when the moon will at most only show a faint crescent. The next conjunction should occur on the day after *hui*, a day that would in its turn be labelled *shuo*, marking the first day of another lunar month.

An eclipse on a day marked *hui* in the calendar thus seems anomalous to a modern reader—who takes it for granted that a solar eclipse can only happen at a conjunction of the sun and moon. Frequent eclipses on *hui* would suggest that the lunar months are running about a day behind the moon, and so the calendar needs adjusting. However, this situation frequently occurred throughout the Western Han—as can be seen by inspecting the list of Western Han eclipses in *Han shu* 27b2, 1500–1506. On the last page the editors note that out of 53 recorded solar eclipses during this period,[6] only 14 actually fell on the day of *shuo*, while 36 fell a day earlier on *hui*, and three even fell the day before that. There is an implied contrast with the eclipses from an earlier run of records, from the late eighth to the early fifth century BCE,[7] where the *Han shu* editors note that most of the 36 eclipses recorded fell on *shuo*, and only a few on *hui*.[8] Despite this, we have no statements from the Western Han that express any sense that the calendar needs adjustment because eclipses are falling earlier than they should—even in the course of the major reform of the astronomical system that took place in 104 BCE, which we shall review in the next chapter.[9] It is not until the Eastern Han that any reference is made to this factor as a possible motivation for change; in chapter 6 we shall look at possible reasons for this.

[5] The history of this word is complex; amongst other questions, it is not clear whether it may have been used for the first day of the month because it had 'first' as one of its root senses, or vice versa. It can also mean 'north', or 'dark'—the latter sense bearing an obvious relation to the moon when it is in conjunction with the sun. See Axel Schuessler (2007) *ABC etymological dictionary of Old Chinese*. Honolulu, University of Hawai'i Press, 399 for a discussion

[6] The editors appear to have miscounted, since in fact there are 54 eclipses listed, of which 3 fell on the day before *hui*, 37 on *hui* and 14 on *shuo*.

[7] From the *Chun qiu* annals, on which see note 53.

[8] See *Han shu* 27B2, 1479–1500. For a modern discussion of these records, see F. R. Stephenson and K. K. C. Yau (1992) 'Astronomical Records in the Chun-Chiu Chronicle.' *Journal for the History of Astronomy* (23): 31.

[9] In the entry for an eclipse in 188 BCE that fell on the day before *hui*, Jing Fang 京房 (fl. 40 BCE) is quoted in terms that imply that eclipses might normally be expected on either *shuo* or *hui*: *Han shu* 27B2, 1500.

2.1.2 Years and reign titles

Next, let us look at the year given in the date we are considering. In the name Yuanding 元鼎 'Epochal Tripod'[10] we see an early example of the use of *nian hao* 年號, literally 'year titles', specially created terms designating a given period within the reign of a ruler. Until the second century BCE, a year was typically labelled according to its number in the reign of a ruler—so, for instance, under the Qin a given year might be referred to as the 15th year of the first emperor. However, during the first century of the Han the custom began of re-starting the count of regnal years at various points during an emperor's time on the throne in order to commemorate some important event or omen, and to give a suitable title to the period thus begun. The consensus is that the *nian hao* system in the full sense of a new year-count, plus a reign title, began with the reign of Wudi 武帝 (r. 141–87 BCE),[11] although there is discussion as to whether the system began at the start of his reign or later.[12] We do know, however, that Wudi's Yuanshou 元狩 'Epochal Hunt' reign period began in 122 BCE to commemorate the capture of an auspicious animal on an imperial hunt,[13] followed by the Yuanding 元鼎 'Epochal Tripod' period which began in 116 BCE to commemorate the discovery of an ancient tripod—an object which will enter the story of this book at a later stage.[14]

In the Ming 明 (1368–1644) and Qing 清 (1644–1912) dynasties, it became the custom for emperors to adopt a single *nian hao* that lasted their entire reign. For that reason, emperors of those dynasties are frequently referred to by their

[10] To distinguish the Chinese names of reign periods I have taken the decision to represent them in roman rather than italic form with elements concatenated, as is my practice for place-names and two-character elements in the names of persons.

[11] This is the form in which the names of Han emperors will be given. The first element is proper to each emperor, and is a posthumous title, *shi* 謚, that was not actually used during their reign. In this case the title Wu 'martial' probably refers to the great expansion of Han territory during his reign. The second element, *di* 帝, has connotations of divinity, and is common to all Han emperors after the first. It is a little like the Roman 'Augustus', and has hence been rendered by some as 'Thearch'; 'Lord' is also possible. Wudi had a personal name—Liu Che 劉徹, but this was never used to refer to him; the character for the given name of an emperor, in this case Che 徹, would normally not be used for any other purpose for a considerable time after his reign, if not for the rest of a dynasty.

[12] See for instance Li Chongzhi 李崇智 (2001) *Zhong guo li dai nian hao kao* 中國曆代年號考 *(An investigation of reign names in China through the ages)*. Beijing, Zhonghua Shuju, 1–2.

[13] *Shi ji* 12, 460–461.

[14] The problems of providing non-misleading translations for *nian hao* without access to explanations from those who chose them are set out clearly in Mary C. Wright (1958) 'What's in a Reign Name: The Uses of History and Philology.' *The Journal of Asian Studies* 18 (1): 103–106. I give translations of some of the reign titles most obviously relevant to the topic of this book, but do not feel obliged to attempt all of those that have to be mentioned in dating.

reign title, such as Qianlong 乾隆 ('Supernal Prosperity') for the emperor who reigned from 1736 to 1795, as if that was their personal name, which it was not. A less misleading usage in such a case would be 'the Qianlong emperor'. But in the period we are discussing, it was normal for each ruler to proclaim a number of successive reign periods, and so no such confusion arises.

But to return to our example—how do we actually know what year the fifth year of the 'Epochal Tripod' period corresponds to in the BCE/CE system of year numbering nowadays used internationally for historical purposes?[15] It is rarely appreciated what a large demand this is: to make the conversion from scratch, one would have to go through the whole of Chinese history up to the present day, adding the lengths of reigns of successive rulers (allowing for interregnums and overlaps), until some date where a Chinese and a Western scholar were able to say 'My reckoning calls this year X. What does yours make it?' and the two sequences can be matched. Such an encounter actually took place when Jesuit missionaries began work in China during the seventeenth century.[16] By that time, Chinese scholars had already written a considerable number of historical works that set the principal datable events from the distant past up to recent times in a unified sequence of years, often (for reasons that will become clear shortly) grouped into successive 60-year periods. It was thus possible for the Jesuits to translate Chinese dates rapidly and accurately into the scheme most familiar to them.

The result, in 1687, was the publication of a comprehensive chronological table as an appendix to a widely circulated epitome of Confucian thought edited by the Jesuit Philippe Couplet and his colleagues.[17] In this work, the first 60-year cycle begins with the supposed date of accession of the mythical Yellow Emperor, Huang Di 黃帝, in 2697 BCE. The first year of the 44th 60-year cycle of this table was thus 117 BCE, and Wudi is stated to have died in the 31st year of the cycle, which was therefore 87 BCE. In terms of reign periods, the Emperor died in the

[15] I use CE (Common Era) and BCE (Before Common Era) in this book rather than AD (Anno Domini 'Year of the Lord [Jesus Christ]') and BC (Before Christ); this parallels the modern Chinese usage *gong yuan* 公元 and *gong yuan qian* 公元前.

[16] The story is told in detail in Nicolas Standaerdt (2012) 'Jesuit Accounts of Chinese History and Chronology and their Chinese Sources.' *East Asian Science, Technology and Medicine* 35: 11–87.

[17] Philippe Couplet, Prospero Intorcetta, Chrétien Herdtrich and François de Rougemont (1687) *Confucius Sinarum philosophus, sive scientia sinensis latine exposita: Studio & opera Prosperi Intorcetta (. . .) Adjecta est tabula chronologica sinicæ monarchiæ ab hujus exordio ad hæc usque tempora*. Parisiis, Apud D. Horthemels. The results obtained by Couplet and his colleagues are the basis of more recent works, such as Mathias Tchang (1905 (repr. Taipei 1967)) *Synchronismes chinois: chronologie complète et concordance avec l'ère chrétienne de toutes les dates concernant l'histoire de l'Extrême Orient*. Shanghai, Imprimerie de la mission Catholique.

second year of the Houyuan 後元 'Later Origin' period. The first year of that reign period was therefore 88 BCE. The preceding reign period, Zhenghe 征和,[18] had four years, and hence its first year was 92 BCE. Working backwards in this way through reign periods with varying numbers of years, we can eventually establish that the first year of Epochal Tripod was 116 BCE, so that the fifth year of that period was 112 BCE, the sixth year of the current 60-year cycle.

2.1.3 Days—sexagenary and Julian

Leaving the month to one side for the moment, let us go straight to the day given in the example we are considering. Using widely available astronomical software,[19] we find that there was one solar eclipse during 112 BCE that was visible from Chang'an, and it fell on 18 June. At about 10 a.m. local time it was at maximum, not far from totality. In the record, the day of this eclipse is designated by the binome term *dingchou* 丁丑. Here we meet for the first time one of the most common features of pre-modern Chinese dates. A day would be named using a pair of characters, one chosen from a cycle of ten characters taken in sequence (the 'heavenly stems' *tian gan* 天干; see Table 2.1). The other is taken from a cycle of 12 characters (the 'earthly branches' *di zhi* 地支; see Table 2.2).[20]

Table 2.1 ***The ten heavenly stems,* tian gan 天干**

1	2	3	4	5	6	7	8	9	10
甲	乙	丙	丁	戊	己	庚	辛	壬	癸
jia	*yi*	*bing*	*ding*	*wu*	*ji*	*geng*	*xin*	*ren*	*gui*

Table 2.2 ***The twelve earthly branches,* di zhi 地支**

1	2	3	4	5	6	7	8	9	10	11	12
子	丑	寅	卯	辰	巳	午	未	申	酉	戌	亥
zi	*chou*	*yin*	*mao*	*chen*	*si*	*wu*	*wei*	*shen*	*you*	*xu*	*hai*

[18] The significance of this name is unclear: see Li Chongzhi 李崇智 (2001), 3. It may be that the name was originally written as 延和, which might mean 'Prolonged harmony'.

[19] I have mainly used Starry Night Pro™ for this purpose.

[20] A number of attempts, none wholly convincing, have been made to explain the origins in remote antiquity of the sets of characters used in the sequences of stems and branches: see the discussion in Endymion Wilkinson (2000) *Chinese history: a manual. Revised and enlarged.* Cambridge, Mass., Harvard University Press, 175–178. In the period discussed in this book, the stems and branches were mere conventional signs and the topic need not therefore be discussed further here.

The result was a pair of characters, often referred to as *gan zhi* ('stem and branch'), that repeated after 60 days. In English, *gan zhi* pairs are often referred to as 'sexagenary day names' or 'cyclical days'. See Table 2.3, which shows the repetition of the first character of the pair after 10 pairs, whereas the second character repeats after 12 pairs. Both the stem and the branch linked with a day might be used separately for divinatory purposes.

The name of the eclipse day, *dingchou*, falls as number 14 in this cycle. And it is not difficult to show that the eclipse did indeed fall on a *dingchou*.14 day—see Box 2.1 for an explanation of how this can be done, using the example of the eclipse of 112 BCE.

The system of sexagenary day numbers was used as far back as we can trace any evidence of dating in China. The earliest examples are found on inscriptions marked on animal bones used for divination by mantic specialists of the Shang

Table 2.3 ***The sexagenary cycle. When referring to sexagenary day names, I normally combine the name with its number in the sequence, as for instance*** **jimao.*16***

甲子	*jiazi*	1	甲申	*jiashen*	21	甲辰	*jiachen*	41
乙丑	*yichou*	2	乙酉	*yiyou*	22	乙巳	*yisi*	42
丙寅	*bingyin*	3	丙戌	*bingxu*	23	丙午	*bingwu*	43
丁卯	*dingmao*	4	丁亥	*dinghai*	24	丁未	*dingwei*	44
戊辰	*wuchen*	5	戊子	*wuzi*	25	戊申	*wushen*	45
己巳	*jisi*	6	己丑	*jichou*	26	己酉	*jiyou*	46
庚午	*gengwu*	7	庚寅	*gengyin*	27	庚戌	*gengxu*	47
辛未	*xinwei*	8	辛卯	*xinmao*	28	辛亥	*xinhai*	48
壬申	*renshen*	9	壬辰	*renchen*	29	壬子	*renzi*	49
癸酉	*guiyou*	10	癸巳	*guisi*	30	癸丑	*guichou*	50
甲戌	*jiaxu*	11	甲午	*jiawu*	31	甲寅	*jiayin*	51
乙亥	*yihai*	12	乙未	*yiwei*	32	乙卯	*yimao*	52
丙子	*bingzi*	13	丙申	*bingshen*	33	丙辰	*bingchen*	53
丁丑	*dingchou*	14	丁酉	*dingyou*	34	丁巳	*dingsi*	54
戊寅	*wuyin*	15	戊戌	*wuxu*	35	戊午	*wuwu*	55
己卯	*jimao*	16	己亥	*jihai*	36	己未	*jiwei*	56
庚辰	*gengchen*	17	庚子	*gengzi*	37	庚申	*gengshen*	57
辛巳	*xinsi*	18	辛丑	*xinchou*	38	辛酉	*xinyou*	58
壬午	*renwu*	19	壬寅	*renyin*	39	壬戌	*renxu*	59
癸未	*guiwei*	20	癸卯	*guimao*	40	癸亥	*guihai*	60

Box 2.1: Finding the sexagenary day number of the eclipse of 18 June 112 BCE

I am writing this on 13 May 2013. According to the traditional Chinese almanac for this year, today has the sexagenary name *jimao*.16.

So if the sexagenary day count used today is consistent with the count used in the Western Han period, we should find that the number of days up to today from the date of the eclipse on 18 June 112 BCE, which we believe was day *dingchou*.14, should be a large number of complete 60-day cycles, plus 2 days – enough to get us from day 14 to day 16. The calculation is simple in theory, but it is not obvious how all those days are to be counted up, taking account of leap years and calendar reforms. Fortunately, astronomers have a way of avoiding such problems, and that is the system of Julian Dates (JD). This is a simple way of giving a number to every instant in past or future, and it works as follows (Smart, W. M. and Green, Robin M. 1979 (reprint of 6th edition 1977): 146).

Noon at Greenwich (on the zero meridian of longitude) on 1 January 4713 BCE is defined as having Julian Date 0. This instant is chosen because it marks the starting point of a number of important cycles in western traditions of calendrical reckoning. Noon on 2 January thus was designated JD 1, and noon on 31 December was JD −1. Midnight at the end of 1 January would be JD 0.5, and JD 0.6 would be 2.4 hours into 2 January, i.e. 2:24 a.m., all at Greenwich time. JD at any instant might once have been found by using tables in an astronomical almanac, but can nowadays be found directly by the use of astronomical software such as Starry Night Pro™.

Using such software, we find that the date and time of the eclipse at 10 a.m. Chang'an local time 18 June 112 BCE correspond to Julian Date 16,80,683.619. We also find that 10 a.m. Chang'an local time 13 May 2013 has Julian Date 2,456,425.619. That makes the number of days elapsed between the two events:

$$2{,}456{,}425.619 - 1{,}680{,}683.619 = 775{,}742,$$
$$\text{and } 775{,}742 = 12{,}929 \times 60 + 2$$

So the remainder of 2 days in excess of complete 60-day cycles is just as required to get from *dingchou*.14 in June 112 BCE to *jimao*.16 in May 2016 CE, and we have thus verified (at least in the given instance) that the sexagenary day numbering system used today is consistent with that used over two thousand years ago.

continued

Box 2.1: *Continued*

Julian Dates, and differences in local time

In deciding on what day an astronomical event occurred, we may need to take account of differences in local time between Greenwich and sites in China. Local time at Chang'an is about 7.2 hours ahead of Greenwich, so the day designated as *dingchou*.14 began and ended 7.2 hours before the corresponding times at Greenwich. In the example above, the time that the solar eclipse occurs ensures that it falls on the same day of the month at both Chang'an and Greenwich, though at different times of day. An event timed around 6 a.m. at Chang'an would, however, be timed at around 10:48 p.m. the previous day at Greenwich.

dynasty in the late second millennium BCE, who worked by interpreting cracks on the bones made by applying a hot iron point. One typical example reads, transcribed into modern characters:

> 癸亥卜. 賓貞: 旬亡咎.
>
> Crack-making [on day] *guihai*.60. Bin divined: 'In [the next] ten days, [there will be] no disaster.' (Wen Shaofeng 温少峰 and Yuan Tingdong 袁庭栋 1983: 87)

In such a case, all we can do is to note the use of the sexagenary cycle, without being able to correlate the day in question with any dating system now in use. But by the time we reach the early Eastern Zhou it becomes possible to make such correlations with a high degree of assurance, because they refer to datable astronomical events, including 37 records of solar eclipses from 720 BCE onwards.[21] Take, for instance, this record from the eighth regnal year of Duke Xuan 宣 of Lu 魯, as given in the *Chun qiu* 春秋 'Spring and Autumn' annals, one of Sima Qian's major sources for the early Eastern Zhou period:

> 八年 [. . .] 秋. 七月 甲子. 日有食之: 既.

[21] These eclipses have been intensively studied for the last two thousand years—early on by imperial astronomers testing the power of their systems to retrodict eclipses centuries before, and most recently by modern astrophysicists studying dynamic changes in the Sun–Earth–Moon system. See for instance F. R. Stephenson and L. V. Morrison (1995) 'Long-Term Fluctuations in the Earth's Rotation: 700 BC to AD 1990.' *Philosophical Transactions of the Royal Society of London. Series A: Physical and Engineering Sciences* 351 (1695): 165–202.

> Eighth year [. . .] Autumn. Seventh month.[22] [Day] *jiazi*.1. The sun was eclipsed [lit. 'consumed']: complete. (*Chun qiu zuo zhuan*, Duke Xuan year 8: 22, 7a in (Ruan Yuan 阮元 (1764–1849) 1973 reprint of original of 1815: vol. 6, 379–1))

Using the correlations between Chinese and European chronology first established by Couplet and his Jesuit colleagues in the 17th century (see footnote 125), we find that Duke Xuan's eighth year was 601 BCE. Was a total solar eclipse actually visible that year at the capital of the state of Lu, Qufu 曲阜, where these annals were recorded? Using the same software as before, we can indeed find an eclipse that would have been very close to total in that city, and took place on 20 September 601 BCE, with eclipse maximum falling close to 3.30 p.m. local time. Even if the sky had not been clear, the resulting period of darkness in the middle of the day would have been very striking. And using the technique explained in Box 2.1, we can verify that the eclipse indeed took place on day *jiazi*.1 in a sexagenary sequence continuous with the sequence used in the eclipse date recorded by Sima Tan in 112 BCE, and still in use today. It seems likely, therefore, that the sexagenary cycle of days has operated unbroken since early in the first millennium BCE, and possibly earlier.[23] The continuity of the sexagenary cycle invites comparison with the use of the seven-day week at the other end of Eurasia. The fact that several different cultures used such short periods for dating—and not always seven days long—makes the question more complex than it is in the place and time that concerns us in this book, though it is certainly difficult to imagine those to whom a seven-day week was of religious importance allowing themselves to lose count.[24]

Before leaving the subject of the sexagenary cycle in dating, we may note that although the earliest records use it only for days, from the end of the Western Han period we begin to find evidence of its increasingly frequent use to designate years within a 60-year cycle. The roots of such a sexagenary year count do, however, appear to go back considerably earlier. Thus in one of the documents recovered from a tomb at Mawangdui 馬王堆 closed in 168 BCE,

[22] The month numbering system used here is the so-called Zhou system, which is two months in advance of the Xia system used from early imperial times onwards: see note 35. The division of the seasons is conventional, with 'Spring' occupying the first three months of the year, so that 'Autumn' falls in the seventh, eighth and ninth months.

[23] Edward L. Shaughnessy (1991) *Sources of Western Zhou history: inscribed bronze vessels*. Berkeley; Oxford, University of California Press, 136.

[24] For a convenient review, see Leofranc Holford-Strevens (2005) *The history of time: a very short introduction*. Oxford, Oxford University Press, 64–79. See also Evans, James (1998). *The history and practice of ancient astronomy*. New York/Oxford, Oxford University Press., 165–6, and more generally on the wider context of chronology in ancient Europe see E. J. Bickerman (1980) *Chronology of the ancient world*. London, Thames and Hudson.

there is a table showing the complete cycle of sixty *gan zhi* names, beginning with *jiayin*.51. Three of the names are marked as associated with regnal years, beginning with the first year of the king of Qin who was to become the first emperor, corresponding to 246 BCE, which is marked on the table under *yimao*.52. This (like the other years marked) corresponds precisely with the result obtained by following the first clear description of how sexagenary year names may be calculated, as given in the astronomical system promulgated in 85 CE: see *Hou Han shu, zhi* 3b, 3060. It does seem, therefore, that the practice of associating years with sexagenary names may have a longer history than some have thought.[25] For convenience of reckoning in historical dating, this system was often projected back into the remote past, long before it was actually in use—hence the 60-year cycles in the sources used by Couplet.[26]

Finally, let us consider the number of the month in which the eclipse falls—if *hui* (the last day) of the fourth month falls on *dingchou*.14, then the fifth month must begin next day, on *wuyin*.15, which is a day of *shuo* and would be 19 June. Counting back four lunar months' worth of days (118 days, reckoning two months of 29 days and two of 30 days) should take us back to the first day of the first month, which turns out to be 21 February.[27] Since that is close to the time when 'Chinese New Year' commonly falls nowadays, it appears that this first lunar month as understood by Sima Tan was similarly defined—and this is indeed the case. This numbering system follows the so-called Xia 夏 [dynasty] count, used for month numbers in the Qin and Han despite the fact that under Qin and for the first century of Western Han the civil year actually began in the month numbered tenth in this count; from 104 BCE onwards the civil year began with the first Xia month, a practice continued by the traditional calendar in modern Chinese almanacs.[28] The relation of the Xia month-count to other systems is discussed below.

[25] See Marc Kalinowski and Phyllis Brooks (1999) 'The Xingde 刑德 texts from Mawangdui.' *Early China* 23/24: 125–202, 135–136 and 145–154. I am grateful to Marc Kalinowski for reminding me of this material. As he points out, the purpose served in the Mawangdui table by the sexagenary terms attached to years was 'not chronological but hemerological' in that it served mainly to show the annual shift of *Tai yin* 太陰, the entity sometimes known as 'counter-Jupiter'.

[26] Using the system described in *Hou Han shu*, the year of accession of the Yellow Emperor, 2697 BCE in Couplet's table, turns out to be labelled *jiazi*.1, which is consistent with it being the starting point for subsequent 60-year cycles.

[27] Readers who wonder whether such calculations do in fact produce historically accurate results may be reassured that they match the data given for the year in question in such a frequently consulted tabulation as Zhang Peiyu 張培瑜 (1990) *San qian wu bai nian liri tianxiang* 三千五百年历日天象 *(An astronomical calendar for 3,500 years)* Henan, Henan jiaoyu chubanshe, 77. In fact such calculations, as well as dates recorded in historical sources, form an important element in the construction of such tabulations as Zhang's.

[28] See section 2.2.1.

2.2 Calendars in early imperial China: from the ground up

2.2.1 Looking at a calendar

For an archaeologist, China is almost an embarrassment. Its soil is so rich in antiquities that the problem is not so much where to find an interesting site to excavate, but rather to decide between the huge number of interesting sites known to exist. For scholars who learn about the past by reading texts, the situation is already difficult enough because of the oceanic quantity of Chinese literature known to us through unbroken textual transmission. For the most ancient periods it was once possible for scholars to enjoy the assurance of having read everything there is left to read. But in recent decades even that limited prop to confidence has been removed, as archaeologists (followed increasingly by commercially motivated tomb-robbers)[29] have uncovered masses of texts in tombs dating from the early imperial age. Most are written in lamp-black ink on bamboo strips, a combination that is particularly resistant to wet conditions that would have destroyed other materials.[30] Some of those texts are early versions of writings still current today, but some represent the work of entirely unknown writers or schools of thought. Additionally, we have many texts that served everyday purposes for ordinary people or low-level officials, the kind of people whose life and thoughts appear only peripherally in the literature preserved and transmitted by the élite scholars who have until recently conditioned our access to China's cultural past.

[29] Some scholars distinguish between texts whose origins lie in archaeological excavation, *fa jue* 發掘, and those 'brought to light' or 'discovered', *fa xian* 發現, which latter term effectively functions as a euphemism for 'stolen'. Typically these 'discovered' documents are transported clandestinely to Hong Kong, where they are purchased for considerable sums by persons who then present them to universities in China of which they are in some cases alumni. As a result, the tomb-robbing industry is rewarded for its criminal activity, which can hardly take place without the connivance of local authorities and of people with at least enough expertise to advise the robbers where to dig next in their search for further valuable finds. On the implications of this for scholarship (including the ever-present risk of fake documents being passed off as genuine), and on the ethical dimensions of using such documents in scholarly research, see Paul R. Goldin (2013) 'Heng Xian and the Problem of Studying Looted Artifacts.' *Dao* 12 (2): 153–160.

[30] On the evolution of writing materials in China, see Tsuen-hsuin Tsien (2004) *Written on bamboo & silk: the beginnings of Chinese books & inscriptions*. Chicago, London; University of Chicago Press. Paper did not come into use as a writing material in China until the beginning of the first millennium CE. Before that time, most common documents were written on bamboo strips (which generally bore a single column of characters each) or on planches of wood, some of which might bear an entire short text. Where cost limitations were less important, texts might be written on a roll of silk.

It would have been hard to live in a society like early imperial China without at least sometimes feeling the need to know what day it was—in the sense of locating the day in question in relation to the start of the current month, and knowing its position in the 60-day cycle. Days do not naturally come with any kind of label attached: in order to put a name to a day human beings need something that corresponds to what 21st-century English speakers call a 'calendar'—a document or device that enables one to give a different label to each succeeding day in a systematic and unique way. In recent decades an increasing number of documents have been excavated from sites of the early imperial age of ancient China that seem to fulfil a function close to that of a modern calendar.[31]

I refer deliberately to a resemblance of function rather than of form, since a number of very different formats of document are known, all of which would have enabled their users to answer the basic question 'what day is it?'[32] Further, I use the English word 'calendar' rather than a Chinese term, despite the risks of importing associated alien concepts into the Chinese past, simply because we are not at all sure what people normally called these documents in the early imperial period. One modern Chinese scholar has adopted a similar stratagem by using the term *li ri* 暦日, literally 'sequenced days' or 'calendrical days', which is well attested in the post-Han period and is used at least once in the period that concerns us.[33]

One common type of calendrical document does, however, follow a pattern that will be easily recognizable to a modern reader once its content is understood. Here in Figure 2.1 is an example, which took a form frequently found in early imperial China—a series of columns of writing, each on its own bamboo strip, with the strips bound together in sequence with string to form a bundle that could be rolled up for storage. The example here has been redrawn using modern-style characters from the original somewhat damaged bamboo slips on which it was written over two thousand years ago.

First, let us look at the strip on the extreme right-hand side, which would have been the first to be read. It bears the four characters *qi nian shi ri* 七年視日 'conspectus of days for the seventh year'.[34] The year in question, corresponding

[31] The reader in search of a more detailed typology may consult Morgan (2013) 184–214.

[32] See the typology set out in Liu Lexian 刘乐贤 (2011) 'Qin Han *li* ri de neirong ji gongyong 秦汉暦日的内容及功用 (On the contents and uses of day-sequences of the Qin and Han).' *Faguo Hanxue* 法國漢學 *'French sinology'* 14: 351–386, 4–6.

[33] See Liu Lexian 刘乐贤 (2011), 6–9.

[34] It may be that the character shown here as *shi* 視 'conspectus' should be read as *zhi* 質, perhaps meaning 'evaluation' in this context. See note 44.

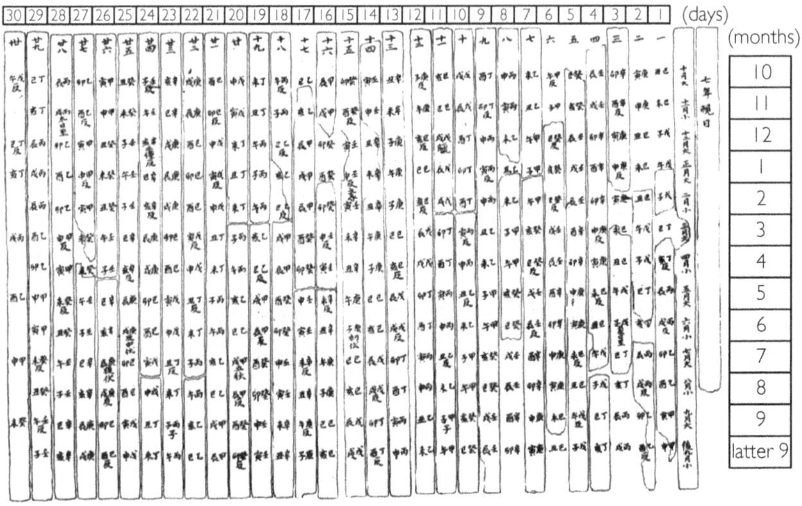

Figure 2.1 Calendar for 134 BCE, from Liu Lexian (2011) figs 1–1 and 1–2; Arabic numbering for days and months has been added.
Reproduced by permission of the editors of *Faguo Hanxue*.

to 134 BCE, followed the sixth year of the Jianyuan 建元 'Establishment epoch' reign period of Wudi, which had begun in 140 BCE, so it was reasonable at the time the document was prepared (presumably near the end of the sixth year) to anticipate that it would be year number seven. But in fact a new reign period, Yuanguang 元光 'Epochal brilliance', was begun that year, apparently in commemoration of the appearance of a comet (*Shi ji* 12, 460), so there was no seventh year after all.

As noted in chapter 1, the term *nian* rendered here as 'year' refers to the civil year, the long-term unit of civil time measurement containing a whole number of months, each containing a whole number of days. The next strip has a series of 13 annotations beginning with *shi yue da* 十月大 'tenth month, large', followed by *shi yi yue xiao* 十一月小 '11th month, small' and *shi er yue da* 十二月大 '12th month, large'. After that, the month-count starts again with the first month (marked *zheng yue* 正月 'standard month') and continues on downwards. It seems therefore that for users of this calendar the year began with the month numbered 10. This is consistent with the pattern we know to have been introduced at the time of the Qin unification in 221 BCE, as part of which, we are told:

改年始, 朝賀皆自十月朔.

They changed the beginning of the year, so that the court ceremonies [for New Year] were held on the first day of the tenth month. (*Shi ji*, 6, 237)

The numbering system for months used here is based on what is commonly called the *Xia zheng* 夏正 'Xia [dynasty] standard', that is the *zheng yue* 正月 'standard month' (effectively 'first month') of the Xia system of month numbers. This is, roughly speaking, the lunar month in which the season of spring is said to begin. A more precise definition will be possible shortly, when we have reviewed the system of seasonal markers known as the 24 *qi* 氣: see the end of section 2.2.1.2. In early imperial China the Xia month-count was the normal reference way of naming months, even when, as in this case, the Xia standard month was not actually the first month of the civil year.[35] After 104 BCE the situation was simplified when the Xia first month was made the first month of the civil year, and it remains the month of 'Chinese New Year' in traditional almanacs published today. A related expression is *zheng shuo* 正朔 'the standard conjunction', which may perhaps be regarded as a contraction of *zheng yue shuo* 正月朔 'conjunction of the standard month', referring to the first day of the lunar month beginning the civil year, on which a luni-solar conjunction would be expected to occur.

Next we may notice the numbers at the top of the strips, which run from 1 to 30. Below those are thirteen rows, consisting of either 29 or 30 sexagenary day names. It may easily be verified that these run in a continuous sequences from row to row, so that (for instance) while the first row ends with 戊午 *wuwu*.55 on the left, the second row begins on the right with 己未 *jiwei*.56.[36] What we have here is in fact a list of the day-names for a 13-month year, whose months mostly alternate between 29 and 30 days, those with 29 days being marked as *xiao* 小 'small', and those with 30 days being marked as *da* 大 'large'. In English, such months are normally called 'short' and 'long'. Let us consider these facts in turn.

[35] According to Sima Qian, 夏正以正月, 殷正以十二月, 周正以十一月. 蓋三王之正若循環, 窮則反本. 'If we take the Xia standard as the standard month, then the Yin standard is the 12th month, and the Zhou standard is the 11th month. So the standards of the three royal lines are like a cycle, which returns to the start when it is completed' (*Shi ji* 26, 1258). The Zhou, Yin and Xia 'standards' were also designated as Celestial *tian* 天, Terrestrial *di* 地, and Anthropic *ren* 人, or labelled with the cyclical characters *zi*.1 子, *chou*.2 丑 and *yin*.3 寅, these being understood in terms of an idealized scheme of the different directions indicated by the tail of the Northern Dipper at dusk throughout the year, as shown on the divinatory devices known as *shi* 栻 '[cosmic] models', 'cosmographs' or 'cosmic boards' (see Cullen (1981)). In imperial times, technical astronomical calculations were frequently based on the Celestial standard *tian zheng* 天正, so that the first month of the count was the Xia 11th month.

[36] In the text of this chapter, sexagenary day-names are written left to right, with the stem followed by the branch. On the calendar we are considering the normal ancient right to left order is used.

2.2.1.1 MONTH AND YEAR LENGTHS

In general, the months on this calendar alternate between short months of 29 days (a 'small month' *xiao yue* 小月) and long months of 30 days (a 'large month' *da yue* 大月), but the second and third months are a pair of successive long months ('joined large months' *lian da yue* 連大月). This is a straightforward consequence of the length of the lunation cited in chapter 1 from the *Huai nan zi* book, which was offered to the throne five years before the year to which this calendar relates. The value there given was 29 $^{499}/_{940}$ days, which is 29.53 days to four significant figures. This interval is a little longer than the average length of 29.5 days that would result from a strict alternation of short and long months, and so it is clear that there should indeed be 'joined large months' from time to time in order to keep months in step with the lunation, as we see in this example with the second and third months.[37]

The calendar has two further features that need more explanation. Firstly, it has 13 months rather than 12. Secondly, the last month in the sequence, which follows the month shown as *jiu yue* 九月 '9th month' is marked as *hou jiu yue* 後九月 'latter ninth month'. What is going on here? Recalling the information given by *Huai nan zi* and cited in chapter 1, a year made up of 12 lunar months will be just under 11 days short of the solar cycle, at which the seasons repeat. As a result, after three years the deficit will have built up to over a month, and, as mentioned in chapter 1, it will be necessary to pause for a month (an intercalary month, *run yue* 閏月) before starting the month-count of the next year. That is evidently the function of the 'latter ninth month' on this calendar.[38]

2.2.1.2 MARKING THE SEASONS

This calendar has four notes showing the passage of the seasons. They are:

'Winter solstice' *dong ri zhi* 冬日至: 11th month, 28th day *bingxu*.23
'Establishment of spring' *li chun* 立春: first month, 15th day *renshen*.9
'Summer solstice' *xia ri zhi* 夏日至: sixth month, 3rd day *wuzi*.25
'Establishment of autumn' *liqiu* 立秋: seventh month, 20th day *jiaxu*.11

[37] Since 29.53 days – 29.5 days = 0.03 day ≈ $^1/_{33}$ day, we should expect 'joined large months' about every 33 months in order to maintain months in step with the phases of the moon.

[38] After 104 BCE, when the beginning of the year was shifted to the first Xia month, it also became the practice to insert an intercalary month after any month of the year, so that, for instance, an intercalary month after the fifth month would be labelled *run wu yue* 閏五月 'intercalary fifth month'. The reasons for choosing a particular month for an intercalation will be discussed later: see section 4.4.5.3.

The term *ri zhi* 日至 rendered here as 'solstice' (from Latin *solstitium* 'standing still of the sun') means literally 'extreme [position of] the sun'. More commonly we find the abbreviated forms *dong zhi* 冬至 and *xia zhi* 夏至, literally 'winter/summer extremes'. Other calendars may show up to eight seasonal markers (the 'eight nodes' *ba jie* 八節), the full set being:

冬至 *Dong zhi* 'Winter solstice'
立春 *Li chun* 'Establishment of spring'
春分 *Chun fen* 'Division of spring (or 'spring equinox')
立夏 *Li xia* 'Establishment of summer'
夏至 *Xia zhi* 'Summer solstice'
立秋 *Li qiu* 'Establishment of autumn'
秋分 *Qiu fen* 'Division of autumn' (or 'autumn equinox')
立冬 *Li dong* 'Establishment of winter'

The terms for the equinoxes, *fen* 分 are, like those for the solstice, abbreviations. Earlier texts refer to *ri ye fen* 日夜分 'The [equal] division between day and night'.[39]

In the early imperial age, all such systems for marking the passage of the seasons were based on the division of the period between winter solstices into equal intervals. Thus, counting from the day of winter solstice to the day of summer solstice on the calendar shown above is an inclusive interval of 183 days from the start of the first day to the end of the last, which allows for the exact interval between the two solstices to be $365\frac{1}{4}$ days $\div 2 = (182\frac{5}{8})$ days. From winter solstice to establishment of spring inclusive is 47 days from the start of the first day to the end of the last, which is consistent with the exact interval between the two markers being $365\frac{1}{4} \div 8 = (45\frac{21}{32})$ days. While this set of eight is the largest that is actually found on calendars, from the second century BCE onwards some texts give a set of 24 seasonal markers, collectively known as the 24 *qi* 氣, of which the *ba jie* are a subset. Each of this set of markers is thus $356\frac{1}{4} \div 24 = 15\frac{7}{32}$ days from the preceding marker. In the early imperial period the full set of 24 is first found in *Huai nan zi*, and is later found in technical contexts, but does not appear on any excavated calendar. The full list is given in Table 2.4 for convenience.

Because of the steady shift of the lunar months relative to the seasons, markers of the kind described here above would fall on different days of the month in

[39] See for instance *Lü shi chun qiu* 呂氏春秋 (*Mr. Lü's Spring and Autumn [Annals]*). (completed 239 BCE, 1922 reprint of Ming woodblock edition,). Lü Buwei 呂不韋 (?–235 BCE), Shanghai, Commercial Press, Si bu cong kan 四部叢刊 collection, 420–424, chapters 2 and 8.

Table 2.4 ***The 24 qi* 氣**

1	冬至	Winter solstice	13	夏至	Summer solstice
2	小寒	Lesser cold	14	小暑	Lesser heat
3	大寒	Great cold	15	大暑	Great heat
4	立春	Establishment of spring	16	立秋	Establishment of autumn
5	雨水	Rain waters	17	處暑	Enduring heat
6	驚蟄	Awakened insects	18	白露	White dew
7	春分	Spring equinox	19	秋分	Autumn equinox
8	清明	Clear and bright	20	寒露	Cold dew
9	穀雨	Grain rains	21	霜降	Frost fall
10	立夏	Establishment of summer	22	立冬	Establishment of winter
11	小滿	Lesser filling	23	小雪	Lesser snow
12	芒種	Grain in ear	24	大雪	Great snow

different years. A properly run system of intercalation would, however, prevent them straying too far from their expected positions. Thus, for instance, the fifth *qi*, Rain Waters, should always fall somewhere within the first Xia month.[40]

2.2.1.3 HEMEROLOGICAL MARKINGS

A calendar is useful because it gives systematic information on when important mark-points in time will be passed. We have noted the presence on this specimen of mark-points that show when certain astronomical events will occur—conjunctions of the sun and moon, and solstices. There were, however, other mark-points that could be highly important for planning human activities, although these refer to events that could not be determined by astronomical observation.

The example before us gives information of two types, the first of which is the position of 'Fan Zhi 反支 days'. These were days when the 'calendar spirit' *shen sha* 神煞 Fan Zhi ('reversed branch') was active, and were in principle ominous. Examples of these on this calendar (marked simply as *fan* 反) are found in the 10th month, days 6, 12, 18 and 24. The six-day intervals result in these

[40] In fact, the system of intercalation used from 104 BCE onwards set out to guarantee that 12 of the full set of 24 seasonal markers would always be found in a given month corresponding to each. These were the odd-numbered *qi* in the list, known as *zhong qi* 中氣 'medial qi', the even-numbered *qi* being distinguished as *jie qi* 節氣 'nodal *qi*' (note, however, the confusing fact that sometimes authors refer to the complete set of 24 *qi* as *jie qi*). Since the interval between two medial *qi* is $365\frac{1}{4} \div 12 = 30\frac{14}{32}$ day, it will eventually be the case that a month does not contain a medial *qi*. This month will be designated as an intercalary month. For more detail, see section 4.4.5.3.

days having one of two earthly branches in their day-names: *wu* 午 and *zi* 子. Each different month has a different pair of branches, all marked as *fan* days. According to a listing in one excavated divinatory handbook of the 'day-book' type that we shall discuss shortly,[41] a month whose first day has the branch *zi*.1 has Fan Zhi days on the days with branches *si*.6 and *hai*.12; a month whose first day has the branch *chou*.2 has Fan Zhi days on the days with branches *wu*.7 and *zi*.1 (as is the case with the 10th month of our example) and so on down the list.[42]

What it could mean in practice when such a day turned up is illustrated by a story about Mingdi (r. 57–75 CE) retold by Wang Fu 王符 (83–170 CE):

> 孝明皇帝嘗問:「今旦何得無上書者?」左右對曰:「反支故.」帝曰:「民既廢農遠來詣闕, 而復使避反支, 是則又奪其日而冤之也.」乃敕公車受章, 無避反支.

> Mingdi once asked [his ministers]: 'At the dawn [audience] today, how is it that there were no petitions submitted?' The ministers answered 'Because of Fan Zhi'. The emperor said 'When the people have abandoned their agricultural work and come to the capital from afar, and on top of that the officials 'avoid Fan Zhi', this amounts to cheating [the people] by stealing a day from them''. Thereupon he decreed that the bureau of complaints should receive documents, and the avoidance of Fan Zhi ceased. (*Qian fu lun* 潛夫論 4,17b (391))

Apart from Fan Zhi days, this calendar marks a few other days:

> 12th month, 11th day *wuxu*.35: La 臘 '[Sacrifice of] Seasoned Meat'
> 12th month, 24th day *xinhai*.48: Chu Zhong 出種 'Putting forth the seed'
> sixth month, 15th day *gengzi*.37: Chu Fu 初伏 'Initial [summer] Retreat'
> sixth month, 25th day *gengxu*.47: Zhong Fu 中伏 'Middle [summer] Retreat'
> seventh month, 26th day *gengchen*.17: Hou Fu 後伏 'Latter [summer] Retreat'

The La sacrifice and its associated Fu days are well known from ancient sources, but there seems no clear agreement when they are to fall. In this case La is marked on the second day with the branch 戌 *xu*.11 following winter solstice, while the three Fu are on the first, second and sixth days with the stem 庚 *geng*.7 following summer solstice. Other sources refer to the third *xu* day after winter solstice for La, and yet other *geng* days for Fu.[43]

[41] Harper and Kalinowski, Eds. (2017, forthcoming), 108.

[42] In the ninth month, the expected *fan* annotation on days 11 *jiazi*.1 and 23 *bingzi*.13 have been replaced by *zi* 子. The reason for this is not clear.

[43] See Liu Lexian 刘乐贤 (2011), 11.

2.2.2 How people used calendars

What did people do with documents such as our bundle of bamboo strips when they had them? In what ways was a calendar useful?

In the first place, calendars were an essential tool for planning and co-ordinating tasks. That is an obvious enough purpose: today we might agree that the budget review meeting for our organization will take place on a date such as Friday 31 May, and everybody can plan to prepare the documents for which they are responsible and be there on time. In ancient China the date would have taken a different form such as 'cyclical day *jiazi*.1, fifth month 17th day', but the effectiveness of the system as a time-planning tool is the same.

In some cases such documents were evidently used to record work already carried out. Another calendrical document of the strip bundle type has entries such as the following:

> 27th year [220 BCE], fourth month:
> Fifth day of month, cyclical day 16: 歸休 'Went on leave'
> Tenth day of month, cyclical day 21: 視事 'Took charge'
> 12th day of month, cyclical day 23: 宿沮陽 'Lodged at Juyang'
> 13th day of month, cyclical day 24: 到介 'Arrived at Jie'
> (Zhu Hanmin 朱汉民 and Chen Songchang 陈松张 2010: 177)

Anybody who has ever had to draw up a report on official activities, or submit an expense account for reimbursement, will recognize the utility of such notes. There seems to have been a well-defined category of documents of this kind, often labelled on the strips themselves as *zhi ri* 質日, an expression whose meaning is not yet completely clear, but may have meant something like 'checking days'.[44]

But a calendar could do more than help an official to list his tasks. Archaeology over recent decades has uncovered a previously unknown kind of calendrical

[44] See the discussion of documents of this type in Su Junlin 苏俊林 (2010) 'Guan yu "zhi ri" jian de ming cheng yu xing zhi 关于'质日'简的名称与性质 (The name and nature of the "zhi ri" bamboo strips).' *Journal of Hunan University (Social Science Edition)* 4: 17–22, and Xiao Congli 肖從禮 (2011) 'Qin Han jian *du* "zhi ri" kao 秦漢簡牘'質日'考 (An investigation of the "zhi ri" texts on Qin and Han bamboo strips and wooden planches).' Fudan University Research Centre for excavated documents and archeology 复旦大学出土文献与古文字研究中心. Chen Wei (2016) "Guan yu Qin Han 'zhi ri' de xin kao cha 关于秦汉'质日'的新考察 (A new examination of the expression *zhi ri* in Qin and Han times)", International Conference on Mantic Arts in China, Erlangen, Germany, has suggested that the use of the expression on these documents may be connected with its usage elsewhere to refer to days on which an official had to be attested as having performed a particular duty; at the time of writing, a version of this article is intended for publication in a future issue of *T'oung Pao*.

literature, a literature that seems to have been intended for use by the middle to lower levels of Han society—people who evidently expected to interact with spiritual entities of lower rank than the exalted powers of heaven and earth with whom the emperor had to deal. These texts are known as 'daybooks' *ri shu* 日書, from the title given to some examples of the genre. The number of such texts that have survived suggest that they were once very widespread indeed, and played an important part in everyday life, although very little sign of their existence is evident from the texts produced by the élite. Looking through one collection of such material, one is struck by the sheer variety of topics on which it was possible to get guidance—marriage, journeys, business deals, house construction, even 'favourable days for sheep' (which differed from those for oxen or chickens). All I can do here is to look at a very small sample of these texts to give an idea of how their contents relate to the way ordinary people used the calendar.

One role that 'daybooks' were clearly intended to play was to help people to choose when to do things, assuming they had a choice. We have already seen one example of how that was done, in the marking of 'Fan Zhi' days by the character *fan* 反 on the specimen calendar discussed above—these were clearly days when one would not want to chance one's luck by doing anything important. On other calendars we may find frequent repetitions of the character 建 *jian* 'Establishment', which typically appears two or three times a month. This is a sign of the use of the so-called *jian chu* 建除 'Establishment and Removal' divinatory system, which assigned twelve different labels to the days of a month, beginning with 'Establishment'.[45] The labels were assigned on the basis of the cyclical day-names—thus for the first month 'Establishment' was marked on the days whose name included the third of the twelve 'earthly branches', *yin* 寅, for the second month on the day whose name included the fourth branch, *mao* 卯, and so on, no provision being made for the intercalary month, which was thus left blank. For the first two days of the sequence of twelve, which are marked successively 'establishment' and 'removal', one excavated daybook text gives the following indications of what to expect:

建日:良日(毆)[也].可為嗇夫,可以祝祠,可以畜(大生)[六牲],不可入黔首.[46]

[45] See John Lagerwey and Marc Kalinowski (2009) *Early Chinese religion*. Leiden; Boston, Brill, 896, Sun Zhanyu 孙占宇 (2010) 'Zhan guo Qin Han *shi* qi jian chu shu tao lun 战国秦汉时期建除术讨论 (Discussion on Jian-Chu Divination of the Warring States and Qin-Han Dynasty).' 西安财经学院学报 *Xi'an Caijing xueyuan xuebao (Journal of Xi'an University of Finance and Economics)* 23 (5): 88–93.

[46] In this and later daybook quotations, obsolete characters in the original have been shown as emended to their modern equivalents, and suggested readings of obscure characters are given, using the usual ()[] convention.

Establishment day: A good day! You may act as a petty official, and conduct prayer and sacrifice; you may raise the six kinds of animal, but you may not bring common people into [the household].

除日: 逃亡不得, 癉疾死. 可以治嗇夫, 可以徹言, 君子除罪.

Removal day: Those who run away will not be caught, those who are sick from exhaustion will die; you may deal with petty officials, you may speak penetratingly; the superior man gets rid of his faults.

(First daybook from Fangmatan 放馬灘日書甲種, strips 13–14, in (Chen Wei 陳偉 2014: vol. 4, p. 18))

And so on for each day of the twelve-fold sequence.

Indications such as those given by the 'Establishment and Removal' system are a useful guide to acting at the right time. On the other hand, things may happen about which we have no choice, and the important question then is what the significance might be of those events happening on one day and not on another. Suppose, for instance, that a house is burgled on a day whose cyclical designation includes the character *zi* 子, the first of the set of twelve earthly branches. Does this give us any clues? Indeed yes, as one text tells us:

子: 鼠也. 盜者(兌)[銳]口, 希(須)[鬚], 善弄, 手黑色, 面有黑子焉, 疵在耳, (臧)[藏]於垣內中糞蔡下.

[Cyclical day] *zi* 子: [day of] the rat, the thief has a pointed mouth, with light moustache, is good at play, with dark hands and with black moles on his face. His ears are flawed. He hides within the wall under the manure stack.

(First daybook from Shuihudi 睡虎地日書甲種, strip 69v (Chen Wei 陳偉 2014: vol. 1, p. 477))

The daybooks give us a rich fund of evidence of the ways in which ordinary people might make use of the calendar as an essential aid to the conduct of life, not only in looking ahead to plan the future, but also in dealing with the unexpected. To those above them, however, the calendar was essentially an instrument of power and authority, and a guarantee of the support of the cosmic order for the imperial state. An emperor saw these matters quite differently from a subject, and it is to the concerns of an emperor that we shall now turn.

CHAPTER 3

The Emperor's Grand Inception, and the defeat of the Grand Clerk

The first two chapters have set the scene for the rest of this book. We now turn to the story of the great reform of the astronomical system that took place in 104 BCE, and to its aftermath.

Nothing in this story turns out quite as expected. In the first place, the decision to carry out a reform was not motivated by any new development in the ways in which the motions of the celestial bodies were observed or predicted, but was driven by Wudi's personal obsession with finding a way to escape death by becoming an 'immortal'. What is more, the reform represented a fundamental cosmic rupture with the preceding dynasty, a rupture that had not been accomplished by the founders of the Han. Then, when we turn to the accounts of the technical aspects of the reform, it becomes clear that strange things were happening amongst the experts at court. As Grand Clerk, *Tai shi* 太史, Sima Qian would seem to have been the natural person to construct the new system, and to tell us how that task was performed. But he seems reluctant to reveal exactly what took place—and appears to have been pushed aside in favour of experts drawn from outside the bureaucracy, and even outside the capital itself. What is more, most of what he does tell us about the background of the new system is found in his monograph entitled Feng *shan shu* 封禪書 'Treatise on the feng and shan sacrifices' (*Shi ji* 28), which centres on ritual matters rather than in his *Li shu* 曆書 'Treatise on the astronomical system' (*Shi ji* 26).

Once the reform was officially concluded, disputes did not cease. One of Sima Qian's successors was dismissed from office after persistently claiming around 78 BCE that the system set up in 104 BCE was fundamentally flawed. The means that his colleagues took to evaluate his allegations were radically different from what had happened in the past. For the first time we see a detailed account of the use of systematic observation of the heavens to provide a basis for deciding whether an astronomical system was correct. Finally, we examine attempts made under the late Western Han to give an account of how such calculations were made in past ages.

3.1 The emperor's winter journey

In late 105 BCE a man called Liu Che 劉徹 made a journey to a mountain. It was a long journey, almost a thousand kilometres, and on some days the temperature in the parts of China through which he travelled would have been below freezing. But this was not a trip that could wait for better travelling weather. He had an appointment to keep, and a task awaited him that he alone could perform. His successful fulfilment of that task was, he believed, a matter of life and death for him personally, as well as being crucial to the peace and prosperity of millions of subjects of the empire that ruled much of the landmass of East Asia.

This was no ordinary traveller: Liu Che was the seventh emperor of the Han dynasty, known to us by his posthumous title of Wudi 武帝 'the martial emperor'. He was travelling from his capital at Chang'an 長安 (modern Xi'an 西安) in the north-west to sacred Mount Tai 泰 in the east of China, modern Shandong 山東 province, to speak in person to the powers that ruled the cosmos. So despite the severe weather, it is likely that this journey was relatively comfortable for the 51-year-old man who sat bundled in furs in the relative warmth of a travelling carriage. His appointment was for the day of the winter solstice, the day when the noon sun is lowest in the sky, which his advisers assured him would happen that year on the first day of the 11th lunar month, or 24 December 105 BCE in the Julian calendar. What is more, in the continuous cycle of sixty day-names that stretched unbroken back to the remotest antiquity that date would be the first day, *jiazi*.1. This was a special moment, whose conditions could not be expected to recur for over a millennium. On that day, which was to be the starting point of a new astronomical system, *li* 曆, he would inaugurate a radical change in the relations of the human world to the cosmic powers.

After his arrival at the foot of Mount Tai, the emperor took his place in the new shrine he had caused to be constructed there, the 'Bright Hall' or 'Hall of

Holiness' Ming Tang 明堂. Its design followed the pattern of the shrines thought to have been built by ancient sage kings as a representation of the cosmic order.[1] After solemn sacrifice, he uttered a prayer, of which these words have come down to us:

> 天增授皇帝太元神策，周而復始．皇帝敬拜 太一.
>
> Heaven has bestowed on me, the Sovereign Emperor, the numinous reckoning of the Grand Origin: the cycle has returned to its starting point. The Sovereign Emperor makes reverent obeisance to the Grand Unity. (*Shi ji* 28, 1401)

The deity 'Grand Unity', Taiyi 太一, to whom the emperor prayed, was an appropriate one to worship at a time of astronomical innovation: he was thought to reside at the pivot-point of the sky near the Pole Star, and to preside over all the lesser powers in charge of the stellar world.[2] In these words, the emperor spoke as only he might do, addressing the supreme powers of the cosmos on behalf of all his subjects, proclaiming and thereby actualizing a new cosmic order. We shall shortly consider in some detail what that meant. But now another dimension of the emperor's journey becomes clear, for our source continues without a break to recount what he did next:

> 東至海上，考入海及方士求神者，莫驗，然益遣，冀遇之.
>
> He travelled east to the seashore, and questioned seafarers and the fang shi 方士 'recipe gentlemen' who sought the spirits,[3] but none of them could produce any proof. So he sent out more [searchers], for he longed to meet with them. (*Shi ji* 28, 1401).

[1] The structure of the original Ming Tang is not specified in any pre-imperial source, and later ritualists were therefore free to propose imaginative reconstructions. Mencius, c. 320 BCE, was, however, quite sure that the possession of a Ming Tang was essential to the proper practice of kingly government, and therefore urged the Duke of Qi 齊 not to destroy the Ming Tang that lay in his territory. This was said to have been built by the ancient Kings of Zhou to give audience to their feudal subordinates. See *Meng zi* 孟子 2a, 13b–14a in *Shi san jing zhu shu*, vol. 8, 35. The commentators suggest that this hall lay at the foot of Mount Tai, so that Wudi's hall would have been a replacement for the lost ancient building. Wudi's Ming Tang was constructed according to an alleged plan of the mythical Yellow Emperor's Ming Tang, providentially brought to light by a local scholar just in time to make up for the complete lack of information on this topic: *Shi ji* 28, 1401. On the connection of Wudi's projects with the Yellow Emperor, see section 3.2.3.

[2] See *Shi ji* 27, 1289 and commentary (note 3) 1290, also the translation in Pankenier (2013), 458. For wider discussions of this deity, see Li Ling (1995) 'An Archaeological Study of Taiyi (Grand One) Worship.' *Early medieval China* 2: 1 and Sarah Allan (2003) 'The Great One, Water, and the Laozi: New Light from Guodian.' *T'oung Pao* 89 (4–5): 237–85, particularly 246–253.

[3] These were specialists in a wide range of recondite arts, some of which might nowadays be classified under the modern concept of 'magic', and some of which might seem to us more religious or even technical in nature. There seems to be no instance of anyone using this term to describe

This time the motivation was purely personal: as we shall see, the emperor hoped that what he had done at Mount Tai would make it possible for him (and him alone) to meet with spirits who would grant him eternal life. To understand better the significance of all this for the emperor and for the state he ruled, we need to go back a century, to the beginning of the Han.

3.2 An emperor in search of immortality

3.2.1 Empire and cosmos at the founding of the Han

Sima Qian's fourth monograph in his great history, the *Li shu* 曆書 'Treatise on the [astronomical] system' (*Shi ji* 26) gives us few details of how the Qin dynasty carried out the calculations that produced the luni-solar calendar, apart from the fact that Qin began its civil year with the tenth Xia month. He does, however, make it quite clear that the Han did not introduce any innovations after its first emperor came to the throne:

> 漢興，高祖曰「北時待我而起」，亦自以為獲水德之瑞．雖明習曆及張蒼等，咸以為然．是時天下初定，方綱紀大基，高后女主，皆未遑，故襲秦正朔服色．
>
> At the rise of Han, the High Ancestor [Liu Bang 劉邦, first emperor of the dynasty] said 'It was the northern borders that rose in my support', so accordingly he took himself to have received an omen of the Power of Water. Even brilliant practitioners of [astronomical] systems, with Zhang Cang and others, all took it to be so. At that time, the empire had only just been settled, and they were only just about to knit together the foundational structures; when [later] the High Empress ruled [though] a woman, there was still no leisure [for dealing with such matters]. So they continued using the standard conjunction[4] and ritual colours of Qin. (*Shi ji* 26, 1260)

themselves; it is only applied to third parties, and not always in a complimentary way. On the place of the *fang shi* in intellectual and religious life, see for instance Joseph Needham and Wang Ling (1956) *Science and civilisation in China ... volume 2: History of scientific thought*. Cambridge, CUP, 132–9, and Michael Loewe and Edward L. Shaughnessy (1999) *The Cambridge history of ancient China: from the origins of civilization to 221 B.C.* Cambridge, UK; New York, Cambridge University Press, 644, 818 and 827.

[4] The expression *zheng shuo* 正朔 translated here refers to the day on which the civil year begins, the first day of the *zheng yue* 正月 'standard month', the first month of the year. Since this is the first day of a lunar month, a luni-solar conjunction would be expected to occur. In the Qin and early Han this was first day of the tenth month of the Xia count: see chapter 2, section 2.2.1. From 104 BCE onwards it became the first day of the first Xia month.

The reference to *shui de* 水德 'the Power of Water' draws on a rich vein of correlative cosmological thought that developed during the Warring States period and was well known by the early imperial age. In this way of thinking, all important entities in the human and natural world were correlated with one of five cosmic 'powers' *de* 德 or 'phases' *xing* 行.[5] These phases were named after, but were not identical with, five important natural kinds—Earth, Wood, Metal, Fire and Water. Many sets of five were correlated by means of this scheme; three sets relevant to the present discussion are shown in Table 3.1.

Table 3.1 ***The Five Phases with correlated colours and regions of the world***

Phase/Power	Colour	Region	Number
Earth	Yellow	Centre	5
Wood	Green	East	8
Metal	White	West	9
Fire	Red	South	7
Water	Black	North	6

The order given here is the so-called 'mutual conquest' *xiang ke* 相克 sequence, based on the notion that earth is 'overcome' by wood, which is in turn 'overcome' by metal and so on. Another possible order in which each phase may lead on to the next is the 'mutual production' *xiang sheng* 相生 sequence, which would run in the order Earth, Metal, Water, Wood, Fire.[6] Clearly the statement attributed to Liu Bang is based on the fact that north is correlated with the phase Water. A few decades before the rise of Han, the *Lü shi chun qiu* 呂氏春秋 'Spring and Autumn [annals] of Mr Lü' completed in 239 BCE under the patronage of the Qin minister Lü Buwei 呂不韋 had correlated the rulers of the past with the phases in the conquest order, beginning with the Yellow Emperor (Earth), of whom we shall shortly hear more, and going on to the Xia (Wood), the Shang (Metal) and the Zhou (Fire). As to what might come next:

[5] On the Five Phases, see Graham (1989), 340–56, also Harper (1999), 809–10. Graham prefers the rendering 'five processes', which underlines the fact that Water, Fire, Metal, Wood and Earth in this context are not to be taken as the names of substances (and certainly not 'elements' in the ancient Greek sense), but as convenient labels for five categories of activity. 'Earth' here is a translation of *tu* 土, with the sense of 'soil' rather than the counterpart of heaven, the earth on which we dwell, which would be *di* 地, a quite different word.

[6] The conventional explanation of these orders is that in 'conquest' terms a wooden shovel can dig earth, a metal knife can cut wood, fire can melt metal, water can extinguish fire, and earth can dam up water. As for 'production', metals 'grow' in the earth, water can condense on cold metal, plants (wood) grow from water, and burning wood produces fire, which in turn produces ashes (earth).

代火者必將水. 天且先見水氣勝. 水氣勝, 故其色尚黑.

> What replaces Fire must put Water in the lead. Moreover Heaven will first make it manifest that the *qi* of water is in the ascendancy. The *qi* of water being in the ascendancy, [the new dynasty] will give precedence to black. (*Lü shi chun qiu* 13, 4b)

That was precisely what the Qin dynasty did when it came to power (*Shi ji* 6, 237–8): its official robes were black, while in accordance with the number associated with Water, official tallies and headgear all measured six inches, the axles of carts measured six feet, the double-pace contained six feet, and an imperial team had six horses. By its choice of Water as its governing phase, Han was in effect accepting the role of continuing the Qin dispensation rather than replacing it.

But as the dynasty became more firmly established, pressure mounted in favour of giving recognition to the fact that Han had replaced Qin and ruled in its own right, so that it had no need to borrow the cosmic symbolism of its predecessor. In the 13th regnal year of Wendi 文帝, which ran from late 167 BCE, someone who had so far not held an official role laid an important proposal before the emperor:

魯人公孫臣上書曰:「始秦得水德, 今漢受之, 推終始傳, 則漢當土德, 土德之應黃龍 見. 宜改正朔, 易服色, 色上黃.」

> Gongsun Chen from Lu submitted a writing, saying 'In the beginning Qin obtained the Power of Water. Now Han has inherited from it. Predicting in accordance with the succession of the end and beginning [of the Five Phases], then Han corresponds to the power of Earth. In response to the power of Earth, a yellow dragon will be seen. It would be appropriate to change the standard conjunction, and change the ritual colour, so as to exalt yellow. (*Shi ji* 26, 1260)

Gongsun Chen made his prediction in accordance with the 'conquest order' in which Earth overcomes Water. As we have seen, yellow is the colour corresponding to Earth. But Zhang Cang, who had supported the continuation of Qin practices at the start of the dynasty, was still in office, and would have none of this amateur interference in his specialty:

是時丞相張蒼好律曆, 以為漢乃水德之始, 故河決金隄, 其符也. 年始冬十月, 色外黑內赤, 與德相應. 如公孫臣言, 非也. 罷之.

> At that time the Chancellor Zhang Cang excelled in harmonics and [astronomical] systems, and he held that Han had its origin in the power of Water, and that the Yellow River having overflowed at Gold Dyke was an omen of that. Beginning the year in the tenth month, with the colours black outside and red within,

> was in correspondence with the [governing] power. As for what Gongsun Chen said, it was wrong. So [Gongsun Chen] was dismissed. (*Shi ji* 26, 1260)

But two years later, it was reported that a yellow dragon had indeed been seen in the county of Qinzhou 秦州 (in modern Gansu), and Zhang Cang thereupon decided to resign on grounds of illness (*Shi ji* 96, 2682). What had actually happened in Qinzhou (and who was responsible for the report) we have no way of knowing, but as a result Gongsun Chen was summoned to court, given the rank of *bo shi* 博士 'scholar of wide learning', and tasked with leading discussions to draft plans for changing the astronomical system and ritual colours. The emperor also accepted proposals to make a number of innovations in religious ritual, including making a imperial visit to the shrines at Yong 雍, which had been a place of worship since the Dukes of Qin took it for their residence in the seventh century BCE (*Shi ji* 28, 1360).

Over the next two years the pressure for change mounted, stirred in part by the arrival of another outsider, Xinyuan Ping 新垣平 from the kingdom of Zhao 趙, who claimed to be able to detect increasing activity in the spirit world by 'discerning the *qi*'. He persuaded the emperor to set up near the capital a new shrine to the *Wu de* 五德 'Five Powers' of the Five Phases (*Shi ji* 28, 1382). The year after his visit to Yong the emperor worshipped at the new shrine, and gave Xinyuan a large sum in gold. The next year, 164–163 BCE, Xinyuan maintained his influence by secretly arranging for someone to present the emperor with a mysteriously inscribed jade cup, whose arrival he predicted by detecting the *qi* supposedly emitted by this ancient treasure. Added to this he claimed that the sun had climbed back to its noon height after passing it once in a single day. As a result, and no doubt also on the advice of Gongsun Chen, the emperor began a new count of regnal years to mark the auspicious event, and the 17th year of the Emperor's reign was counted as the first year of the Houyuan 後元 'Later Origin' period (*Shi ji* 28, 1383).

Xinyuan Ping's next claim was that he had discerned the *qi* of the ancient sacred tripods which had embodied the power of the kings of the Zhou 周 dynasty, long sunk in the Si 泗 river in Shandong, and a shrine was set up on the river bank at Fenyin 汾陰, in the hope that the tripods would be recovered. But Xinyuan's growing influence had evidently made enemies, for shortly after this he was denounced as a fraud, and after investigation he was executed together with his entire clan. The wave of confidence on which the emperor had ridden for the last few years was broken, and he lost interest in making further innovations in calendrical or ritual matters. No doubt the major incursion of the Xiongnu 匈奴 nomads from the northern steppe in the following year helped to fix the court's attention on other matters.

3.2.2 Wudi: empire and emperor in a new cosmic order

The reign of Wendi was followed by that of Jingdi 景帝 (r. 156–141 BCE), who made no significant innovations in state policy in relation to the cosmic powers. But in the spring of 141 BCE, Wudi came to the throne at the age of sixteen. In accordance with the practice inherited from Qin, the first full civil year of his reign began in the tenth month of that year, when he held a great New Year audience to take the advice of princes, court officials and scholars as to the course he should pursue. The country had enjoyed relative peace and security for a number of years; the people were thought to be increasing in prosperity and the government's reserves of cash and grain were said to be straining the available storage capacity (*Shi ji* 301, 1420). There was a general feeling in the court that the Han empire was at last firmly established enough to proclaim its independence of Qin institutions by breaking with the calendrical and ritual practices it had taken over from its predecessor. Those who exerted most influence on the young emperor in this direction were a group of so-called *Ru* 儒 scholars,[7] whose teaching had been out of favour in preceding reigns. One of the emperor's first acts was to approve a memorial presented by the Lieutenant Chancellor Zhao Wan 趙綰, in which those who supported the tradition of 'legalist'[8] statecraft which had come down from Qin times were condemned as 'disturbing the good order of the state' (*Han shu* 6, 156). Zhao and other literati drafted proposals for the emperor to begin his reign with a ritual tour of the country, during which he would perform the ancient *feng* 封 and *shan* 禪 ceremonies on Mount Tai,[9] and carry out a reform of the astronomical system and the dominant ritual colour, thus marking a definitive rupture with Qin practice.

At this point, however, the Empress Dowager Dou 竇 intervened. According to Sima Qian, she was an adherent of the Huang-Lao 黃老 teaching, which is now thought to have been a synthesis of the mystical statecraft of the *Lao zi* 老子 book with the principles of 'legalist' realpolitik.[10] She certainly had no liking for the literati group advising the young emperor, who had already moved

[7] As mentioned previously in this book, section 1.1, this term has conventionally been translated as 'Confucian', but in the present context this may be misleading: see Graham (1989), 31. We can, however, say that the *Ru* laid particular stress on the value of ancient texts as the basis of all worthwhile knowledge and social practice.

[8] This term was coined by later historians, who have tended to group together in retrospect various ancient thinkers who might not necessarily have seen themselves as having a common purpose.

[9] On the *feng* and *shan* rituals see Loewe (1982), 130 ff.

[10] Graham (1989), 170–1.

against men who must have been close to her party. Perhaps Zhao and his supporters were anticipating a countermove from her when they suggested in the tenth month of 140 BCE that the Empress Dowager should no longer take an active part in affairs of state. But it was the Empress Dowager who won: Zhao was soon charged with a number of serious offences and shortly afterwards he was said to have committed suicide in prison (*Shi ji* 28, 1384 and 107, 2843; *Han shu* 6, 157).

Although the emperor continued to show his support for the literati by setting up professorships *bo shi* 博士 in the Five Classics in 136 BCE (*Han shu* 6, 159), as mentioned in chapter 1, by the time the Empress Dowager died in the following year it was too late to pursue the idea of beginning his reign with the symbolism of cosmic change, and all such projects, including astronomical reform, were laid aside for some years.

The posthumous title 'Wu' (The Martial) by which the emperor is known to history is certainly justified by the vigour and success of Chinese military policy during his reign, which saw imperial power exerted deep into central Asia. But unlike his ancestor Liu Bang, who founded the Han, Emperor Wu took no direct part in fighting. Much of his personal energy was devoted to establishing relations with the unseen world of spiritual powers. It is possible to distinguish two main strands in Emperor Wu's religious activity:

(a) The use of intermediaries to make contact with the spirits. Some of these intermediaries were *wu* 巫, 'mediums' whom the spirits would possess and through whom they would speak. Others claimed to have means of contacting the mysterious *xian* 僊 (also written 仙) 'immortals', beings who lived in eternal felicity on the islands of Penglai 蓬萊 far beyond the eastern sea, but might sometimes be found closer at hand. With proper preparation it was possible to induce *xian* to visit the world of men, and a few favoured human beings might even hope to be allowed to join their number and obtain immortality.

(b) The further development of the ancient imperial worship of the cosmic powers, a process already begun by Wendi.

Neither of these motivations should be seen as marking out Wudi as in any way eccentric. For an emperor to be concerned that he should play his religious role to best advantage could be seen as entirely proper—though not everybody would necessarily agree with his decisions in this area. And as for the wish to attain the happy status of an immortal, he was certainly not alone in his time in hoping for this—even though as emperor the resources he could devote to this pursuit were vastly greater than those available to others.[11] At a crucial stage in the emperor's reign the two strands became linked together, so that Wudi's obsession with personal immortality was thrown in on the side of establishing a new relation between the dynasty and the powers which controlled the universe. It was as a necessary part of this process that reforms in celestial calculation were finally introduced. And now another emperor comes onto the scene.

3.2.3 The Yellow Emperor intervenes

Two years after the death of the Empress Dowager, in the regnal year 134–133 BCE the emperor paid his first visit to the shrines at Yong 雍, and performed the ceremony of *jiao* 郊 'The Bounds' in honour of the Five Powers (*Shi ji* 28, 1384). Thereafter this ritual was to be performed every three years, although the emperor did not always attend in person. At this time the emperor came under the influence of Li Shaojun 李少君, the first of a series of spiritual intermediaries to whom imperial favour was extended. Li claimed to be several hundred years old, and told the emperor that during a sea voyage he had met Master Anqi 安期, one of the immortals of Penglai. According to Li:

> 祠竈則致物, 致物而丹沙可化為黃金, 黃金成以為飲食器則益壽, 益壽而海中蓬萊僊者乃可見, 見之以封禪則不死, 黃帝是也.
>
> Through worshipping the god of the stove you will become able to command all creatures, and thus you will be able to turn cinnabar to gold. By the use of eating and drinking vessels made of this gold you will lengthen your days, and having done so you will be able to visit the immortals of Penglai in the midst of the sea. When you have visited them, you may perform the *feng* and *shan* rituals and obtain immortality. This is what the Yellow Emperor did. (*Shi ji* 28, 1385)

Li Shaojun's suggestion was a brilliant piece of statecraft, since it brought together all the elements that could impel Wudi into action—his personal obsession with immortality, the use of ritual to underline the power and legitimacy of his dynasty, and the emulation of a figure who was a model for any Han ruler—the 'Yellow Emperor' *Huang di* 黃帝. But who was the Yellow Emperor?

[11] On the general question of the search for immortality and the *xian*, see Michael Loewe (1979) *Ways to paradise: the Chinese quest for immortality*. London, Allen & Unwin, also Needham and Wang Ling (1956), 8–12.

The importance of the Yellow Emperor in the time of Wudi is shown by the decision of Sima Qian to take his reign (rather than that of the sage emperor Yao revered by the *Ru* scholars and canonized in the *Book of Documents*) as the start of his history of mankind.

黃帝者,少典之子,姓公孫,名曰軒轅.生而神靈,弱而能言,幼而徇齊,長而敦敏,成而聰明.

The Yellow Emperor was the son of Shaodian, of the line of Gongsun, and his personal name was Xuan Yuan. At his birth he [showed] miraculous power, and could speak when still an infant. He developed rapidly as a child, and as he grew up he was careful and diligent, while as an adult he was of brilliant intelligence. (*Shi ji* 1,1)

Xuan Yuan organized armed resistance to the forces of disorder and violence, and after their defeat he was recognized as ruler by his grateful subjects. As a result of his tireless work, universal peace came to the world he ruled, and its effects spread beyond the world of the merely human:

萬國和,而鬼神山川封禪與為多焉.獲寶鼎.迎日推筴[12] ... 順天地之紀,幽明之占,死生之說 ... 旁羅日月星辰水波 ... 有土德之瑞,故號黃帝.

The myriad states were in harmony, and marvelled at his [skills in relation to] the ghosts and spirits, the hills and streams, and the *feng* and *shan* [rituals]. He obtained a precious tripod, and predicted the days in accordance with the reckoning. ... He followed the Eras of heaven and earth, [comprehending] the prognostications of dark and light, and the doctrines of life and death ... He expounded on the sun, moon, stars and celestial markers [above] and on the movements of the waters [beneath] ... He received favourable omens of the power of Earth, so he was called the 'Yellow Emperor'. (*Shi ji* 1, 6)

In other accounts, the Yellow Emperor was said to be responsible for such things as the invention of weapons, clothing and houses, as shown in the caption of this image from the second century CE Wuliang 武梁 tomb shrines (Figure 3.1).

The caption to the left of the figure reads:

黃帝多所改作.造兵,井田,垂衣裳,立宮宅.

The Yellow Emperor made many innovations. He invented weapons, the well-field system [of taxation],[13] draped [people in] clothing and built palaces and houses

[12] This character serves here as a variant of the more common *ce* 策, which is used when the same phrase is written in *Shi ji* 281,393.

[13] According to traditional accounts, the 'well-field' system involved dividing a square plot centring on a well into nine smaller square plots. The eight outer plots were allocated to families for their own use, but the central plot was cultivated by all of them in common, and provided revenue for the ruler.

Figure 3.1 The Yellow Emperor, as depicted in a relief from the Wuliang tomb shrines (second century CE). (Feng Yunpeng 馮雲鵬 and Feng Yunyuan 馮雲鵷 1929 (first edn. 1823): vol 9: 3—10); cf. (Wu, Hung 1989: 250).

A text recovered from a tomb closed in the second century BCE contains writings that give another account of the Yellow Emperor, under the title of Huang Zong 黃宗 'The Yellow Ancestor', and tell us of his acts after he ascended the throne. First, he put in place the offices of the *San gong* 三公 'Three Excellencies', who were the principal counsellors of the ruler; next he set up feudal states with their princes and ministers. Immediately after this

> 數日, 曆[14]月, 計歲, 以當日月之行.
>
> He counted off the days, sequenced the months and reckoned the solar cycles,[15] so as to correspond to the movements of the sun and moon.[16]

For a young emperor as full of ambition and self-confidence as Wudi, all this constituted an image of a model ruler that he found entirely congenial. And as we shall see, he did his best to live up to it in all its details.

[14] This is simply a variant way of writing the word *li* more commonly written 曆.

[15] *Sui* 歲 can often simply be translated 'year'. But in this case it clearly has its more precise meaning—the annual cycle of the sun from winter solstice to winter solstice, which is not a whole number of days or lunar months (see section 1.3).

[16] See Leo S. Chang and Yu Feng (1998) *The four political treatises of the Yellow Emperor: original Mawangdui texts with complete English translations and an introduction*. Honolulu, University of Hawai'i Press, Chapter 2, Jing 經 'Classics', 145–6; my translation attempts to follow the structure of the original more closely than that of Chang and Feng.

The emperor's faith in Li Shaojun was so great that even Li's death shortly afterwards failed to shake him; in his view Li had simply been 'transformed' into an immortal himself. He set in motion a great search amongst all the *fang shi* 方士on the sea coasts of the ancient eastern states of Yan 燕 and Qi 齊 in an effort to find someone else who could contact Master Anqi for him, and help in his quest to imitate the Yellow Emperor. In view of the rewards offered there was no lack of applicants, and the emperor's reign must have been a golden age for *fang shi*. The problem was that whatever their initial sincerity may have been they were inevitably tempted to promise more than they could perform, and the emperor had no mercy when they were detected in fraud.

In 120 BCE Shao Weng 少翁 from Qi was given high rank for summoning the spirit of a dead concubine of the emperor's at a seance in the palace. He promised more direct communications with the unseen world, but after a year nothing had materialized. In desperation he faked a 'spirit message', but was secretly executed when the emperor noticed the suspicious similarity of the calligraphy to Shao's own handwriting (*Shi ji* 28, 1387–8). A few years later, in 113 BCE, when Wudi was beginning to regret having executed Shao before his art had had time to produce results, an old associate of Shao, Luan Da 欒大, came forward to take his place. Like Li Shaojun he claimed to have met Master Anqi of Penglai, but he had been rebuffed as being a mere commoner. If only the emperor would give him high rank and treat him with honour, perhaps his next trip to Penglai might result in establishing successful relations with the immortal court. The emperor responded enthusiastically, marrying him to a princess with a rich dowry. During the months when Luan was ostensibly preparing for his expedition he was treated as almost more than the equal of the emperor. There was an understandable outbreak of enthusiasm amongst the *fang shi* of the eastern seaboard, all of whom now claimed to have secret methods for contacting the immortals. All, no doubt, also claimed to be in need of a substantial amount of research funding.

It was not, however, from expeditions over the seas that contact with the Yellow Emperor finally came, but as part of the programme of 'restoration' of what were presumed to be ancient cosmic cults. In 133 BCE Wudi had been persuaded to set up a new shrine to the south-east of the capital, in honour of the Grand Unity, Taiyi 太一, who was said to rule the sovereigns of the Five Powers worshipped at Yong (*Shi ji* 28, 1386). In 121 BCE it was claimed that a 'unicorn' had been captured after the emperor had performed the ceremony of the Bounds. Encouraged by this omen the emperor took personal possession of the territory round the sacred Mount Tai, and likewise of the land around Mount Chang. All the five sacred mountains of the kingdom were now in his hands, and the feeling grew that the time for the great imperial rites of *feng* and *shan* was near (*Shi ji* 28, 1287). The last time an attempt had been made to perform these was in 219 BCE, when the first emperor of Qin had climbed Mount Tai to announce his accession to the cosmic powers; according to the story current under the Han, he had been met with a great storm, a clear omen of cosmic disapproval (*Shi ji* 281, 367).

Perhaps with this example in mind, as well as because it was doubtful how the ceremonies were to be performed, the emperor hung back for several years. The events which led up to the consummation of the *feng* and *shan* ceremonies had a more precise connection than has previously been thought with the choice of 104 BCE as a suitable year for reform.

The story begins in the summer of 113 BCE, when Luan Da was preparing for his embassy to Penglai. A *wu* 巫 spirit medium was conducting invocations at the shrine of *Hou tu* 后土 'Sovereign Earth', which the emperor had recently set up at Fenyin 汾陰. This was the site where the emperor's grandfather, Wendi, had commissioned ceremonies in 164 BCE in an attempt to recover the lost sacred tripods of the Zhou dynasty. On seeing a glint of metal, the *wu* dug away the soil and recovered a strange tripod vessel, *ding* 鼎, inscribed with indecipherable writing.

Wudi was doubtful what the significance of this discovery might be, especially in view of a recent series of evil omens such as bad harvests and flooding on the Yellow River. His ministers nevertheless recounted the stories of the great bronze tripods that had been in the possession of the sage rulers of high antiquity, and assured him that the discovery at Fenyin was highly complimentary to the wisdom of the emperor's rule. The tripod was given an honoured place in the hall of imperial audience (*Shi ji* 28, 1392). The tripod itself was probably a genuine enough object from the Shang or early Zhou dynasties, perhaps buried at the site of an earlier shrine. However, spurred on by the rich success of Luan Da and hopeful of avoiding the fate of his predecessor, a certain Gongsun Qing 公孫卿 from Qi 齊 decided that this was an opportunity not to be missed.

His chance to get the emperor's attention came in the autumn of 113 BCE, when there were discussions at court about whether the emperor himself ought to carry out the worship of the deity *Tai yi* 太乙 Grand Unity that had been inaugurated 20 years earlier in 133 BCE. Gongsun Qing announced that:

> 今年得寶鼎,其冬辛巳朔旦冬至,與黃帝時等.
>
> This year a precious tripod has been discovered, and this winter the solstice will fall on the first day of the month, on day *xinsi*.18 [of the sexagenary cycle]; things are as in the time of the Yellow Emperor. (*Shi ji* 28, 1393)

He had a document on this subject, that he attempted to present to the emperor, but at first he was blocked by a court official, who told him 'That business about the tripod is all over now—why raise it again?' Perhaps he was a little tired of gentlemen from Qi with mystic and rather expensive secrets. Gongsun simply sent in his document by another route; it read as follows:

> 黃帝得寶鼎 宛朐, 問於鬼臾區. 鬼臾區對曰: 『(黃)帝得寶鼎神策, 是歲己酉朔旦冬至, 得天之紀, 終而 復始.』 於是黃帝迎日推策, 後率二十歲復朔旦冬至, 凡二十推, 三百八十年, 黃帝僊登于 天.
>
> The Yellow Emperor discovered a precious tripod at Wanqu. He asked Guiyu Qu about it, who replied 'Your Majesty has obtained this precious tripod, [with its] numinous reckoning. This year the winter solstice will fall on the first day of the month, on a *jiyou*.46 day. To accord with the Era of heaven, things must complete [the cycle] and return to their starting point.' Thereupon the Yellow Emperor predicted the days in accordance with the reckoning; in the 20th year, the winter solstice fell once more on the first day of the month, and after 20 such reckonings, making 380 years, the Yellow Emperor ascended to heaven as an immortal. (*Shi ji* 28, 1393)

From what we have already seen, it is no surprise to find the Yellow Emperor involved in calculations about the calendar—nor would it have been a surprise to Wudi. Let us examine the implications of what Gongsun Qing's account. We may understand the statements he ascribes to Guiyu Qu in terms of the standard 'quarter remainder' calendrical periods already explained in chapter 1 of this book. Beginning with the *ce* 策'reckonings', since twenty reckonings make 380 years, we may deduce that one reckoning is:

$$380/20 \text{ years} = 19 \text{ years}$$

In quarter remainder terms, this is one *zhang* 章 'Rule'. If we begin with a day during which the moment of winter solstice coincides with luni-solar conjunction (and hence falls on the first day of a month), then 19 years later (i.e. at the start of the 20th year of the sequence), they will coincide once more, but at a different time of day. After four Rules, 76 years, making one *bu* 蔀 'Obscuration', the coincidence will recur at the same time of day as at the start of the first Rule. We may note that if the sequence of lunar months has its usual relation to the seasons, maintained by appropriate intercalations, that month will be the eleventh of the usual Xia count.

A quarter remainder *ji* 紀 'Era' is 1,520 years, and begins with the coincidence of winter solstice and the moment of luni-solar conjunction at midnight beginning the first day of the month, usually specified as a *jiazi*.1 day. Now supposing that, as stated in the story told by Gongsun Qing, new moon and winter solstice fall together on day *jiyou*.46, 380 years before the beginning of a new Era cycle. Since 380 years is not only 20 Rules but also 5 Obscurations, then since winter solstice and conjunction coincide at midnight at the start of the Era, they will also coincide at midnight 380 years earlier. But since that midnight begins a *jiyou*.46 day, on what day must they fall at the start of the next Era? Again assuming that each cycle from one winter solstice to the next is 365 ¼ days as in all quarter remainder systems, then 380 such solar cycles amount to:

$$380 \times 365\tfrac{1}{4} \text{ days} = 138{,}795 \text{ days},$$

$$\text{and } 138{,}795 = 2{,}313 \times 60 + 15.$$

Thus the cyclical day number of the start of the next Era Cycle is found by counting fifteen days forward from the *jiyou*.46 day from which the reckoning began. Since 46 + 15 = 61, this takes us through to the beginning of the next cycle of 60, a *jiazi*.l day. On such a day, when winter solstice and the new moon of the 11th month coincided at midnight beginning the first day of a new 60-day cycle, *jiazi*.l, the Yellow Emperor ascended to heaven.

The striking thing about these conditions are that they are precisely those which were specified in the calendrical reform which was eventually carried out in 104 BCE, and in fact, as we shall see, Gongsun Qing's name headed the list of those who proposed the change when it was finally made. Thus, in 113 BCE, Gongsun Qing gave the Emperor Wu warning of the approach of a new Era Cycle according to the Yellow Emperor's reckoning, but whereas the Yellow Emperor had 380 years before favourable conditions for his ascent to heaven occurred, Emperor Wu had only eight clear years' notice of what would happen on a *jiazi*.1 day near the end of 105 BCE.

For Gongsun Qing it was important that the discovery of the tripod at Fenyin had taken place in a year which was followed by a winter solstice on the first day of the month, just as things had been in the year when the Yellow Emperor made his discovery. We can be sure that Gongsun Qing's statement of conditions in 113 BCE represented the actual date assigned to winter solstice in that year, since Sima Qian precedes his description of the ceremonies conducted on that day—on which see below—as follows:

> 十一月辛巳朔旦冬至…
>
> The 11th month, [day] *xinsi*.18, conjunction, winter solstice … (*Shi ji* 28, 1395)

But what does such a date imply about the day on which winter solstice should fall eight years later? The interval from local Chang'an midnight on the *jiyou*.46

day near the end of 113 BCE (24 December 113 BCE) to midnight beginning the day *jiazi*.1 near the end of 105 BCE (25 December 105 BCE) is 2,923 days. Taking a mean lunation as the usual quarter remainder value of $29\ ^{499}/_{940}$ days, we see that this is close to a whole number of lunations:

$$99 \times 29\ ^{499}/_{940}\ \text{days} = 2{,}923\ ^{521}/_{940}\ \text{days}$$

Now $^{521}/_{940}$ day is 0.5543 day, or 13 hours and 18 minutes. Thus, so long as the conjunction of the 11th month on the *xinsi*.18 day in 113 BCE falls no later than about 10:41 a.m. local time, the conjunction of the 11th month will fall before the end of the *jiazi*.1 day in 105 BCE, as it should to match up with the conditions at the time of the Yellow Emperor's ascent. But what about the winter solstice? If we look for the nearest multiple of 365 ¼ days to 2,923 days, we find:

$$8 \times 365\ ^1/_4\ \text{days} = 2{,}922\ \text{days}$$

So if a solstice falls on the *xinsi*.18 day in 113 BCE, then it can fall no later than the day before *jiazi*.1 in 105 BCE, i.e. on *guihai*.60, 24 December. Perhaps Gongsun did not do that calculation, or if he did, he did not bring it to the emperor's attention. There is, however, an alternative: if the solar cycle were to be just a little longer than 365 ¼ days, then eight such cycles would slightly exceed 2,922 days, so that a winter solstice falling very near the end of *xinsi*.18 in 113 BCE would imply a winter solstice in 105 BCE that might just edge into the beginning of *jiazi*.1. And as we shall see, the new system adopted eight years later did in fact make the solar cycle a small amount longer than the quarter remainder value, an amount which over eight years would have added up to just under two minutes, making a solstice on *jiazi*.1 at least marginally possible on the basis of the 113 BCE *xinsi*.[17] We cannot tell whether Gongsun Qing already had such a change in mind when he made his statement, or whether he simply ignored the difficulty about the solstice, and nobody else noticed the problem.

Wudi reacted favourably to Gongsun's document, and summoned him for consultation. Gongsun explained that he had received the document from Shen Gong 申公 'Sire Shen' of Qi. From references elsewhere, we know that Shen Gong was a *Ru* scholar who had come to the capital at the beginning of the Wudi's reign 27 years earlier, when he was already into his eighties, and had taken part in the abortive discussions aimed at recognizing the Han dynasty as partakers of a new cosmic dispensation (*Shi ji* 121, 3120–2). Shen Gong was now dead, but during his lifetime, said Gongsun:

[17] See Box 3.1. The increase in the length of the solar cycle was equivalent to 14 seconds.

> [...] 與安期生通, 受黃帝言, 無書, 獨有此鼎書. 曰『漢興復當黃帝之 時』. 曰『漢之聖者在高祖之孫且曾孫也. 寶鼎出而與神通, 封禪. 封禪七十二王, 唯黃 帝得上泰山封』. 申公曰: 『漢主亦當上封, 上封能僊登天矣. [...]』
>
> He was in contact with Master Anqi, from whom he heard the sayings of the Yellow Emperor. [He left] no writings; all we have is this text about the tripod. It says 'The rise of Han takes us back to the time of the Yellow Emperor' and 'The holy sage of Han will be amongst the sons and grandsons of the High Ancestor (i.e. Liu Bang). When the precious tripod appears he will establish contact with the spirits and perform the *feng* and *shan* ceremonies. Of the seventy two kings who have [attempted] the *feng* and *shan*, only the Yellow Emperor [really] succeeded in climbing Mount Tai to perform the *feng* ceremony.' Shen Gong told me that 'The ruler of Han must likewise ascend [the mountain] and [perform] the *feng*. When he has ascended [the mountain] and [performed] the *feng*, he will be able to rise up to heaven as an immortal.' (*Shi ji* 28, 1393)

Gongsun went on to stress the significant parallels between Wudi and the Yellow Emperor; like Wudi, the Yellow Emperor had anciently performed the Bounds ceremony at Yong, which was also the site of the tomb of the Yellow Emperor's minister Guiyu Qu, who had advised on the finding of the tripod. Wudi's palace at Ganquan 甘泉 was where the Yellow Emperor had met with a great concourse of divinities. After casting a new tripod with bronze from Mount Jing 荊, the Yellow Emperor and more than seventy of his ministers and concubines had been borne up to heaven by a dragon. The place where this happened was none other than Tripod Lake, where in 118 BCE Wudi had been cured of a severe illness by the intervention of a female medium, through whom he had received mysterious messages from the spirits (*Shi ji* 28, 1388).

All this delighted the emperor, who was eager to repeat his predecessor's ascent to heaven, and he appointed Gongsun Qing to the rank of Court Gentleman. Unlike his predecessors, Gongsun did not die at the hands of the executioner. Although in years to come he was adept at feeding the emperor's appetite for stories of occult manifestations, he never seems to have put himself forward as an intermediary with the spirit world in the style of Li Shaojun, Shao Weng and Luan Da. From now on it was the emperor himself who was to be responsible for contacting the unseen powers, and if the results were unsatisfactory there would be no doubt where the blame lay. Shortly after the rise to power of Gongsun Qing the emperor tired of the unfortunate Luan Da, who was evidently never going to summon up the courage to depart for Penglai, and had him executed (*Shi ji* 28, 1395).

The emperor now set in motion the process of recognizing the new cosmic dispensation under which his dynasty ruled, thereby, as he hoped, ensuring that, like the Yellow Emperor, he would achieve immortality. At the winter solstice following the discovery of the tripod the emperor, clad in yellow robes, personally worshipped the Grand Unity for the first time. In his sacrificial prayer he stated:

> 天始以寶鼎神策授皇帝, 朔而又朔, 終而復始, 皇帝敬拜見焉.
>
> Heaven began by granting the precious tripod and the numinous reckoning to [me,] the Sovereign Emperor; now after new moon succeeding upon new moon the cycle has reached its end and will begin again. [Now as] Sovereign Emperor I reverently perform obeisance. (*Shi ji* 28, 1395)

Sima Qian's father Sima Tan, who at that time held the office of *Tai shi* 太史 'Grand Clerk', was amongst those who sent in memorials congratulating the emperor on the favourable omens that had accompanied the sacrifice (*Shi ji* 28, 1395). From this time on, although the emperor never lost interest in making direct contact with Penglai, his main attention was devoted to preparing for the *feng* and *shan* ceremonies, by performing which he hoped to emulate the Yellow Emperor. There was perhaps a moment of scepticism on the emperor's part when after sacrificing at what was said to be the Yellow Emperor's tomb at Mount Qiao 橋 (in modern Shaanxi 陝西 province) he asked:

> 吾聞黃帝不死. 今有冢, 何也?
>
> I heard that the Yellow Emperor did not die. But now here is his tomb—how can that be? (*Shi ji* 28, 1396)

But his attendants had a reply ready—after the Yellow Emperor had ascended to heaven, his ministers had buried his hat and robes at this site. And that was apparently enough to reassure Wudi that he could continue to believe.

Having made the necessary preparations, the emperor performed the *feng* and *shan* ceremonies on Mount Tai in the fourth Xia month of the regnal year of 111–110 BCE. For the *feng* ritual an earthen mound nine feet high and twelve feet broad was built, beneath which were buried jade tablets with a secret inscription.[18] The ceremony was that used in the worship of Grand Unity, and took place on the eastern side of the mountain. Afterwards the emperor climbed to the summit with a single companion, and performed

[18] Since one meaning of *feng* is 'an earthen mound', it seems likely that the name of the ritual is derived from this.

secret rites. Next day he descended and the *shan* ceremony took place, the form of worship being that used for Sovereign Earth. Throughout he was clad in yellow robes (*Shi ji* 281, 398). On completion of the ceremonies the emperor's ministers gathered to offer felicitations in a ritual 'Hall of Holiness', *ming tang* 明堂 at the foot of the mountain. The emperor was convinced that the ritual had been a success, and travelled hopefully to the shores of the eastern sea to see if the immortals of Penglai would show themselves at last. Although he was disappointed yet again in this respect, the appearance of what may have been two comets in the heavens that autumn was taken as an omen of cosmic approval (*Shi ji* 28, 1399).

Finally, on 25 December 105 BCE the long anticipated moment arrived, and as recounted at the start of this chapter, the emperor marked with solemn sacrifice the repetition of the conditions that had occurred in the time of the Yellow Emperor, when winter solstice and the first day of the 11th month both fell on day *jiazi*.1.

Although the emperor sent numerous messengers to the eastern coast to watch for immortals, none appeared. Later in the month he went himself, but saw nothing. Not at all discouraged, however, he returned home and on the advice of Gongsun Qing and various *fang shi* he began a number of building projects designed to follow the example set by the Yellow Emperor and attract visits from immortals. In 104 BCE the new astronomical system was inaugurated:

> 夏, 漢改曆, 以正月為歲首, 而色上黃, 官名更印章以五字. 因為太初元年.
>
> That summer, the Han reformed the astronomical system, making the first [Xia] month into the head of the [civil] year, precedence being given to the colour yellow, and official titles and seals being revised so as to contain five characters. This was taken to be the first year of the Grand Inception [reign period]. (*Shi ji*, 28, 1402).[19]

Despite the fact that the emperor neither subsequently saw any immortals nor became immortal himself, his belief remained strong. Sima Qian concluded his chapter by noting somewhat drily that even though more and more *fang shi* brought news of divine beings to court, 然其效可睹矣 *ran qi xiao ke du yi* 'Their results remained to be seen' (*Shi ji* 28, 1404).

[19] The number five and the colour yellow pertain to the phase Earth, the newly adopted patron power of the dynasty. One commentator on a parallel passage in the *Han shu* notes that the titles of posts on official seals were, where necessary, modified slightly so they would contain five characters: *Han shu* 6, 200, note 2. The *Han shu* passage says that the new system was promulgated in the fifth month, presumably for similar reasons.

3.3 The outcome of it all: the events of 105–104 BCE and their aftermath

3.3.1 The Grand Clerk's account

Despite Sima Tan's encouragement of Wudi's worship of Grand Unity three years earlier, failing health prevented him from accompanying the emperor for the *feng* and *shan* rituals of 110 BCE. In his deep disappointment at being unable to be present at what he saw as an epoch-making event, he laid on his son Sima Qian the solemn obligation to follow him in taking up the hereditary post of Grand Clerk, which, he said, had been a responsibility of the family from as far back as the Zhou dynasty. Not long after, he died, and as he had wished, his son took office as Grand Clerk in 108 BCE (*Shi ji* 130, 3295–6). Sima Qian was therefore the official with principal responsibility for observing, interpreting and (as far as was possible) predicting celestial phenomena at the time that the great reform of 104 BCE took place.

Most of the story told so far in this chapter comes from Sima Qian's account given in the sixth of his monographs on special topics, the *Feng shan shu* 封禪書 'Account of the *feng* and *shan* [rituals]' (*Shi ji* 28). From that account we know why the 104 BCE reform was launched—but what did the reform look like from the point of view of the technical experts who had the responsibility of watching the heavens, and of actually carrying out the calculations that that underpinned the calendar? For that aspect of the story, we may turn back to Sima Qian's fourth monograph, the *Li shu* 曆書 'Account of [astronomical] systems' (*Shi ji* 26). This monograph consists of two sections, the first of which is a historical narrative, and the second of which consists of a calendrical tabulation.[20] Of the historical narrative, only the last six out of 33 columns of main text in a modern edition deal with the events of Emperor Wu's reign. Before that, we are given an outline of the development of astronomical systems from the time of the Yellow Emperor up to the early Han.

The theme of Sima Qian's historical outline is simple, and is summed up at the end of the introductory passage of his narration:

[20] It is doubtful whether the second section is really by Sima Qian. Certainly some of it cannot be by him—since it gives basic calendrical data for the 76 years (an Obscuration) from 104 BCE onwards, and in doing so uses reign titles that were not created until long after Sima Qian's death in 86 BCE. There is the further point that this tabulation is made using quarter remainder methods, which produce different results from the '81 factor' methods actually in use from 104 BCE—which Sima Qian would have had to use as part of his normal duties.

王者易姓受命, 必慎始初, 改正朔, 易服色, 推本天元, 順承厥意.

When a king changes the [ruling] clan-name and receives the Mandate [of Heaven], it is essential for him to take care of the way he begins, changing the standard conjunction, altering the colours of ritual dress, finding his basis in the celestial origin, and harmoniously upholding its meaning. (*Shi ji*, 26, 1256)

The problem, however, is that things always fall away from the excellent models set up by a succession of idealized rulers who bring order from chaos, and chaos inevitably returns in the hands of incompetent successors. Thus, after the Yellow Emperor's founding work in establishing an orderly calendar, disorder broke out amongst the 'Nine Li [tribes]' *jiu li* 九黎, so that 'people and spirits mingled', and the Yellow Emperor's grandson Zhuan Xu 顓頊 had to bring things back to a proper state. But this did not last, and the job had to be redone by Yao 堯 and Shun 舜. The succession of the Three Dynasties of Xia, Shang and Zhou saw an orderly shift in the month of New Year, which moved from the first Xia month (where it is today) back to the preceding 12th month and then the 11th month as the dynasties succeeded each other. But order broke down with the later Zhou kings:

史不記時, 君不告朔, 故疇人子弟分散, 或在諸夏, 或在夷狄, 是以其禨祥廢而不統.

The clerks did not record the seasons, and the lords did not announce new moons. Thus the Specialists[21] and their disciples were scattered, some into civilized lands, but others amongst the barbarians. So all [that brings] good fortune was abandoned, and there was no continuity. (*Shi ji*, 26, 1258–9)

Worse still, he tells us, in the 26th year of King Xiang 襄 of Zhou (626 BCE), the Spring and Autumn annals record that an intercalary month was inserted after the third month, rather than at year end as it should have been. Chaos continued under the Warring States, and the reunification under Qin brought no resolution, despite the adoption of the new patron phase of Water, with consequent changes in the ritual colour to black, and to a New Year in the tenth Xia month. As we saw earlier in this chapter (at the start of section 3.2) Sima Qian tells us that the unsettled state of affairs under the first Han rulers left no time to attend to such matters, and the practices of Qin were continued. Then, as we have seen,

[21] This passage is the *locus classicus* for the expression *chou ren* 疇人, which is not widely found in ancient literature. The early commentators on this text take *chou ren* as a reference to persons exercising hereditary expertise. Its clear connection with calendrical skills in the context where it first occurs led to it being interpreted later as 'someone skilled in calculation', a sense possibly related to the similarly written and pronounced word *chou* 籌 'calculate'.

Wendi at first followed the advice of Gongsun Chen to consider reforms in the calendar, but later abandoned his plans.

But finally Sima Qian comes to the time of Wudi. The striking thing here, apart from the brevity of the discussion, is that his own role as Grand Clerk is not mentioned at all. Nor are any other office-holders mentioned by name. Two names are mentioned at the outset, but neither of them has a government post:

> 至今上即位，招致方士唐都，分其天部；而巴落下閎運算轉曆，然後日辰之度 與夏正同．乃改元，更官號，封泰山．

> When it came to the accession of the present emperor, he summoned the *fang shi* Tang Du, who distinguished the divisions of the heavens, while Luoxia Hong from Ba carried out calculations to revise the astronomical system. After that, the *du* of the chronological markers were in accordance with the Xia new year. Thereupon they began a new count of years, reformed the names of offices, and the *feng* ritual was performed at Mount Tai. (*Shi ji* 26, 1260)

Of these two men, the first had been the teacher of Sima Qian's father (*Shi ji* 130, 3288); Ba was in Sichuan, far from the capital. These were non-official consultants. When officials are referred to, there are no names, and it is evident that things did not go well so far as they were concerned. The passage just cited continues:

> 因詔御史曰: 「乃者, 有司言星度之未定也, 廣延宣問, 以理星度, 未能詹也 … 」.

> Thereupon an edict was issued to the Imperial Clerks, saying 'Now, the responsible officials have said that the degrees of the stars have not yet been determined. A broad enquiry for advice has been widely promulgated, so as to set the degrees of the stars in order, but they have not yet been able to reach a conclusion … ' (*Shi ji* 26, 1260)

In the rest of the edict, the emperor cites the Yellow Emperor as an example of a ruler supremely successful in setting such matters, beginning with what was the most important aspect of the question from the emperor's point of view:

> 蓋聞昔者黃帝合而不死，名察度驗，定清濁，起五部，建氣物 分數．然蓋尚矣．書缺樂弛，朕甚閔焉．朕唯未能循明也，紬績日分，率應水德之 勝．今日順夏至，黃鐘為宮，林鐘為徵，太蔟為商，南呂為羽，姑洗為角．自是以後，氣復正，羽聲復清，名復正變．

> Now [We] have heard that formerly the Yellow Emperor accorded [with the cosmos] and did not die. He checked the names and observed the *du* [of the stars], he corrected the pitch [of musical notes], he drafted the order of the Five Categories [*sc.* the Five Phases], he established the numerical basis of [seasonal]

> *qi* phenomena. So [such matters] evidently go back to the remote past. But to Our great sorrow, the writings are lost and the musical tradition is debased. We have not been able to follow [the Yellow Emperor's] brilliance, and weave together the day fractions, and so far [We] have conformed with the [notion of] the dominance of the power of Water [inherited from the Qin]. Now the days will follow the solstices of Xia. The pitchpipe Yellow Bell sets *gong* [the fundamental note of the scale], Forest Bell sets [the note] *zhi*, Grand Budding sets [the note] *shang*, Southern Regulator sets [the note] *yu*, and Maiden Purity sets [the note] *jue*.[22] So from now on, [seasonal] *qi* return to order, the notes are again pure, and the names are once more in their correct order. (Shi ji 26, 1260)

In view of what we have seen that Wudi had been led to expect from following the Yellow Emperor's example, it is not surprising that he sees the Yellow Emperor as providing the way out of the problem posed by his official's lack of capacity. The emperor then announces his intention to choose for the origin of the new astronomical system the precise date that in 113 BCE Gongsun Qing had stated to be in accord with the practice of the Yellow Emperor:

> 十一月甲子朔旦冬至已詹, 其更以七年為太初元年. 年名『焉逢攝提格』月名『畢聚』, 日得甲子, 夜半 朔旦冬至.

> [The occurrence of] the winter solstice on a *jiazi*.1 day on the first day of the 11th month has already been determined; let the seventh year [of the present reign period] be the first year of the Taichu 'Grand Inception' period. The year is named *yanfeng shetige*, and the month is named *biju*.[23] The day is placed at *jiazi*.1, and at midnight beginning the first day of the month, it is winter solstice. (*Shi ji* 26, 1260–1)

[22] These are the notes of a pentatonic scale. They are not given in ascending order of pitch, which would have been *gong, shang, jue, zhi, yu*, but in the order in which they would (in theory, and ignoring end-effects) have been generated by multiplying the length of the Yellow Bell pitch-pipe alternately by ⅔ and ⁴⁄₃. On this see the account given by Sima Qian in *Shi ji* 25, 1249, and the discussion in Joseph Needham, Ling Wang and K. G. Robinson (1962) *Science and civilisation in China, volume 4: Physics and physical technology. Part 1: Physics*. Cambridge, Cambridge University Press, 171–6.

[23] As the commentators point out, *yanfeng shetige* is a term made up from the elaborate forms of the names for the branches and stems of the sexagenary cycle listed in the (third century BCE) glossary *Er Ya* 爾雅 'Approaching Elegance'. *Yanfeng* stands for *jia* 甲, and *shetige* for *yin* 寅, so the year is named as *jiayin*.51 in the sexagenary sequence. In the term for the month, *bi* indicates that the month began on a day corresponding to the stem *jia* 甲, and *ju* that it is the first of the count (in this case the Celestial count, two months ahead of the Xia count): *Er Ya* 6, 95–96 in *Shi san jing zhu shu*. The reference to the month is reasonable enough, but as we shall see it is highly anomalous to designate 104 BCE as a *jiayin*.51 year, since in the sequence in use after 104 BCE it would have been *dingchou*.14. An attempt at a solution is found in Liu Tan 劉坦 (1955) *Lun xing sui ji nian* 論星歲紀年 *(On dating years by the Jupiter cycle)* Beijing, Science Press, 22–25, which argues that the *jiayin*.51 year-name was a proposal by Sima Qian and his colleagues that was never actually put into use. Liu also deals with the one-year shift forward in the year-name compared with the system apparently in use before 104 BCE.

And that is all we have from Sima Qian's monograph on astronomical systems about the execution of the Grand Inception reform. As expected, the text of the edict places the start of the new system as Gongsun Qing had predicted in 113 BCE, and it is made clear that Wudi wished to follow the example of the Yellow Emperor. But the rest of the account is so brief as to be puzzling. Did the Grand Clerk himself play no role in the proceedings? What was it that 'the officials' *you si* 有司 were unable to do that required the intervention of outsiders like Tang Du and Luoxia Hong? Was there any change in the fundamental constants of the system, such as the lengths of the mean lunation and solar cycle? And what is the astronomical significance of introducing the reference to pitchpipes? Sima Qian was in an excellent position to answer all these questions, but says nothing. However, as we shall see, there was much more that could have been said, but was not mentioned in the *Shi ji* account.

3.3.2 The account in the *Han shu*

Ban Gu 班固 (32–92 CE) wrote the *Han shu* monograph on mathematical harmonics and astronomical systems *Lü lu zhi* 律曆志 in the late first century CE. Unlike Sima Qian, who participated personally in the Grand Inception reform, Ban Gu had to rely on what he could find in official files and other documents from nearly two centuries before his day. On reading his account, it is clear that he is able—or perhaps we should say 'willing'—to tell us a much more detailed story than the one that Sima Qian decided to give us. For a start, he tells us that Sima Qian, as Grand Clerk, was a principal mover in the technical aspects of the reform, something that the Grand Clerk himself elides from his story:

> 至武帝元封七年，漢興百二歲矣，大中大夫公孫卿，壺遂，太史令司馬遷等言「曆紀壞廢，宜改正朔」.
>
> When it came to the seventh year of the Yuanfeng 'Epochal *Feng* [ritual]' period of Wudi [104 BCE], it was 102 years since the rise of Han; the Grand Counsellors of the Palace Gongsun Qing, and Hu Sui, the Grand Clerk Sima Qian and others, stated: 'The guiding threads of the astronomical system have been destroyed and abandoned. It would be fitting to correct the Standard Conjunction.' (*Han shu* 21a, 974–5; Cullen 2017, 367)

So, according to Ban Gu, Sima Qian joined with Gongsun Qing (who as mentioned in section 3.2.1 was the person who had originally put the idea of change into Wudi's head in 113 BCE) to propose, presumably in a memorial to the emperor, that a reform of the astronomical system should be set in motion. The text goes on to recount that—just as had been the case for the *feng* and *shan* sacrifices—the emperor ordered the learned Ni Kuan 倪寬 to give advice on what should be done. And, as mentioned by Sima Qian on that occasion, Ni Kuan adroitly replied that the emperor was the only person truly qualified to make such decisions:

> 今二代之統絕而不序矣, 唯陛下發聖德 …
>
> Now the continuity with the Concordances of the two preceding dynasties has been broken; so that the only solution is for Your Highness to manifest his sagely virtue … (*Han shu* 21a, 975; Cullen 2017, 368)

Next the *Han shu* quotes substantially the same edict as recorded by Sima Qian, in which the emperor refers to the failure of his officials to settle matters, and then, following the precedent set by the Yellow Emperor, announces the conditions from which the new astronomical system will begin. That having been said, we are given details unmentioned by Sima Qian:

> 遂詔卿, 遂, 遷與侍郎尊, 大典星射姓等議造漢曆. 乃定東西, 立晷儀，下漏刻, 以追二十八宿相距於四方, 舉終以定朔晦分至, 躔離弦望. 乃以前曆上元泰初四千六百一十七歲, 至於元封七年, 復得閼逢攝提格之歲，中冬十一月甲子朔旦冬至, 日月在建星, 太歲在子, 已得太初本星度新正.
>
> Then an edict was issued to [Gongsun] Qing, [Hu] Sui, [Sima] Qian and the Attendant Gentlemen Zun, the Senior Star Observer She Xing and others, that they should consult on the construction of a Han astronomical system. So they fixed east and west, set up gnomons [lit. 'instruments for observing shadows'], and set water clocks running, in order to find the extents of the 28 lodges in the four quarters,[24] and to seek a conclusion [of cycles] so as to fix the first and last days of months, with the equinoxes and solstices, and to trace the orbits of crescents and full moons. Thereupon they found that counting back the system to High Origin Grand Inception, it was 4,617 years[25] to the seventh year of the Yuanfeng period, when it was once more a *yanfeng shetige* [= *jiayin*.51][26] year, and in the second month of winter the winter solstice fell on a *jiazi*.1 day on the first

[24] On the connection between waterclocks and the 28 lodges, see chapter 5, section 5.4.

[25] In his major textual analysis of the *Han shu* Wang Xianqian suggests that this figure is in error, and should be restored to 4,560 years, which is the period at which system origin conditions repeat in systems of the quarter remainder type (see note 83): Wang Xianqian 王先謙 (1900, reprinted 1959) *Han shu bu zhu* 漢書補注 *(The Han shu, with addenda and comments)*. Beijing, (reprint) Commercial Press, p. 1667. I concur with him that the figure given here is influenced by later practice based on the system that was, as we shall see, adopted in preference to that proposed by Sima Qian and his colleagues.

[26] The anomalous nature of this year-name has already been noted: see note 177.

> day of the 11th [Xia] month, the sun and moon were at the Establishment Star,[27] and the Great Year was at *zi*.1,[28] so that they had obtained the new standard for the basic stars of the Grand Inception. (*Han shu* 21a, 975; Cullen 2017, 368–9)

To find Sima Qian working with Gongsun Qing is no surprise, and the participation of Hu Sui evidently took place with Sima's approval, since elsewhere in the *Shi ji* he speaks well of what he learned of Hu Sui's character while working with him 'to correct the pitchpipes and astronomical system' *ding lü li* 定律曆 (*Shi ji* 108, 2865). We may note that for the first time we actually find a name given to the new system—*Han li* 漢曆 'the Han astronomical system'. We shall shortly see this term used later in the Western Han, although the more common term in later writing is *Tai chu li* 太初曆 'the Grand Inception system'.

But what follows next is surprising:

> 姓等奏不能為算, 願募治曆者, 更造密度, 各自增減, 以造漢太初曆. 乃選治曆鄧平及長樂司馬可, 酒泉候宜君, 侍郎尊及與民間治曆者, 凡二十餘人, 方士唐都, 巴郡落下閎與焉. 乃選治曆鄧平及長樂司馬可, 酒泉候宜君, 侍郎尊及與民間治曆者, 凡二十餘人, 方士唐都, 巴郡落下閎與焉. 都分天部, 而閎運算轉曆.

> But [She] Xing and his colleagues submitted a memorial saying they had been unable to perform the calculations, and they wished to recruit specialists in [astronomical] systems, so as to make the *du* more accurate, with each proposing their own additions and subtractions, in order to make a Grand Inception system for the Han. So they chose the system specialists Deng Ping, Sima Ke from Changle, Yi Jun the Watcher from Jiuquan,[29] the Gentleman in Attendance Zun, in all more than twenty specialists in astronomical systems from amongst the people, amongst whom were Tang Du, and Luoxia Hong from Ba prefecture. [Tang] Du distinguished the divisions of the heavens [i.e. the lodges], while [Luoxia] Hong carried out calculations to revise the [astronomical] system. (*Han shu* 21a, 975; Cullen 2017, 369–70)

Sima Qian's own account did mention Tang Du and Luoxia Hong—but he made no reference to the sudden recruitment of so many others, or to the

[27] The Establishment Star *jian xing* 建星 is in the lodge Dipper. However, the Grand Inception system did not finally use this as the winter solstice position, but placed it instead at the start of the next lodge, Ox; see *Han shu* 21b, 1005, and Cullen (2017), 112. It appears that the change of personnel referred to in the next paragraph led to significant changes in the astronomical basis of the new system—as we are told, the new team 'observed new *du* for the stars, and the motions of sun and moon'.

[28] The term *tai sui* 太歲 'Grand Year' here refers to the sequence of twelve cyclical characters (the 'earthly branches') used as part of the designation of years in a sexagenary cycle: see note 79. In the Triple Concordance system as set out by Liu Xin based on the Grand Inception system, the civil year beginning in 104 BCE was 丙子 *bingzi*.13.

[29] Jiuquan, located in modern Gansu province, was a frontier prefecture under the Han. Despite the temptation to translate *hou* 候 as 'observer' in this context, the word seems to have been a military title in this case (compare *Han shu* 69, 2980), and hence I translate as 'watcher'.

problem that led to this decision. What exactly was it that the first group of experts had found so difficult to deal with? That is by no means obvious—although it might have been the strange choice of *jiayin*.51 for the year-name—but whatever it was, it is clear that the skills relevant to solving such problems were expected to be widely distributed outside the group of those officials in the capital bureaucracy, such as Sima Qian, whose jobs gave them responsibilities related to the heavens, and might even be found in distant provinces. But next we come to specifics, and amongst other things we find the solution to the mysterious reference to the Yellow Bell pitchpipe in Sima Qian's account:

> 其法以律起曆，曰：「律容一龠，積八十一寸，則一日之分也．與長相終．律長九寸，百七十一分而終復．三復而得甲子．夫律陰陽九六，爻象所從出也．故黃鐘紀元氣之謂律．律，法也，莫不取法焉．」與鄧平所治同．於是皆觀新星度，日月行，更以算推，如閎，平法．法，一月之日二十九日八十一分日之四十三．[...]

> Their method derived the astronomical system from the regulating [pitchpipes]. They said: 'The [Yellow Bell] regulating pitchpipe has a capacity of 1 *yue*, which is a volume of 81 *cun*. So [that number] is the divisor for one day. It is commensurable with the length. The regulator is 9 *cun* long, and a 171-fold division takes us back to the conclusion. After three repeats, one [again] obtains *jiazi*.1 [as the day number]. The regulators are yin and yang, [corresponding to] nine and six, from which the lines of the hexagrams of the *Book of Change* are generated.[30] Thus the Yellow Bell is called a "regulator" since it gives the guiding thread for the original *qi*. "Regulator" means "pattern/divisor", and there is nothing that does not take its pattern from it.' This agreed with what Deng Ping had determined. Thereupon they observed new *du* for the stars, and the motions of sun and moon, and recalculated in accordance with the pattern of [Luoxia] Hong and [Deng] Ping. The pattern was that the days of one lunation were 29 and $^{43}/_{81}$ days. [...][31] (*Han shu* 21a, 976; Cullen 2017, 370)

[30] In the divinatory process that generates the hexagrams of the *Book of Change*, nine corresponds to an unbroken (yang) line, and six to a broken (yin) line.

[31] At this point the *Han shu* has the following text, which seems out of place. No commentators offer an explanation of its significance, but it evidently relates to when the next first crescent appears in relation to the end of a month that may have either 29 or 30 days. It may be an intruded note, or displaced text from elsewhere. 先藉半日，名曰陽曆；不藉，名曰陰曆．所謂陽曆者，先朔月生；陰曆者，朔而後月乃生．平曰：「陽曆朔皆先旦月生，以朝諸侯王羣臣便．」

'If it is in advance and borrows half a day, that is called a yang sequence; if there is no borrowing, it is a yin sequence. What is meant by "yang sequence" is that the moon appears before the [calculated] conjunction. It is a "yin sequence" if the moon only appears after the [calculated] conjunction. [Deng] Ping said "On the first day of a yang sequence, the moon always appears before dawn, for the convenience of the nobles, princes and all the ministers who attend court."' These Yin and Yang sequences seem to have nothing to do with the terms used by Liu Hong nearly three centuries later to define the latitude cycle of the moon: see section 8.3.1.2.

This new method evidently followed radically different principles from the *si fen* 四分 'quarter [remainder]' methods described in chapter 2, which appear to have been the basis of all astro-calendrical calculations before 104 BCE. The basic numbers of the new system are set out in Box 3.1. They are the same as those given in the detailed system description given by Liu Xin c. 10 AD, and recorded in *Han shu* 21b, to which we shall turn later.

Box 3.1: Basic constants of the system introduced in the Grand Inception reform of 104 BCE

The text gives a clear statement of the length of a lunation (synodic month), 29 $^{43}/_{81}$ days. This exceeds the quarter remainder value of 29 $^{499}/_{940}$ days by only $^{0.012}/_{940}$ day, 1.1 seconds, and the modern mean value (see 1.3) by 23 seconds.

No explicit statement is made about the length of the solar cycle. In fact, from the fuller information given by Liu Xin c. 10 CE and recorded in the second part of the *Han shu* monograph on harmonics and astronomical systems (*Han shu* 21b), we learn that solar cycle length is derived by combining this lunation length with the familiar equivalence that we have already seen in quarter remainder methods:

19 solar cycles = 235 lunations

Thus the length of a solar cycle will be 235 × (29 $^{43}/_{81}$)/19 days

= 235 × 2,392/(19 × 81) days = 562,120/1,539 days = 365 $^{385}/_{1,539}$ days

This exceeds the quarter remainder value of 365 ¼ days by only $^{0.25}/_{1,539}$ days, 14 seconds, and the modern value (see 1.3) by 10 minutes.

1,539 solar cycles contain a whole number of days:

365 $^{385}/_{1,539}$ days × 1,539 = 562,120 days

They will also contain a whole number of lunations, since there are 235 lunations in 19 solar cycles, and:

1,539 × $^{235}/_{19}$ = 19,035, which is a whole number.

Since 1,539 = 9 × 171, we can see the significance of the references to the numbers 9 and 171 in the *Han shu* text.

Since 562,120 days = (9,368 × 60 + 40) days, it is plain that 1,539 solar cycles will not take us back to the same sexagenary day that we started from, but to

continued

Box 3.1: *Continued*

40 days later. Since 3 × 40 = 120, after 1,539 × 3 = 4,617 solar cycles we shall get back to the same sexagenary day—exactly as the text told us:

'After three repeats one [again] obtains *jiazi*.1 [as the day number].'

In the full account of the Grand Inception system given in *Han shu* 21b the periods above are named:

1,539 years: a *tong* 統 'Concordance'; 4,617 years: a *yuan* 元 'Origin'.

In the next part of the Ban Gu's account, it becomes clear that Sima Qian himself no longer had a voice in the technical decisions that were being made:

> 乃詔遷用鄧平所造八　十一分律曆，罷廢尤疏遠者十七家，復使校曆律昏明. 宦者淳于陵渠復覆太初曆晦朔弦望，皆最密，日 月如合璧，五星如連珠. 陵渠奏狀，遂用鄧平曆，以平為太史丞.
>
> Then [Sima] Qian was ordered to use the system that Deng Ping had devised based on the regulators, with a factor of 81, and to discard seventeen methods that were less accurate, so as to use dawn and dusk timings that were in accord with the astronomical system and the regulators. The Palace Eunuch Shunyu Lingqu reworked the last days of lunations, conjunctions, crescents and oppositions for the Grand Inception system, and they were all extremely accurate, with 'the sun and moon like joined jade discs, and the five planets like a string of pearls.' [Shunyu] Lingqu memorialized accordingly, whereupon Deng Ping's system was put into use, and [Deng] Ping was made Deputy Grand Clerk. (*Han shu* 21a, 976; Cullen 2017, 371)

So the Grand Clerk was *instructed* to use a new method devised by a person who had been called in as one of the specialists 'from among the people' in order to deal with the fact that the original group of three officials—himself, Hu Sui and Gongsun Qing—could not get the figures to work out as required. What is more, the principal innovator, Deng Ping, appears to have been imposed on him as a deputy. It is perhaps not at all surprising that Sima Qian did not choose to record his own discomfiture in detail.

3.3.3 The justification of the new system

It is now time to examine a number of important aspects of the events of 104 BCE, and their consequences.

The records of what happened in 104 BCE are without precedent in the history of celestial calculation in China. In the accounts given by the *Shi ji* and the *Han shu*, we can for the first time trace in detail the work of a group of named specialists tasked with creating a new astronomical system as the basis of calendrical calculations. The *Han shu* text is also the earliest text to give a name to such a system—the *Han li* 漢曆 'Han system', although in later times it was more normally referred to as the *Tai chu li* 太初曆 'Grand Inception system'. Moreover, as we shall shortly see, we are given enough information to enable us to begin to run calculations as specified by the new system, and to test whether the results of those calculations are in accordance with the calendar actually in use after 104 BCE.

Another striking feature of the Grand Inception reform is that (so far as the record goes) we have only two pieces of evidence as to why the reform was undertaken:

1. We know that eight years earlier, in 113 BCE, Gongsun Qing had persuaded Wudi that inaugurating a new system based on coincidence of winter solstice and the conjunction of the 11th Xia month in 105 BCE might enable him to emulate the Yellow Emperor, and ascend to heaven as an immortal.
2. We know that the figure of 81 was ostensibly chosen as the denominator of the day fraction in lunation length on the grounds that this was supposedly the volume of the Yellow Bell pitchpipe. But as shown in Box 3.1, another effect of this change is to produce the slight increase in solar cycle length that makes it possible for a winter solstice at the start of day *jiazi*.1 in 105 BCE to be consistent with the fact that the calendar of 113 BCE, referred to by Gongsun Qing, placed the winter solstice on day *xinsi*.18: see section 3.2.3.

Did winter solstice and the actual moment of luni-solar conjunction (the 'true conjunction') actually coincide at the instant chosen for the new system—midnight beginning the *jiazi*.1 day which was the first day of the 11th Xia month in 105 BCE (25 December in the proleptic Julian calendar)? A check with laptop astronomical software (Starry Night Pro™) yields the following results, expressed in local time at Chang'an, to the nearest ten minutes:[32]

Winter solstice: 23 December, 19:35
True conjunction: 24 December, 07:10

So the actual solstice is over 28 hours earlier than assumed, while the true conjunction is 17 hours earlier than assumed; the mean conjunction, which is more important for prediction future conjunctions, did not fall until some time after

[32] Sivin (1969), 23 interpolates in the tables of Tuckerman (1962), producing results that are (reduced to Chang'an time) about two hours earlier than that given here for the solstice, and three hours later for the conjunction.

the true conjunction.[33] We noted in section 2.1.1 that there were clear signs during Western Han that conjunctions were being systematically predicted about a day too late, as shown by the frequent occurrence of solar eclipses on the last day of the lunar month, *hui*, rather than on the first day, *shuo*, when conjunction (and hence any eclipse of the sun) should ideally fall. In fact, out of 29 Han eclipses recorded before 104 BCE, only five fell on the first day of the month. The rest fell on *hui*, except for three cases where they fell on the day before *hui*. As suggested by the significant lag in the predicted conjunction in December 105 BCE, the situation did not improve much after the Grand Inception reform: of the 25 Western Han solar eclipses recorded after the reform, only 9 fell on *shuo*, with all the rest on *hui*. However, nobody associated with the Grand Inception reform gives any sign of seeing eclipses not falling on *shuo* as being anomalous, nor do we have any record of anybody else in the Western Han making any remark on eclipses being a useful tool for evaluating the performance of a *li*. We shall consider possible reasons for this in chapter 6.

Apart from the use of eclipses to check on conjunction predictions, the means available for checking on actual astronomical conditions at the time of the Grand Inception reform were limited. Solstices could only be determined by watching for the length of the noon shadow of a gnomon of height 8 *chi* 尺 (about 2 metres) to reach its maximum length (winter solstice) and minimum length (summer solstice). Shadow length varies only slowly close to the solstices, and calculations by Nakayama Shigeru referred to previously suggest that an error of 1 cm in estimating the length of the shadow would have created an error of 4 to 5 days in fixing the winter solstice.[34] Unless a solar eclipse occurs, luni-solar conjunction is not susceptible to direct observation (since the moon is too close to the sun to be seen), and its occurrence can only be deduced by the first visibility of a thin crescent moon soon after sunset around two days later, as the moon shifts eastwards away from the sun. Observations made around 25 December 105 BCE would therefore have been unlikely to point clearly to any discrepancy between prediction and observation, especially if the question

[33] We may estimate the timing of the mean conjunction near 24 December 105 BCE by listing the instants of true conjunction from June 105 BCE to June 104 BCE, and then finding the best fit to those of a series of equally spaced events at intervals of a mean synodic month (29.53059 days); best fit is defined by minimizing the root mean square error between the two series of events. The mean conjunction on 24 December 105 BCE thus defined was about 9 hours later than the true conjunction on that date. During the period considered the interval between conjunctions varies from the mean value by about five or six hours either way, so any such value should only be taken as a general indication that prediction is substantially behind observation; the actual discrepancy will vary considerably from month to month.

[34] Nakayama (1969), pp. 242–3.

asked was not 'what is our best estimate of the instants of solstice and conjunction?' but instead (and more probably) 'do the observed phenomena flatly contradict the conditions supposed by the new system?'.

In fact, the reasons for choosing midnight beginning 25 December 105 BCE as the starting point of the new system did not depend crucially on contemporary observation, since as we have seen this date had already been laid down in 113 BCE, on the basis of Gongsun Qing's account of the Yellow Emperor. It was the emperor's wish to emulate the Yellow Emperor in attaining immortality that was the real driving force behind the reform.

3.3.4 The adoption of the new system

As we have seen, we have two accounts of the events of 104 BCE from the technical point of view. The only account by someone who actually took part, Sima Qian, appears to be deliberately sketchy, and gives us no hint of the radical changes in technical practice and introduction of new personnel that are revealed in the *Han shu* account. The latter, however, was mostly written by Ban Gu 班固 two centuries after the events described. We have to ask how far Ban Gu's description may have read back into the past the thought and practice of his own time—particularly the work of Liu Xin 劉歆 (c. 50 BCE–23 CE), large parts of whose writings on cosmology and astronomical systems he appears to have transcribed.[35] Thus, for instance, *Han shu* 21b begins with a detailed description of an astronomical system whose basic constants are identical to those laid out in the brief *Han shu* references to the 104 BCE events. Can we be sure that Deng Ping's was really the system that was put into effect?

Fortunately, we need be in no doubt that the new methods of calculation said in the *Han shu* to have been introduced in 104 BCE were actually in use not long after that time. Table 3.2 shows a transcription of the principal data from an excavated specimen of a bamboo strip calendar dated to 69 BCE. Figure 3.2 shows a photograph of the original.

As with the earlier specimen discussed in chapter 2, the top of each strip is marked with the numbers of days in the month, running from right to left. The rows below give the sexagenary day names of the corresponding days of each month, transcribed in the table as the corresponding numbers. The calendar is missing its first few strips on the right, which would have given the numbers of the months and the sexagenary day names for the first three days of each month; the strip for the eleventh day is also missing. It is, however, possible to restore

[35] See the comment by Yan Shigu 顏師古 (581–645 CE) in *Han shu* 21a, 956 note 7.

Table 3.2 ***Transcription of principal data from excavated calendar of 69 BCE (Liu Lexian 刘乐贤 2011: 23, fig. 2). Restored data shown in italics***

30	29	28	27	26	25	24	23	22	21	20	19	18	17	16	15	14	13	12	*11*	10	9	8	7	6	5	4	*3*	*2*	*1*	Day	*Month*
	5	4	3	2	1	60	59	58	57	56	55	54	53	52	51	50 LC	49	48	*47*	46	45	44	43	42	41	40	*39*	*38*	*37*		*1*
35	34	33	32	31	30	29	28	27	26	25	24	23	22	21	20	19	18	17	*16*	15	14	13	12	11	10	9	*8*	*7*	*6*		*1 int*
	4	3	2	1	60	59	58	57	56	55	54	53	52	51	50	49	48	47	*46*	45	44	43	42	41	40	39	*38*	*37*	*36 CF*		*2*
34	33	32	31	30	29	28	27	26	25	24	23	22	21 LX	20	19	18	17	16	*15*	14	13	12	11	10	9	8	*7*	*6*	*5*		*3*
4	3	2	1	60	59	58	57	56	55	54	53	52	51	50	49	48	47	46	*45*	44	43	42	41	40	39	38	*37*	*36*	*35*		*4*
	33	32	31	30	29	28	27	26	25	24	23	22	21	20	19	18	17	16	*15*	14	13	12	11	10	9	8	*7* XZ	*6*	*5*		*5*
3	2	1	60	59	58	57	56	55	54	53 LQ	52	51	50	49	48	47	46	45	*44*	43	42	41	40	39	38	37	*36*	*35*	*34*		*6*
	32	31	30	29	28	27	26	25	24	23	22	21	20	19	18	17	16	15	*14*	13	12	11	10	9	8	7	*6*	*5*	*4*		*7*
2	1	60	59	58	57	56	55	54	53	52	51	50	49	48	47	46	45	44	*43*	42	41	40	39	38 QF	37	36	*35*	*34*	*33*		*8*
	31	30	29	28	27	26	25	24 LD	23	22	21	20	19	18	17	16	15	14	*13*	12	11	10	9	8	7	6	*5*	*4*	*3*		*9*
1	60	59	58	57	56	55	54	53	52	51	50	49	48	47	46	45	44	43	*42*	41	40	39	38	37	36	*35*	*34*	*33*	*32*		*10*
	30	29	28	27	26	25	24	23	22	21	20	19	18	17	16	15	14	13	*12*	11	10 DZ	9	8	7	6	5	*4*	*3*	*2*		*11*
60	59	58	57	56	55	54	53	52	51	50	*49*	48	47	46	45	44	43	42	*41*	40	39	38	37	36	35	*34*	*33*	*32*	*31*		*12*

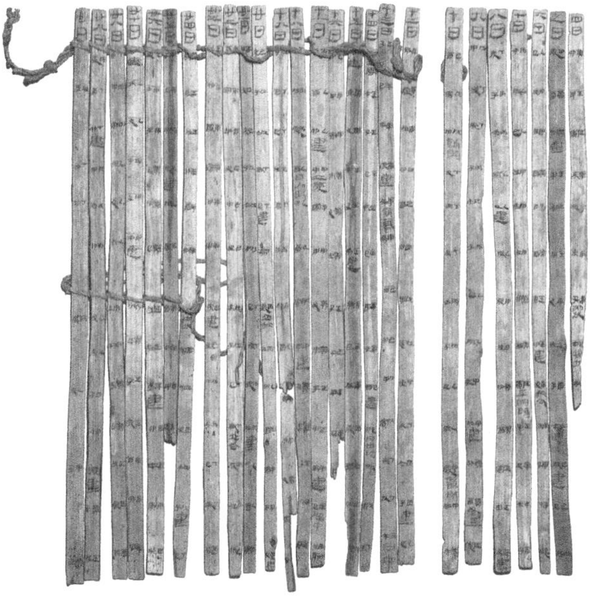

Figure 3.2 Bamboo strip bundle type calendar, 69 BCE. From Liu Lexian 2011, p23 fig 2, modified to show the strip for day 7 in the correct position between the strips for days 6 and 8. The strips for days 1, 2, 3 and 11 are missing so the first strip on the right corresponds to day 4. The last strip on the left corresponds to day 30.

the missing data (shown in italics). But even without any detailed deciphering of the writing on the strip, two features are already obvious:

(a) The calendar has thirteen rows corresponding to months, and this year therefore has an intercalary month.
(b) As expected, only some months have a day on the last (leftmost) strip, which marks the 30th day—except that the third and fourth month-rows both have a 30th day, and are thus an example of successive long months.

Looking more closely, we can see that day 14 of the month in the row just below the row of day numbers has the annotation *li chun* 立春 'Establishment of Spring' (LC in the transcription). This is the fourth of the 24 *qi*, implying that the fifth *qi* falls just at the end of the month, making this the first Xia month (see 2.2.1.2). This calendar therefore follows one new feature of the Grand Inception reform, which made this month the first of the civil year, rather than beginning with the tenth Xia month as had been the practice in Qin and early Western Han. Further confirmation is given by the fact that the annotation *dong zhi* 冬至 'winter solstice' (transcribed DZ), the first of the 24 *qi* appears in the row corresponding to the 11th month, as expected if this is the 11th Xia month. Other *qi* annotations are:

Chun fen 春分 'Spring Equinox' (transcribed CF) second month, day 1, restored.
Li xia 立夏 'Establishment of Spring' (transcribed LX) third month, day 17.
Xia zhi 夏至 'Summer Solstice' (transcribed XZ) fifth month, day 3, restored.
Li qiu 立秋 'Establishment of Autumn' (transcribed LQ) sixth month, day 20.
Qiu fen 春分 'Autumn Equinox' (transcribed QF) eighth month, day 6.
Li dong 立冬 'Establishment of Winter (transcribed LD) ninth month, day 22.

But do the details of this calendar really fit what the new system requires for the year in question? Detailed calculations show that this is precisely the case: see Box 3.2 for examples of how this can be done.

Box 3.2: Using the new system to calculate dates in the 69 BCE calendar

Winter solstice

The calendar marks the winter solstice as falling on sexagenary day *guiyou*.10, and we know that the winter solstice of late 105 BCE fell at the midnight beginning day *jiazi*.1.

The total number of solar cycles elapsed from the 105 BCE solstice to this one is:

$$105 - 69 = 36.$$

Each solar cycle contains 365 $^{385}/_{1,539}$ days.

So the total number of days elapsed is:

$$36 \times (365\ ^{385}/_{1,539}) \text{ days} = 36 \times\ ^{562,120}/_{1,539} \text{ days}$$
$$= 13,149\ ^{9}/_{1,539} \text{ days}$$

Now 13,149 = 219 × 60 + 9

Therefore if the 105 BCE solstice fell at midnight beginning day *jiazi*.1, the solstice of 69 BCE should fall $^{9}/_{1,539}$ day after midnight on the day with sexagenary number:

1 + 9 = 10, i.e. *guiyou*.10, just as marked on the calendar for the ninth day of the 11th month.

First days of months

Since we know the number of days from midnight at the start of the 11th month of 105 BCE, and we know that the length of a lunation is 29 $^{43}/_{81}$ days, we can calculate the number of whole lunations elapsed between that date and the winter solstice of 69 BCE. This will be the whole number result of:

continued

Box 3.2: *Continued*

$(13{,}149\ {}^{9}\!/\!_{1{,}539}\ \text{days})/(29\ {}^{43}\!/\!_{81}\ \text{days}) = (20{,}236{,}320/1{,}539)/(2{,}392/81)$
$= (20{,}236{,}320/2{,}392) \times (81/1{,}539)$
$= (20{,}236{,}320/2{,}392)/19$
$= 445$ to the nearest whole number

So the number of days elapsed is:

$$445 \times (29\ {}^{43}\!/\!_{81})\ \text{days} = 13{,}141\ {}^{19}\!/\!_{81}\ \text{days}$$

And $13{,}141 = 219 \times 60 + 1$ day
So since the 11th month of 105 BCE begins on a *jiazi*.1 day, the lunation just before the 69 BCE winter solstice should fall on a *yichou*.2 day, beginning the 11th month. Since the first to fourth day names of that month are missing, we can only check the name for the fifth day, which is *jisi*.6, as expected if the first day was *yichou*.2.

Similar calculations verify that all the other dates on this calendar are as predicted by the Grand Inception system.

3.4 The affair of 78 BCE

The basic solar and lunar constants of the Grand Inception system remained in use until the late first century CE. It was not, however, immune to criticism. In 78 BCE, 27 years after the Grand Inception reform, the Grand Clerk Zhang Shouwang 張壽王 submitted a memorial to Zhaodi 昭帝 (r. 87–74 BCE), at that time only 15 years old, in which he stated:

> 曆者天地之大紀，上帝所為．傳黃帝調律曆，漢元年以來用之．今陰陽不調，宜更曆之過也．

> The astronomical system is the great thread guiding heaven and earth, made by the Supernal Lord. From the beginning of the Han [up to the Grand Inception reform], the dynasty made use of the Yellow Emperor's *Tiao lü li* 調呂曆 'astronomical system harmonizing the standard pitchpipes'. Now [under the new system] Yin and Yang are no longer in harmony, and it is time to put right the faults in the system. (*Han shu* 21a, 978; Cullen 2017, 371)

Once more we see the Yellow Emperor as an obvious reference point in calendrical matters. The memorial was referred for further investigation to the calendrical expert Xianyu Wangren 鮮于妄人. At first he is said to have questioned

Zhang Shouwang, without receiving a satisfactory response. His reaction to that is very revealing of the way he thought that disputes about astronomical systems ought to be settled:

> 妄人請與治曆大司農中丞麻光等二十餘人雜候日月晦朔弦望，八節二十四氣，鈞校諸曆用狀．奏可．
>
> Wangren requested that he should be allowed to work with the expert in *li* Ma Guang, [holder of the office of] Palace Assistant to the Grand Supervisor of Agriculture, and more than twenty other persons to make varied solar and lunar observations, including last and first days of lunations, crescents and oppositions, the eight main divisions of the solar cycle and the 24 *qi*, and to make comparisons of the merits in use of various *li*. His request was approved. (*Han shu* 21a, 978; Cullen 2017, 372)

The results of this programme, which was continued for two years on the 'Pure Terrace of the Shanglin [park]' *Shang lin qing tai* 上林清臺,[36] were reported as having vindicated the predictions of the Grand Inception system and thus to have shown that Zhang's astronomical system was seriously in error.

Zhang did not confine his unorthodox views to calendrical matters. He was also condemned for holding that the time of the Yellow Emperor was more than 6,000 years before the time of his memorial, whereas the official view placed his reign 3,269 years before the time Zhang wrote. He also believed that two sovereigns had held the throne who were not in the accepted sequence, one in succession to the founder of the Xia dynasty, and another 'between the Shang and Zhou dynasties'. This second, he held, was a woman. Zhang was arraigned on a charge of 'Great Disrespect [to the Emperor]', roughly equivalent in later European terms to blasphemy combined with *lèse majesté*, but he was eventually pardoned. Despite this good fortune he could not keep his ideas to himself and was eventually re-arrested, after which we hear no more of him.

3.4.1 The question of the 'six systems'

At one point in the narrative we followed in the previous section, Xianyu Wangren reports one of his group's conclusions as follows:

> 壽王曆乃太史官殷曆也．
>
> The system used by [Zhang] Shouwang is in fact the Yin system [recorded in] the Grand Clerk's office. (*Han shu* 21a, 978; Cullen 2017, 373)

[36] This appears to have been the term used for the observation site of official skywatchers at this period. From the Eastern Han onwards, the term 'Numinous Terrace' *Ling tai* 靈臺 is used.

This is our first encounter with the assumption that there were a number of named 'alternative' astronomical systems in circulation, and it is worth pausing to look at this situation in more detail. By the second century of the common era, the idea seems to have become widely accepted that before the imperial age there had existed different *li*, each with its own characteristics and its own name. The first sign of this view is found in the text of the *Han shu*, which states:

> 故自殷周，皆創業改制，咸正曆紀，服色從之，順其時氣，以應天道，三代既沒，五伯之末史官喪紀，疇人子弟分散，或在夷狄，故其所記，有黃帝，顓頊，夏，殷，周及魯曆.
>
> So from the Yin and Zhou [dynasties] all established their duties and reformed their regulations, comprehensively corrected the sequence of their systems, and the ritual colours followed this, according with the *qi* of the seasons, so as to respond to the way of Heaven. When the three dynasties collapsed, then at the end of the time of the Five Hegemons, the Clerks' officials missed their reckonings, and the masters and their disciples were scattered, some amongst the outer barbarians. Thus in their records, there are the systems of the Yellow Emperor, Zhuan Xu, Xia, Yin, Zhou and Lu. (*Han shu* 21a, 973; Cullen 2017, 366)[37]

And a few pages later:

> 至孝成世，劉向總六曆，列是非，作五紀論.
>
> In the time of Chengdi (r. 33–7 BCE), Liu Xiang gave a complete account of the six [astronomical] systems, setting out the rights and wrongs of them, and composed his *Wu ji lun* 'Discussion of the five sequences of time'. (*Han shu* 21a, 979; Cullen 2017, 374)

Since Liu Xiang 劉向 (79–8 BCE) and his son Liu Xin 劉歆 (c. 50 BCE–23 CE) were responsible for the bibliographical work that the *Han shu* editors reproduced in abbreviated form, it is not surprising to find the following entries in the *Li pu* 歷譜 '[astronomical] systems and listings' section of the *Han shu* bibliography, each followed by the number of rolls of bamboo strips composing the given text:

> 黃帝五家曆三十三卷.
> 顓頊曆二十一卷.
> 顓頊五星曆十四卷.
> 日月宿曆十三卷.

[37] A short reference to the role of political disorder in the origins of the six systems, referred to as the *liu jia* 六家 'six schools' is also found in the editorial essay on the composition of the *Han shu* in *Han shu* 100b, 4241.

夏殷周魯曆十四卷.
天曆大曆十八卷.
漢元殷周諜曆十七卷.
The systems of the Yellow Emperor [and the other] five schools: 33 rolls.
The Zhuan Xu system: 21 rolls.
The Zhuan Xu system for the 5 planets: 14 rolls
The system for the lodges of the sun and moon: 13 rolls.
The Xia, Yin, Zhou and Lu systems: 14 rolls.
The Celestial system and the Great system: 18 rolls.
Tables of the Yin and Zhou systems [from] the first year of Han: 17 rolls.
(*Han shu* 30, 1765–6)

All these writings, together with Liu Xiang's *Wu ji lun*, are now lost. We do know, however, that the latter text must have survived into the time of the Tang scholar Kong Yingda 孔穎達 (574–648 CE), since he notes that the account of the Yin system that it contained only gave methods for the calculation of the *qi* (the subdivisions of the solar cycle, mentioned in section 3.3.4) and conjunctions, and specified no means for calculating planetary motions.[38]

No doubt Liu Xiang gave a complete list of when each of these systems had their system origin *li yuan* 曆元—that is, the moment when all elements such as the solar and lunar cycles, and the cycle of sexagenary days, were at their initial states, which may be used as a convenient starting point for predicting later conditions. But in default of his work we have no document from early imperial times that gives us this information in complete form.[39] His son Liu Xin did, however, write a major chronological study, the *Shi jing* 世經 'Canon of the Ages' in which he compared the dates in ancient texts with the predictions of his *San tong li* 三統曆 Triple Concordance system, which was based on the Grand Inception system: see chapter 4, section 4.5. He frequently gives comparisons with dates according to the Yin system; for example, he tells us that according to the Yin system the winter solstice of late 48 BCE coincided with conjunction on day *jiazi*.1, making it a *ji shou* 紀首 'Era Head' for the Yin system (*Han shu* 21b, 1024). As the Era

[38] See *Shi san jing zhu shu, Shi jing, Da ya, Da ming* 大明. (Edn of Ruan Yuan, 543–2): 劉向五紀論載殷曆之法. 唯有氣朔而已. 其推星在天黿, 則無術焉. 'The methods for the Yin system contained in Liu Xiang's *Wu ji lun* only have [calculations for] the *qi* and the conjunctions. As for calculating whether a [given] planet was in [the asterism] Celestial Turtle [as was said to have been the case when the Zhou conquered the Shang dynasty], it has no such procedure.'

[39] Several centuries later, we find a list of the systems with the years elapsed from their system origins to 714 CE in ch.105 of the *Kai yuan zhan jing* 開元占經 of Qutan Xida 瞿曇悉達, c. 725 CE. In the preface to his monograph on calendrical astronomy now incorporated in the *Hou Han shu*, Sima Biao (c. 240–c. 306 CE) gives us a list of the sexagenary years of the origins for each system, but no actual dates (*Hou Han shu, zhi* 3, 3082).

Cycle used with quarter remainder systems such as the Yin system is 1,520 years, the preceding Era Head for the Yin system must have been in late 1,568 BCE, and the winter solstice of later 105 BCE could not thus have been an Era Head according to the Yin system. Since this was the solstice that began the Grand Inception system, Zhang Shouwang's rejection of the Grand Inception is understandable, if, as his colleagues stated, he was in fact using the Yin system.

3.4.2 Identifying the Qin and Han systems: textual evidence

Thus, in the last few decades of the first century BCE, when Liu Xiang did his work, it seems to have been accepted by some that the 'six systems' had actually existed, presumably since some time in the pre-imperial period. So if we wish to identify the system in use before 104 BCE, it seems reasonable to ask which one of them it could be.

There is in fact is no statement dated to before 104 BCE that actually tells us the name of a system in current use, or even gives us the name of any system.[40] The *Huai nan zi* does not do so. In his account of relevant issues from early times up to 104 BCE in the *Li shu* 曆書 'Treatise on the calendar', chapter 26 of the *Shi ji*, Sima Qian says that the Han dynasty continued the practice of the Qin, but he does not mention a named system. Nor does he do so in the quite frequent references to controversies on calendrical matters in the *Feng shan shu* 'Treatise on the *feng* and *shan* sacrifices', chapter 28 of the *Shi ji*. When the *Han shu* covers the same ground nearly two centuries later, things are quite different. Consider for instance the contrast between two very similar passages, first the *Shi ji* and then the *Han shu*. The first has already been cited earlier in this chapter:

> 漢興, 高祖曰「北時待我而起」, 亦自以為獲水德之瑞. 雖明習曆及張蒼等, 咸以為然. 是時天下初定, 方綱紀大基, 高后女主, 皆未遑, 故襲秦正朔服色
>
> At the rise of Han, the High Ancestor [Liu Bang 劉邦] said 'It was the northern borders that rose in my support', so accordingly he took himself to have received an omen of the Power of Water. Even brilliant practitioners of [astronomical] systems, with Zhang Cang and others, all took it to be so. At that time, the empire had only just been settled, and they were only just about to knit together the foundational structures; when [later] the High Empress ruled [though] a

[40] In one famous passage in the Analects, Confucius advises a disciple who asks about how to run a state *Xing Xia zhi shi* 行夏之時 'Follow the seasons of Xia' (*Lun yu* 論語 15, 10). But seems likely to be no more than an injunction to use the Xia first month as the reference point for the year.

woman, there was still no leisure [for dealing with such matters]. So they continued using the standard conjunction and ritual colours of Qin. (*Shi ji* 26, 1260)

So all that is said to have been carried on from Qin is the correlation with the phase Water, the use of the Xia tenth month for the New Year, and the ritual colour black. Nothing is said about methods of calculation. The *Han shu* account is significantly different:

漢興, 方綱紀大基, 庶事草創, 襲秦正朔. 以北平侯張蒼言, 用顓頊曆, 比於六曆, 疏闊中最為微近. 然正朔服色, 未覩其真, 而朔晦月見, 弦望滿虧, 多非是.

At the rise of Han, they were only just about to knit together the foundational structures, and many matters were dealt with provisionally, so they continued using the standard conjunction of Qin. In accordance with the words of the Beiping Marquis Zhang Cang, they used the Zhuan Xu system, since in comparison with [the rest of] the Six Systems, it was the closest approximation out of those inaccurate [systems]. They had not yet seen the real [situation in relation to] the standard conjunction and the ritual colours. Thus the moon might be visible on the first and last days of a month, and the crescents and oppositions and waxings and wanings were mostly in error. (*Han shu* 21a, 974)

Three elements absent from the *Shi ji* version appear in the *Han shu* account:

(a) There is a reference to a named system, the Zhuan Xu system.
(b) The 'six systems' are mentioned.
(c) The notion of accuracy is introduced.

Events at the start of the Han dynasty are an area where it is clearly very unlikely that the *Han shu*, completed around 110 CE, had access to sources additional to those of the *Shi ji*, completed around 90 BCE. We may reasonably conclude that the *Han shu* editors are reading back later views into their modification of the *Shi ji* material.

It seems, on the basis of what we have seen so far, difficult to trace the idea of the 'six systems' any further back than the reign of Chengdi (r. 33–7 BCE), when Liu Xiang wrote. There is, however, one possible reference to a named system that may be earlier. As we have seen the main discussions of calendrical matters in the *Shi ji* do not contain the references to the *Zhuan xu* system that appear in the account given in the *Han shu*. However, after the section of the *Shi ji* that contains the biography of Zhang Cang, presumably written some time before 90 BCE, there is the following editorial entry:

太史公曰:「張蒼文學律曆, 為漢名相, 而絀賈生, 公孫臣等言正朔服色事而不遵, 明用 秦之顓頊曆, 何哉？

> His Honour the Grand Clerk observes: Zhang Cang's literary learning [embraced] harmonics and the calendar, and he was a famous minister of the Han. But he treated with contempt Jia Sheng, Gongsun Chen and others who spoke of the matter of the standard conjunction and ritual colours, and did not follow them. How could it be that he clearly used the Qin's Zhuan Xu system?
>
> (*Shi ji* 96, 2685)

Sima Qian's last words are worth comparing with a similar expression in the *Han shu*:

> 贊曰: 張蒼文好律曆, 為漢名相, 而專遵用秦之顓頊曆, 何哉？
>
> Encomium: Zhang Cang's literary learning excelled in harmonics and the calendar, and he was a famous minister of the Han. But how could it be that he confined himself to using the Qin's Zhuan Xu system?
>
> (*Han shu* 42, 2103)

I am frankly unsure what to make of this situation. Did Sima Qian really write the reference to the Zhuan Xu system in the *Shi ji* passage, or was it later inserted on the model of the *Han shu*? As is well known, parts of the *Shi ji* have suffered alteration or are later 'restorations': immediately following the editorial entry cited above, the narrative continues with accounts of events after Sima Qian's death, which he clearly cannot have written himself. If the last sentence is omitted, the passage is in no way inconsistent with what is said about calendrical matters before 104 BCE in major sections of the *Shi ji*, where nothing is said about the Zhuan Xu system. But even if it is included, it is still not obvious what it means. The significance of *ming* 明 in this sentence is not clear, and as a result one cannot tell whether the sentence is blaming Zhang Cang for using the Zhuan Xu system, or criticizing others for saying that he did so. Nearly two centuries later, the *Han shu* on the other hand is clear—Zhang Cang is being criticized for allegedly using the Zhuan Xu system, which it has already stated (see quotation above) that he adopted from the Qin.

Even taking account of this passage, we can make the following provisional conclusions:

(a) It does not seem that any of the 'six systems' are referred to in any document from before 104 BCE.
(b) The first evidence of the 'six systems' being thought of as a group is not found until the last decades of the first century BCE.

Can we find one of the 'six systems' that fits the surviving evidence of calendar dates from before 104 BCE? That is easier said than done, as shown by the ever-lengthening list of scholars who have attempted to find a solution over past centuries.[41] Simply by way of illustration of what is involved, let us try to apply to the excavated 134 BCE calendar the data we have for the Yin system, which was, according to the *Han li* editors, the real basis of the system that Zhang Shouwang claimed in 78 BCE was the one used in early Han.

As we have seen, according to Liu Xin, an Era Head for the Yin system fell at the winter solstice of late 1,548 BCE, on day *jiazi*.1. That implies that conjunction and the solstice coincided at midnight beginning the day. Now, from that day to the winter solstice of later 135 BCE, preceding the civil year 134 BCE, there are:

$$1{,}568 - 135 = 1{,}433 \text{ solar cycles, each of } 365\tfrac{1}{4} \text{ days.}$$

$$\text{Now } 1{,}433 \times 365\tfrac{1}{4} \text{ days} = 523{,}403\tfrac{1}{4} \text{ days, and}$$

$$523{,}403\tfrac{1}{4} = 8{,}723 \times 60 + 23\tfrac{1}{4}$$

So since the 1,548 BCE solstice fell on sexagenary day *jiazi*.1, the solstice of 135 BCE that we are calculating should fall on day 1 + 23 = 24, day *dinghai*. But the calendar itself marks winter solstice on day *bingxu*.23 in the 11th month, so the Yin system prediction is a day too late. If we do similar calculations using the lunation length, we find that the first days of most months predicted by the Yin system match the calendar, but those for the second, fourth and sixth months are a day too early.[42] So whatever system was used in 134 BCE, it cannot have been the Yin system as it was known to Liu Xin over a century later.

[41] Zhang Peiyu 张培育 (2007) 'Gen ju xin chu tu li ri jian du shi lun Qin he Han chu de li fa 根据新出历日简牍试论秦和汉初的历法 (An investigation of the calendrical systems of the Qin and early Han, based on recently discovered strip and planche calendars 中原文物).' (5): 62–77, 62–3 points out that efforts to give systematic reconstructions of the Qin and Han calendars go back at least as far as Liu Xisou 劉羲叟 (1018–1060). His article addresses the same problem in its turn. The most recent listing and critical overview of the considerable number of attempts to solve such problems in recent decades on the basis of an ever-growing amount of evidence from excavated texts is given in Li Zhonglin 李忠林 (2012) 'Qin zhi Han chu (qian 246 zhi qian 104) li fa yan jiu—yi chu tu li jian wei zhong xin 秦至汉初（前 246 至前104）历法研究 — 以出土历简为中心 (Researches on astronomical systems from Qin to early Han (246 BCE to 104 BCE)—centring on excavated calendrical bamboo slips).' *Zhong guo shi yan jiu* 中國史研究 *'Studies in Chinese history'* 2 (134): 17–70, 17–23.

[42] For those interested in checking these results, we may calculate as follows: if the winter solstice falls 523,403 ¼ days after system origin, and if the lunation is $(29 + \frac{499}{940})$ days long then the number of lunations since system origin up to the winter solstice is: $(523{,}403\tfrac{1}{4})/(29 + \frac{499}{940}) = 17{,}723$ to the nearest whole lunation. Now, $17{,}723 \times (29 + \frac{499}{940}) = 523{,}375$ to the nearest whole number, which is the number of days from the winter solstice at system origin to the conjunction preceding this winter solstice. And 523,375 = 8,722 × 60 + 55. So the predicted day of conjunction is number 1 + 55 = 56, *dingsi* 丁巳, as marked on the calendar. The conjunction of the next month is 17,724 lunations from system origin, and calculations proceed accordingly for that and subsequent months.

What about the Zhuan Xu system? The problem here is that unlike the case of the Yin system, we do not have an early complete statement of the conditions at its system origin. A quotation from a lost work ascribed to Cai Yong 蔡邕 (132–192 CE) tells us:

> 顓頊曆術曰: 天元正月己巳朔旦立春 …
>
> The procedures of the Zhuan Xu system state 'The celestial origin is in the first month, day *jisi*.6, first day of the month, Establishment of Spring … ' (*Hou Han shu, zhi* 2, 3038, commentary)

We are not, however, told in what year the system origin fell. For that we may turn to the eighth century divinatory compendium *Kai yuan zhan jing* 開元占經 'Divination manual of the *Kai yuan* period' compiled c. 725. According to chapter 105, the conditions of the Zhuan Xu system may be represented as follows:

> 古今曆積年及章率: 古今曆上元已來, 至今開元二年甲寅歲積 … 顓頊曆上元乙卯, 至今二百七十六萬一千一十九算外.
>
> The accumulated years and Rule constants of ancient and modern *li*: From the High Origins of ancient and modern *li* up to the present, second year of the *Kai yuan* period, the year [name] being *jiayin*.51 … from the High Origin of the Zhuan Xu system, [year] *yimao*.52 up to the present is outside the count of 2,761,019 years. (Qutan Xida 瞿曇悉達 c. 725 CE: 105, 1a–2a, vol 807, 943–4)

The second year of the *Kai yuan* period was 714 CE. We must note that this year is 'outside the count', meaning if we count the first year of the system as number 1, then 713 CE is number 2,761,019. Thus to find the High Origin of the Zhuan Xu system, we must count back 2,761,019 years from 714 CE, which (allowing for the absence of a year 0) takes us to 2,760,306 BCE. This is indeed a *yimao*.52 year in the system used from Eastern Han onwards. Since, however, we are only interested in the dates of *qi* such as *li chun* 立春 'Establishment of Spring' (the fourth *qi* counting from winter solstice) and the beginnings of lunar months we can remove whole multiples of the Era period of 1,520 years, and instead use the more convenient origin of 1,506 BCE, another *yimao*.52 year.

Following a similar procedure to the preceding calculation, we may find the number of solar cycles elapsed from Establishment of Spring in 1,506 BCE to Establishment of Spring in 134 BCE as follows:

$$1{,}506 - 134 = 1{,}372$$

So the days elapsed are:

$$1{,}372 \times 365\tfrac{1}{4} = 501{,}123$$

and

$$501{,}123 = 8{,}352 \times 60 + 3$$

So the day of Establishment of Spring this year should be number 6 + 3 = 9, *renshen* 壬申. This is in fact precisely the day shown on the calendar for this instant in the solar cycle. Similarly we may calculate the dates of the other three *qi* shown on this calendar: the preceding winter solstice, which is correctly predicted on day *bingxu*.23, the subsequent summer solstice, again correctly predicted on day *wuzi*.25, as is Establishment of Autumn on day *jiaxu*.11. So the seasons seem to work out quite well. But if we turn to the months, the match is worse than with the Yin system: the 11th, second, fourth, sixth, eighth and latter ninth months all have their first days predicted a day too early.[43] So it seems that neither the Yin system nor the Zhuan Xu system as later known can have been the system in operation in 134 BCE. The same can be said of the other four of the 'six systems', at least as they are described in the *Kai yuan zhan jing*.

Under these circumstances, there seems good reason to be sceptical whether the 'six systems' as they are known from around the beginning of the Common Era are real survivals of ancient practice, rather than later reconstructions. This is not a new point of view. In the early seventh century CE, Kong Yingda 孔穎達 (574–648 CE) stated:

> 然古時真歷，遭戰國及秦而亡．漢存六歷，雖詳於五紀之論,皆秦漢之際假託為之．
>
> The real [astronomical] systems of ancient times were lost during the Warring States and the Qin dynasty. [As for] the 'six systems' preserved by Han, even though their details were given in [Liu Xiang's] *Wu ji lun* 'Discussion of the five sequences of time', all of them were faked around the time of the transition from Qin to Han. (*Shang shu* 2, 18a (commentary) in (Ruan Yuan 阮元 (1764–1849) 1973 reprint of original of 1815: vol. 1, 25b)).

Zu Chongzhi 祖沖之 (429–500 CE) had set out the evidence for this point of view in detail two centuries earlier. Out of the six points he makes against the genuineness of the 'six systems', perhaps the most effective are his two final ones:

[43] A solution proposed by some of the first scholars to study the 134 BCE calendar was, in essence, that conjunctions predicted by the *Zhuan Xu li* for times after noon on a given day would be taken as falling on the next day. See Chen Jiujin and Chen Meidong (1989) 'Cong Yuan Guang li pu ji Ma Wang Dui boshu tianwen ziliao shitan Zhuan Xu li wenti' (An investigation of the Zhuan Xu system on the basis of the Yuanguang almanac and the Mawangdui silk manuscript), in *Zhongguo tianwen wenwu lunji* [Collected articles on ancient Chinese astronomical relics], Wenwu chubanshe, Beijing 83–103.

春秋書食有日朔者凡二十六, 其所據曆, 非周則魯. 以周曆考之, 檢其朔日, 失二十五, 魯曆校之, 又 失十三. 二曆並乖, 則必有一偽 … 古之六術, 並同四分, 四分之法, 久則後天. 以食檢之, 經三百年, 輒差一日. 古曆課今, 其甚疏者, 朔後天過二日有 餘. 以此推之, 古術之作, 皆在漢初周末, 理不得遠. 且却校春秋, 朔並先天, 此則非 三代以前之明徵矣

There are twenty-six instances in which the *Spring and Autumn Annals* record a solar eclipse [together with] the [sexagenary] day, [and noting that it fell on] the day of conjunction [at the start of the month]. The system this was based on must have been that of either [the kings of] Zhou, or [the Dukes of] Lu [the state where the records were made]. But if we check against the [so-called] 'Zhou' system it gets twenty-five of them wrong, and if we check against the [so-called] 'Lu' system it gets thirteen of them wrong. Both the two systems are off, so they must both be fakes. … The 'six ancient systems' are all of the quarter remainder type, and all such systems lag behind the heavens if [used for] a long period. Checking against eclipses, they have an error of one day after 300 years.[44] If we try using them today, they are extremely inaccurate, and conjunctions [are predicted] a bit over two days late behind the heavens. Working on this basis, the creation of these 'ancient methods' must have taken place some time between the end of Zhou and the start of Han, and they cannot reasonably be far [from that time]. Moreover, if we go back to the time of the *Spring and Autumn Annals*, the conjunctions [are predicted] in advance of the heavens. So it is quite clear that these systems' commentary cannot be from the time of the Three Dynasties or earlier. (*Hou Han shu, zhi* 2, 3030, commentary)

3.4.3 Was there any system in Qin and early Han?

In our attempt to identify the system in use before 104 BCE, we used one document only, the 134 BCE calendar. But in fact the number of data that have to be satisfied by any solution—or solutions—is much greater than the thirteen conjunction dates and the three seasonal *qi* listed on that document. In addition to all the dates listed in the received literature of Qin and Han times, we now have to deal with an increasing mass of dates from excavated documents. A recent study by Li Zhonglin 李忠林 lists a very large number of items, comprising 305 data points giving the sexagenary days of conjunctions, 136 data points giving the sexagenary day names of days within given months, and 7

[44] This is a conventional figure, but may be checked against the actual value of the quarter remainder lunation length, $29\ ^{499}/_{940}$ = 29.530851 days, and the modern value of 29.530587 days. The difference, 0.000264 days, mounts up to 1 day after 3,788 lunations, which is close to 306 years.

items giving the sexagenary days of seasonal *qi*, and on that basis attempts a reconstruction of Qin and early Han calendrical practice.[45] The earliest date in his repertoire is the conjunction of the 10th month beginning the second year of the reign of King Zheng of Qin, which fell towards the end of 246 BCE on day *guiyou*.10, and the last is the conjunction of the 11th month that fell in late 105 BCE, on day *jiazi*.1.

Let us suppose that there does exist a single quarter remainder type system that is consistent with, at a minimum, the 305 conjunction dates in Li's collection. Then, given that such a system will predict conjunctions at fixed intervals of $29\ ^{499}/_{940}$ day, there must exist at least one point in time from which one instant during each recorded conjunction day is distant from our starting point by a whole multiple of $29\ ^{499}/_{940}$ days. Li's conclusion is that no such single starting point can be found, even making reasonable allowances for possible copying errors in the original records. It is therefore pointless for us to argue whether the system in use before 104 BCE was the Yin system, the Zhuan Xu system, or another of the so-called 'six systems', or indeed some other hitherto unknown system: there simply was no single system of the kind known from 104 BCE onwards.

Li's solution to this situation is to suggest that calendrical practice from the Qin to 104 BCE can be divided into three main periods, during each of which the reckoning is continuous:

(A) From 246 BCE, the first year of the reign of King Zheng 政 of Qin, who became the first emperor, to the 12th Xia month of the year corresponding to 202 BCE, the fifth year of the first ruler of Han—which was in fact the year when he accepted the imperial title, when a change of practice might have been expected (*Shi ji*, 8, 379).

(B) From the first Xia month of 202 BCE to the 'latter 9th month' of 164 BCE—which was followed by the year when Wendi, under the influence of Gongsun Chen 公孫臣 after having dispensed with the services of the aged Zhang Cang, decided to begin a new year-count (*Shi ji* 28, 1381–3; 10, 430)—an incident already discussed earlier in the present chapter.

(C) From the first Xia month of 163 BCE to the 5th Xia month of 104 BCE, when we know that the new Grand Inception system was put into practice.

[45] See Li Zhonglin 李忠林 (2012), 24–36.

None of the sequences of dates constructed by Li Zhonglin is based on a single 'system origin' in the sense known from the 104 BCE reforms onwards. Instead, he identifies a starting date for calculations, when a conjunction begins at midnight on a named day some time before the period of currency of the date sequence, and calculates conjunctions on that basis. The discontinuities between periods identified by Li consist essentially in the decision to make the timing of conjunctions slightly earlier: from period A to period B the shift is identified as $^{144}/_{940}$ day, and from B to C as $^{18}/_{940}$ day.[46] The overall shift of $^{162}/_{940}$ day will leave many conjunctions on the same day as originally predicted, but those conjunctions that fall close to the start of a day may end up being shifted to the day before. One important example is the final date in Li's repertoire, the conjunction of the 11th month falling in late 105 BCE, which we know to have been assumed to fall on day *jiazi*.1; if system A had continued to operate, this conjunction would have been placed $^{100}/_{940}$ day into day *yichou*.2, instead of at $^{878}/_{940}$ day into day *jiazi*.1, which is the day on which system C places it. The *qi* divisions of the solar cycle are taken to follow the pattern seen in the Zhuan Xu system, and intercalations are added after the Xia ninth month on a fixed pattern within each 19-year Rule, although the starting point of this pattern shifts as we change systems. Looking back to earlier in this chapter, these prescriptions do fit in with the statement made by Gongsun Qing in 113 BCE—that the conjunction of the 11th month near the end of that year fell on day *xinsi*.18, when winter solstice also fell.[47]

It would be premature to claim that Li's results have settled the problem—this is a debate that seems likely to continue. But whatever solution may be proposed as the basis for dates before 104 BCE, it seems unlikely that any one of the 'six systems' studied by Liu Xiang and his successors—or indeed any other single system—can be taken as representing actual practice during the entire Qin and

[46] One obvious reason for this tendency to shift the calendar forward in this way is that in the early Han there was clear evidence that conjunctions were being predicted too late, as shown by the fact that in the period the majority of solar eclipses (which can only occur at the instant of conjunction, supposedly the first day of a month) were mostly recorded as happening on the last day of the months in which they fell: see *Han shu* 27c2, 1500–03.

[47] See Li Zhonglin 李忠林 (2012), 69 for the conjunctions in question. Li calculates solstices according to the later reconstruction of the Zhuan Xu system, which is said to have begun its reckoning from coincidence of luni-solar conjunction and the 4th *qi* 'Establishment of Spring' *li chun* 立春 at midnight beginning a *jisi*.6 day corresponding to 9 February 1506 BCE. Although that system does predict that the 113 BCE winter solstice fell on day *xinsi*.18 as was stated to be the case by Gongsun Qing and Sima Qian, for reasons explained earlier no quarter remainder system can make that prediction and also predict that the winter solstice of late 105 BCE will fall on a *jiazi*.1 day In fact the Zhuan Xu system places the 105 BCE solstice on the preceding day, *guihai*.60.

early Han period. Indeed, we may well ask whether the concept of a single integrated calendrical system with a name was even known before the Grand Inception reform.[48]

In the next chapter, we shall look at the first named astronomical system embodied in a detailed text with full instructions for its use—the *San tong li* 三統曆 'Triple Concordance system' constructed by Liu Xin at the end of Western Han.

[48] My present position is thus considerably more sceptical than the one set out in some of my previous writing, from Cullen (1993) up to Cullen (2011b), in which I accepted the consensus that the Zhuan Xu system was in use in Qin and early Han.

CHAPTER 4

The Triple Concordance system and Liu Xin's 'Grand Unified Theory'

The last chapter took us from the Grand Inception reform of 104 BCE to the final years of the Western Han. In 9 CE, the powerful courtier Wang Mang deposed the infant prince of Han and took power as first emperor of his own dynasty, proclaiming that he would restore the power and legitimacy once exercised by the most ancient rulers of China. Part of the justification for Wang Mang's rule was his claim to have mastery of the interface between the human world and the cosmic powers. To help him sustain this claim, Liu Xin created what (in the terms of his day) amounted to a 'Grand Unified Theory', in which all the important regularities of nature were explained with reference to numbers. Part of Liu Xin's programme was a new rationale and expansion of the system of celestial calculation that had been constructed for Wudi in 104 BCE, tying it into a single frame of reference that included metrology and musical harmony in a system based upon the fundamental numbers of the cosmos thought to be encoded in the ancient Book of Change. He also added means for predicting planetary motions. We shall look in detail at the Triple Concordance system that resulted from this initiative, and follow through some examples of calculation based on it. Here we shall meet for the first time certain common features of all such systems.[1]

[1] A complete translation of the Triple Concordance system, with commentary and worked examples, will be found in Cullen (2017), which also includes translations of two other systems created later in the Han dynasty.

Liu Xin also undertook an ambitious project to use his new system to reconstruct the records of astronomical phenomena in a number of revered historical texts. His work gives us an in-depth view of how one of the central figures in the early development of Chinese mathematical sciences saw the world and tried to make sense of it.

4.1 The historical background of Liu Xin's work

Liu Xin 劉歆 (46 BCE–23 CE) was a fifth generation descendant of a younger brother of Liu Bang, founder of the Han dynasty. As we shall see, however, he was closely associated with movements that led to the humiliating displacement of the Liu clan from power for 14 years from 9 to 23 CE. Since most of our information about him comes from sources written after the restoration of the Han to power, we must treat those sources—principally the *Han shu* written by Ban Gu—with some caution. But Liu Xin's reputation as a scholar does not seem to have been unduly affected by what many must have seen as a betrayal of his allegiance to the dynasty founded by his ancestors.

Neither he nor his father Liu Xiang 劉向 (79–8 BCE) had close connections with those members of the imperial clan who held noble rank. Their influence both in their own day and in later centuries was due to their striking intellectual and literary achievements rather than to the holding of high offices of state, or to family relationships. They worked together on a great bibliographical project, launched in 26 BCE, which involved them in collating and cataloguing the immense imperial collection of texts that had accumulated since the reign of Wudi. This collection had been increased by a recent order to collect texts from throughout the empire. An abbreviated version of the listing they eventually produced around 6 BCE now forms chapter 30 of the *Han shu*, and gives us a tantalizing picture of the literature available in one great Chinese collection around the beginning of the common era.[2]

[2] For an account of the Lius' bibliographical activities, see the outline in P. van der Loon (1952) 'On the Transmission of Kuan-tzŭ.' *T'oung Pao* 41 (⅕): 357–393. A recent collection of studies of the intellectual and cultural context within which they worked is Michael Nylan (ed.), Griet Vankeerberghen (ed.) and Michael Loewe (ed.) (2015) *Chang'an 26 BCE: an Augustan Age in China*. Seattle, University of Washington Press. A penetrating study of the father and son as individuals, emphasizing the differences in their intellectual stances, is Michael Loewe (2015), 'Liu Xiang and Liu Xin' in Nylan (op. cit.) 369–389. The task facing the Lius was much more demanding than that of modern librarians cataloguing a collection of printed and bound books, who can usually expect that each item will come with an author's name and title, contents that are common to all copies in a given edition, and a publication date. Their work as collators and editors of texts existing in a large number of variant or duplicate manuscript versions seems quite frequently to have given them an effectively authorial role. Much of the ancient Chinese literature we have today must have been modified or even shaped by their labours.

In his youth, Liu Xin made friends with a certain Wang Mang 王莽 (c. 45 BCE–23 CE), with whom he had served as a *huang men lang* 黃門郎 'Gentleman of the Yellow Gates'—a respectable post for a young man at court.[3] Their relationship was to endure until a little before both of them met their end within a few weeks of each other, Liu by suicide to avoid execution for plotting against Wang, and Wang at the hands of soldiers who slew him in his burning palace. Wang came from a family that was not prominent in national affairs until in 54 BCE a young woman of the Wangs, Wang Zhengjun 王政君 (71 BCE–13 CE) was selected for the harem of Xuandi 宣帝 (r. 74–49 BCE). This would normally have brought a certain amount of imperial favour to her family, but the combination of her longevity and eventual promotion to the rank of Dowager Empress ensured that many of her male relations, including Wang Mang, attained high office.

It eventually became clear that Wang aimed at the highest office of all, an office that he finally attained when in 9 CE he secured the abdication of a child emperor of the Liu clan and took the throne as the first and last emperor of what he called the Xin 新 'New' or perhaps 'Renewing' dynasty.[4] His rise to power had been carefully prepared, and one important element in that preparation lay in his self-presentation to scholar officials, a group without whose support the empire simply could not be governed. The role he took in the state was designed to recall the glories of the revered founders of the Zhou dynasty, whose practices were the model most admired by Confucius. In 1 CE, it was even proposed that Wang Mang's guardianship of the young emperor Pingdi should be recognized by the conferral of the title Zhou Gong 周公 'Duke of Zhou', the title that had been held by Dan 旦, the loyal guardian and preceptor of the young King Cheng 成 (r. 1042–1021 BCE). When Pingdi died in 6 CE, Wang Mang ensured that the throne passed to the two-year-old Liu Ying 劉嬰, and his role as acting emperor continued until 9 CE. Finally, a well-orchestrated succession of reports of favourable omens, reinforced by memorials from many officials urging him take over supreme power, led to his assumption of the imperial title. In a carefully staged ceremony, he held the child's hand, weeping and lamenting 'for a long time' *liang jiu* 良久 as he explained that the time for the end of Han had come, and that Heaven had called him to rule. He expressed his sense of cosmic destiny in words which must have had a deep historical resonance for all those

[3] *Han shu* 26, 1972.

[4] On the possible meanings of the dynastic name, see Chauncey S. Goodrich (1957) 'The Reign of Wang Mang: Hsin or New?' *Oriens*: 114–18. Wang Mang's dynasty did not endure, and the Han was eventually restored. Inevitably, therefore, traditional Chinese historiography has treated him as a villainous usurper without any positive personal qualities apart from a talent for plotting and manipulation. More balanced accounts of his reign will be found in Twitchett, Loewe and Fairbank (1986), 223–251 and Loewe (2000), 536–545.

present, but which may also prepare us to understand why a renewed astronomical system would play a part in his assumption of power:

> 咨爾嬰, 昔皇天右乃太祖, 歷世十二, 享國二百一十載, 曆數在于予躬.
>
> Oh Ying! Formerly August Heaven supported your Great Ancestor [Liu Bang, founder of the Han], so his descendants succeeded for twelve generations, and possessed the state for two hundred and ten years, [but now] the sequence of reckonings rests on my person. (*Han shu* 99b, 4009)

The expression *li shu* 曆數 'the sequence of reckonings' (or perhaps 'sequence of numbers') had already been used in the Empress Dowager's edict of 5 CE, discussed in chapter 1; she had claimed that the 'sagacious emperors and perspicacious kings' of the past had all followed it. And the *Book of Documents* recorded that when in high antiquity Emperor Shun 舜 had called Yu 禹 to ascend the throne, he had said:

> 天之歷數在汝躬.
>
> Heaven's sequence of reckonings rests on your person. (*Shang shu* 4, 8b in (Ruan Yuan 阮元 (1764–1849) 1973 reprint of original of 1815: vol. 1, 55–2))

Nobody present would have failed to realize what was intended—nor could they have been unaware of the astronomical implications of the word *li*, which as we have seen can also mean '[astronomical] system'.

Had anybody doubted that Wang Mang intended to make such an astronomical connection, they had only to listen to the proclamation he issued shortly afterwards, in which he listed the titles and duties of the seven new great officers of state he intended to create. The list began:

> 歲星司肅, 東(獄)[嶽]太師典致時雨, 青煒登平, 考景以晷.
>
> Jupiter has charge of conscientiousness; [under its patronage] The Grand Master of the Eastern Peak sees to it that timely rain comes about, and [by its] caerulean brilliance he promotes peace, examining the shadow using a gnomon. (*Han shu* 99b, 4101)

And so he continued for three other planets,[5] the sun, the moon and the Northern Dipper, in each case naming an office and its responsibilities, and ending by

[5] The only planet omitted from the list is Saturn. It may be that it was left out because it was the planet associated with the phase Earth, under whose patronage Wang Mang thought his new dynasty was to rule. One planet had to be omitted in order to make room for the Dipper, which for Wang Mang represented a major source of cosmic power, occupying as it did a central place in the heavens: see chapter 6, section 6.5, for an account of its representation on models of the cosmos used for hemerological divination. For the story of how Wang Mang relied on the symbolism of the Dipper, and invoked it against his enemies in his last extremity, see *Han shu* 99c, 4190 and Needham, Wang and Robinson (1962), 271–3.

specifying an activity of cosmic significance—examining the sounds by the pitchpipes (Mars), examining the measures of capacity by weighing (Venus), examining the stars with the clepsydra (Mercury), examining the square by taking the setsquare as a pattern (the moon), examining the circle by taking the compasses as a pattern (the sun) and examining the measures of length with the plumbline (the Northern Dipper, *bei dou* 北斗, made up of the seven bright stars of Ursa Major). Wang Mang was certainly the first ruler to cite the planets as a source of his power, and as we shall see Liu Xin constructed the first astronomical system to give the planets a fully integrated role in its workings.

Liu Xin was closely involved in the ideological preparations for Wang Mang's rise to power. Thus in 4 CE we find him charged with the construction of ritual buildings that were specifically meant to resemble those of the Zhou (*Han shu* 12, 359); these included a *ling tai* 靈臺 'Numinous terrace', though it is not clear whether this structure yet served as a base for sky-watching. The title Liu Xin held at this time referred back to even more ancient models: he was the Xi He 羲和, a term derived from the names of the sky-clerks commissioned by Emperor Yao in remotest antiquity (see section 1.1). That same year, Wang Mang sponsored a great conference of literati who might possess the kinds of knowledge he wanted to co-opt in support of his growing influence:

> 徵天下通知逸經, 古記, 天文, 曆算, 鍾律, 小學, 史篇, 方術, 本草及以五經, 論語, 孝經, 爾雅教授者, 在所為駕一封軺傳, 遣詣京師. 至者數千人.
>
> He summoned those in the empire who had a comprehensive knowledge of lost classical texts, ancient records, celestial patterns, calculations [pertaining to] astronomical systems, bells and pitchpipes, elementary studies,[6] historical compilations, recipes and procedures, pharmacognosy, and who professed the Five Classics, the Analects [of Confucius], the Classic of Filial Piety, and the Literary Expositor. They were transported in carriages of the official relay system in vehicles bearing a single seal, and sent to the capital. Several thousand of them arrived. (*Han shu* 12, 359)

We are fortunate in having a report of the results of this activity: with some apparent embarrassment, Ban Gu informs us in the introduction to the monograph on harmonics and astronomical systems in the *Han shu* that Liu Xin was charged with summarizing the conclusions reached by these scholars, and that much of what follows is a censored and expurgated version of what he wrote (*Han shu* 21a, 955). Even then, it is possible to see signs of Liu Xin's efforts to

[6] 'Elementary studies' *xiao xue* 小學 are defined in the *Han shu* summary of the Lius' bibliographical work as relating to the analysis and proper understanding of written characters: *Han shu* 30. 1719–23.

boost Wang Mang if one knows how to read the signs. Take this passage, from near the end of a section on the cosmic significance of weights and measures acknowledged to contain Liu Xin's views:

> 書曰:「予欲聞六律, 五聲, 八音, 七始詠, 以出內五言, 女聽.」予者, 帝舜也.
>
> The [Book of] Documents says: 'I wish to hear the six pitch-tubes, the five notes, and the eight kinds of sound, and the seven sources of poetry, in order to send forth and take in the five words [of each line of verse?]. You are to listen [on my behalf].' The 'I' is Emperor Shun. (*Han shu* 21a, 972)

Now the biography that Wang Mang had crafted for himself made him a descendant of Emperor Shun (*Han shu* 99b, 4105), who had accepted the abdication of Emperor Yao 堯, supposedly the ancestor of the imperial clan of Han—the message was obvious enough. Shun's interest in the technical matters discussed by Liu Xin therefore reflected on Wang Mang himself. And when Wang Mang listed the cosmological tasks associated with the great offices of state in the accession proclamation issued in 9 CE that we discussed earlier, what he had to say resonated strongly with the structure and content of Liu Xin's report as recorded in the *Han shu*.

4.2 A new astronomical system? Facing the problem of the planets

Wang Mang's formal accession was accompanied by systematic calendrical and cosmological innovations never before fully recorded at the accession of a dynasty, despite the precedent for some of them in Wudi's reforms of 104 BCE. The 12th Xia month near the end of 8 CE was proclaimed as the first month of the first civil year of Wang's reign, and a number of ritual changes were announced. These included the adoption of yellow as the ritual colour, since the new dynasty was held to be under the patronage of the power of Earth, as well as the change from the standard clepsydra division of a day into 100 *ke* 刻 to 120 *ke*.[7]

[7] The change of patron phase to Earth was in some sense a re-adoption, since the change from the Water phase of Qin to Earth had been a feature of the Grand Inception reform of 104 BCE. The notion that the Han's true phase was Fire later became current, as did the idea that the natural sequence for the cosmic basis of rule was the peaceful 'production' sequence of the phases, which was adopted in preference to the 'conquest' sequence previously used. In the production sequence, Earth followed naturally from Fire, as discussed in chapter 3, section 3.2.1. See *Han shu* 99a, 4094–6 and 99b, 4113, also Twitchett, Loewe and Fairbank (1986), 738 and Michael Loewe (1994) *Divination, mythology and monarchy in Han China*. Cambridge, Cambridge University Press, 55–60. On the division of the day into *ke*, see the discussion of the use of clepsydras in chapter 5, section 5.4.

Liu Xin did not fail to use his expertise in astronomical and calendrical matters to add to Wang's prestige. While preserving the core data laid down by the Grand Inception reform, such as its values for the lengths of the lunar and solar cycles, and use of the winter solstice of 105 BCE as a starting point for calendrical calculations involving the sun and moon, he systematized and extended it in ways that were ambitious enough to merit a new name: the *San tong li* 三統曆 'Triple Concordance system'. The term *San tong* 三統 'Triple Concordance', or 'Three Concordances' may have been adopted from the book *Chun qiu fan lu* 春秋繁露 'Luxuriant Dew of the Spring and Autumn annals', ascribed to Dong Zhongshu 董仲舒. Dong Zhongshu, who was active in the time of Wudi, was a thinker deeply admired by Liu Xin, who said of him:

仲舒遭漢承秦滅學之後,六經離析,下帷發憤,潛心大業,令後學者有所統壹

Zhongshu encountered the consequences of Han following on from the destruction of learning by Qin, when the Six Classics were in disorder; he stirred up his ardour in seclusion, plunging himself into the great task, so that later scholars had something to unite them in concord (*tong*). (*Han shu* 56, 2526)

In the *Chun qiu fan lu* we are told of the three ages of the past, each corresponding to a colour, a different 'standard month' (*zheng yue* 正月) for the start of the year, and a correspondingly different position of the sun and moon at the start of that month.

三代改正,必以三統天下

The Three dynasties [Xia, Shang and Zhou] changed their standards, and thus made all under heaven triply concordant (*tong*). (*Chun qiu fan lu* 7, 5b)[8]

The first references to this term in the context of an astronomical system are found in an encomium on Liu Xin's writings in the *Han shu*, probably written about 60 years after his death. There we read:

三統曆譜考步日月五星之度.有意其推本之也.

[His] Triple Concordance system and Listing[9] examined and predicted[10] the degrees [of motion] of the sun, moon and five stars [*sc.* 'planets']. Its aim was to get to the basis of these. (*Han shu* 36, 1973)

[8] See p. 264 in *Chun qiu fan lu* 春秋繁露 *(Luxuriant Dew of the Spring and Autumn annals).* (1592 woodblock edition, photographic reprint 1978). Dong Zhongshu 董仲舒 (c. 179–c. 104 BCE), Kyōto, Chūbun shuppansha.

[9] On this 'listing' see section 4.5.

[10] Literally 'paced out' *bu* 步; this is a common conventional expression in connection with the motion of the celestial bodies. As we shall shortly see, the five visible planets may be referred to as the *wu bu* 五步, 'Five Pacers'.

Here we see the first mention of a new feature of Liu Xin's astronomical system: as well as providing a means for calculating the motions of the sun and moon, which are the foundations for the construction of the calendar, he also built into his system a means for predicting the motions of the planets. This is something that was not mentioned at all at the time of the Grand Inception reform of 104 BCE, but as we have seen it was quite consistent with the attention paid to the planets in Wang Mang's accession proclamation.

That does not mean, of course, that the basic regularities of planetary motion were not a topic of interest before the time of Liu Xin. The *Huai nan zi*, offered to the throne in 139 BCE, gives explicit information about the time taken for Jupiter and Saturn to make complete circuits of the heavens—12 years for Jupiter and effectively 28 years for Saturn (since it is stated that the planet moves through one lodge out of the 28 each year). For Venus, we are told its periods of visibility and invisibility in days, although there seems to be some confusion in the figures given. Little is said about Mars, and Mercury is said to appear once in each season.[11] In the *Shi ji*, completed by about 90 BCE, Sima Qian gives identical data for Jupiter and Saturn, and says rather more about Mars than *Huai nan zi*. The statements made about the periods of Venus and Mercury are similar to those in *Huai nan zi*.[12]

But the most systematic and ambitious attempt to describe planetary motion known from before the time of Liu Xin is undoubtedly found in a manuscript, possibly of late Qin date, excavated from a tomb that was closed in 168 BCE. The modern editors gave this manuscript the title *Wu xing zhan* 五星占 'Prognostics of the Five Planets', since it is wholly concerned with the motions and times of appearance of the planets, and the prognostications that may be drawn from observation of these.[13] Like the other two sources mentioned, this manuscript says much more about Jupiter, Saturn and Venus than about Mars and Mercury. For the first three planets, it not only describes the basic regularities of their motion, but gives year-by-year tabulations as follows:

(a) The lodges with which Jupiter is seen when it first becomes visible in the east before dawn,[14] based on a 12-year cycle. Five complete cycles are given, together with the first ten years of the sixth cycle.

[11] See *Huai nan zi* 3, in *Huai nan hong lie ji jie*, 89–92, also Major (1993), 73–7.

[12] See *Shi ji* 130, 1312–30, and Pankenier (2013), 473–89.

[13] I have given a full translation of this document in Cullen (2011a), with introductory comments, and have analysed it further in Cullen (2011b). See these articles for background to the description given here. The *Wu xing zhan* is further discussed in Morgan (2013) chapter 2.

[14] Often called the 'apparent morning rising' of the planet, with 'apparent' having the sense 'visible', as opposed to the 'true morning rising', when the planet is in conjunction with the sun and therefore rises with it in the morning, but is invisible to observers on earth because of the sun's intense luminosity.

(b) The lodges with which Saturn first becomes visible in the east before dawn, based on a 30-year cycle. Two complete cycles are given, together with the first ten years of the third cycle.
(c) The phases of Venus, based on an eight-year cycle; within each eight-year period there are five complete phase cycles, beginning with a first morning rising; for each phase cycle we are told the month of first morning rising, and the asterism with which Venus rises, the time interval until last morning rising, the month in which this occurs, and the asterism with which Venus then rises, the time interval until the first evening setting, the month in which this occurs and the asterism with which Venus then sets, the time interval until last evening setting, the month in which this occurs and the asterism with which Venus then sets, and the time interval until the next first morning rising.[15] Eight complete eight-year cycles are given, together with the first six years of the ninth eight-year cycle.

The most striking feature of these tabulations is that they all three begin with the first regnal year of the King of Qin who was to become the first emperor of the Qin dynasty, 246 BCE, and all end in 177 BCE. What is more, all three planets are said to be first visible in the lodge House during the first tabulated year. This was the lodge whose fifth *du* was given by *Huai nan zi* as the starting point for the sun and moon at the *yuan shi* 元始 'Origin Initiation' (for which, however, it does not specify a year): see section 1.3. The fifth *du* of House was also said by later scholars to have been the starting point of the Zhuan Xu system, which has often been assumed to be the system used by the Qin.[16] Further, analysis of the predictions given in the tables suggests strongly that they are based on simple numerical schemes. The use of a common starting point for these planetary reckonings recalls the use of a 'system origin' *li yuan* 曆元, the common starting point for all calculations that lies at the basis of all astronomical systems before the 13th century.

[15] For Venus, as for Mercury, there is a 'first morning rising' when it is first far enough to the west of the sun to be visible just before sunrise, followed by a 'last morning rising' as it approaches the sun again. After a period of invisibility close to the sun, during which conjunction occurs, the planet reappears to the east of the setting sun in the evening, and sets not long after it, its 'first evening setting', and eventually moves back close enough to the sun to have its 'last evening setting', followed by invisibility again, and eventually another first morning rising.

[16] See *Huai nan zi* 3, in *Huai nan hong lie ji jie*, 804 and Major (1993), 83. For the view that this was the starting point of the Zhuan Xu system, see for instance *Xin Tang shu*, 27a, 602–3. I am now more sceptical than I was previously about the reality behind the suggestion that the Zhuan Xu system as later described was actually used by Qin: see section 3.4.1.

Apart from the surviving sources described here, a further sign of systematic interest in the planets before Liu Xin is found in the bibliographical monograph of the *Han shu*, which we know to have been based on work by Liu Xin and his father. The titles of three texts found there are:

> 顓頊五星曆十四卷
> 傳周五星行度三十九卷
> 自古五星宿紀三十卷
>
> The five planet system of Zhuan Xu, 14 rolls.
> Chuan Zhou[17] [on] the *du* of motion of the five planets, 39 rolls.
> The sequence of the five planets through the lodges since antiquity, 30 rolls. (*Han shu* 30, 1765—1766)

The fact remains, however, that Liu Xin's system is the first one known in which the same system origin is explicitly used for the sun, moon and planets, and this had a marked effect on how far from his day that system origin had to be placed, as remarked in the *Hou Han shu*:

> 王莽之際,劉歆作三統,追太初前(世)[卅]一元得五星會庚戌之歲,以為上元.
>
> In the time of Wang Mang, Liu Xin made the Triple Concordance [system]. This pushed back 31 Origin [cycles] before the Grand Inception [system origin], to find a conjunction of the five planets in a *gengxu*.47 year, and took that as its High Origin. (*Hou Han shu, zhi* 3, 3082)

The Grand Inception system origin, the point when the sun and moon were in their initial conditions and from which calculations began, fell at the winter solstice of late 105 BCE, when the conjunction of the 11th Xia month and winter solstice were said to have coincided at the beginning of a *jiazi*.1 day, the conditions specified for an Origin Head *yuan shou* 元首. Since an Origin cycle of the Grand Inception system is 4,617 years in length (see chapter 3), the new High Origin was clearly placed at the winter solstice near the end of the Julian Year calculated as follows:

$$(105 + 31 \times 4{,}617)\ \text{BCE} = 143{,}232\ \text{BCE}$$

The choice of the winter solstice of that year (which on the Triple Concordance reckoning fell on 2 December of the proleptic Julian calendar) for the High Origin of the Triple Concordance system depended on satisfying a number of requirements:

[17] This appears to be the name of a person, although the name is not otherwise known.

(a) The solstice chosen for the sun, moon and planets to be simultaneously at their initial positions had to be a whole number of Origin cycles, 4,617 years, from the 105 BCE winter solstice, so that the system would continue to predict basic calendrical data in the manner laid down in 104 BCE and practised during the rest of the Western Han.
(b) The lengths adopted for planetary cycles had to be consistent with what could be learned about their values from long-term records.
(c) The combination of (a) and (b) had to predict to a satisfactory degree the observations of planetary positions made in his own day.

Assuming that (a) and (b) had been fulfilled, solving (c) is certainly not a trivial problem. It would involve the use of what has been called the 'Chinese remainder theorem', that is, a method for determining a number which, when divided by a series of specified numbers (in this case the synodic periods of the planets), leaves a series of specified remainders (the portion of the synodic cycle of each planet uncompleted at the moment specified in (c)).[18] But although Liu Xin did not leave us a clear record of how he actually chose the moment of his High Origin, there are signs that he put even greater constraints on himself. Rather than simply trying to fulfil the purely observational and mathematical demands of (a), (b), and (c), he appears to have been determined that the basic planetary constants he used should have mathematical properties that he saw as being of cosmic significance, and rather than selecting a single set of observation data to match, he set out to match a whole range of historical records. All this was, however, part of a much wider project to use numbers to describe not just the way the celestial bodies behaved—but the way everything behaved. To this project we now turn.

4.3 Numbers and a theory of everything

In modern times, the cultures of ancient East Asia have largely been studied in the West by scholars with an orientation towards the humanities, who have naturally tended to see the cultures they studied through the lenses of that

[18] On this topic see, for instance Ulrich Libbrecht (1973) *Chinese mathematics in the thirteenth century: the Shu-shu chiu-chang of Ch'in Chiu-shao*. Cambridge, MIT Press, chapter V, particularly 367–9; also Needham and Wang Ling (1959), 119–22 and Yan Li and Shiran Du (1987) *Chinese mathematics: a concise history*. Oxford, Clarendon, 161–6.

orientation. Numbers were not central to their own world-view, and the role of numbers in the cultures they studied was not always considered fully—if it was considered at all. This bias was not always there amongst western scholars: a student wishing to take the degree of Master of Arts in mediaeval Paris or Oxford would have begun his studies with the *quadrivium* of arithmetic, geometry, music theory, and astronomy, subjects seen as an essential foundation for serious philosophical and eventually theological study.[19] The Jesuit missionaries who first went to China in the 17th century had all undergone training that assumed that a high degree of numeracy was an essential ingredient in the formation of a Christian intellectual.[20]

There is evidence that some groups in early imperial China also held that numbers were an important key to understanding the cosmos. In its strongest form, this view may be exemplified in a manuscript, thought to be of Qin date, that was included in the collection apparently robbed from a tomb, sold in Hong Kong, and then presented to Beijing University by an alumnus:[21]

> 魯久次問數于陳起曰: ‘久次讀語, 計數弗能並屰(徹), 欲屰(徹)一物, 可(何)物為急?’陳起對之曰: ‘子為弗能並屰(徹), 舍語而屰(徹)數, 數可語殹(也), 語不可數殹(也).’久次曰: ‘天下之物, 孰不用數?’ 陳起對之曰: ‘天下之物, 无不用數者.’
>
> Lu Jiuci asked Chen Qi about numbers: ‘I have studied discourses,[22] and the reckoning of numbers, and I cannot fully grasp both. If I want to fully grasp one thing, which is more vital?’ Chen Qi replied: ‘If you cannot fully grasp both, leave discourses and fully grasp numbers. With numbers, you can discourse; but with discourse [alone], [dealing with] numbers is not possible.’ [Lu] Jiuci asked ‘Of all the affairs in the empire, to which are numbers not applicable?’

[19] The roots of the *quadrivium* go back to the ancient Mediterranean world: in the seventh book of Plato's *Republic* (c. 380 BCE), that begins with the famous analogy of the cave, we are told that the education of the ‘guardians’ of the ideal city-state must certainly include arithmetic, geometry and astronomy. About the role of music, Plato is less enthusiastic. See *Republic* VII, 521e–531c in *Republic*. (2013). Plato (428⁄427 or 424⁄423–348⁄347 BCE), Cambridge, MA; London, Harvard University Press, 134–169.

[20] Jami (2012), 22–4.

[21] It is perhaps appropriate to warn the non-sinological reader that an increasing number of scholars outside China have begun to raise doubts about the ethics of using such stolen materials, or indeed about their reliability as historical sources. See, for instance Goldin (2013). This quotation is therefore to be taken with due reserve.

[22] The term *yu* ‘discourse’ appears to be an abbreviation of the phrase *bai jia zhi yu* 百家之語 ‘the discourses of the hundred schools’ (*Shi ji* 87, 2546), a reference to the vigorous debates on philosophical issues that took place between groups of pre-Qin thinkers. It might be paraphrased here as ‘philosophy’.

> Chen Qi replied: 'Of all the affairs in the empire, there is none to which numbers are not applicable.'[23]

The version of Liu Xin's conference report that has come down to us makes it clear that he too saw the understanding of numbers as an important key to understanding the world as a whole. Moreover, in introducing his censored version of Liu Xin's material in the *Han shu*, Ban Gu takes care to make it plain that this view was not limited to the supporters of traitors and usurpers:

> 虞書曰「乃同律度量衡」, 所以齊遠近立民信也. 自伏戲畫八卦, 由數起, 至黃帝, 堯, 舜而大備. 三代稽古, 法度章焉. 周衰官失, 孔子陳後王之法, 曰:「謹權量, 審法度, 修廢官, 舉逸民, 四方之政行矣.」漢興, 北平侯張蒼首律曆事,孝武帝時樂官考正. 至元始中王莽秉政, 欲燿名譽, 徵天下通知鐘律者百(餘)餘人, 使羲和劉歆等典領條奏, 言之最詳. 故刪其偽辭, 取正義, 著于篇.

> In the *Book of Yu* [the section of the *Book of Documents* dealing with Emperors Yao, Shun and Yu] it is said 'Then he made uniform the musical scales, and [the measures of] length, capacity and the scales': this was in order to make equal [those from] far and near, and to establish the trustworthiness of the people. From when Fu Xi started from numbers and drew the eight trigrams, we arrive at the time of the Yellow Emperor, Yao and Shun, and all was grandly complete. The Three Dynasties examined antiquity, and patterned their measures and rules on them. When the Zhou dynasty decayed and official posts were lost, then Confucius set out the methods for later kings, saying 'Take care of the weights and quantities, look into the patterns and measures, restore offices which have been cast aside, and reinstate those who have abandoned [public life]. Then the ordering of the four quarters will be set in action.' With the rise of Han, the Beiping Marquis Zhang Cang headed all affairs of musical scales and astronomical systems. In the time of Wudi the Music Office was examined and reformed. When it came to the Yuanshi reign period [1–5 CE], Wang Mang held power, and wished to add lustre to his name. So he summoned over a hundred persons from all over the Empire who had a thorough knowledge of bells and musical scales, and put the Xi He [official] Liu Xin and others in charge of making an orderly report [of their deliberations]. They discussed these matters

[23] See Chinese text in Han Wei 韩巍 (2015) 'Bei Da cang Qin jian "Lu Jiuci wen shu yu Chen Qi" 北大藏秦简《鲁久次问数于陈起》初读 (A Brief Study of the Text Entitled "Lu Jiuci Asks Chen Qi about Numbers" on the Qin Bamboo Slips Collected by Peking University).' *Bei jing da xue xue bao: zhe xue she hui ke xue ban* 北京大学学报：哲学社会科学版》 (*Journal of Peking University: Humanities and Social Sciences*) (2): 29–36. I discuss this dialogue in Christopher Cullen (2015) 'Lu Jiuci yu Chen Qi de dui hua he zao qi de Zhong Guo shu xue shi 鲁久次与陈起的对话和早期的中国数学史 "The dialogue of Lu Jiuci and Chen Qi, and the early history of mathematics in China".' *Zi ran ke xue shi yan jiu* 自然科学史研究 (*Studies in the History of Natural Science*) 34 (2): 254–7.

in great detail; so I have cut out all heretical expressions, and selected the correct meaning, setting it out in a chapter. (*Han shu* 21a, 955)

In an early section of his report, Liu Xin sets out the ambitious nature of his claims for the role of numbers:

夫推曆生律制器, 規圜矩方, 權重衡平, 準繩嘉量, 探賾索隱, 鉤深致遠, 莫不用焉. … 其法在算術. 宣於天下, 小學是則. 職在太史, 羲和掌之.

In deriving astronomical systems, producing pitchpipes and making vessels, encompassing the circle and setting right the square, weighing the heavy and balancing what is equal, levelling the [builder's] line and making fair the capacities, in 'Seeking the profound and drawing out the obscure, hooking up from the deep and attaining to the distant'[24]—there is nothing in which [numbers] are without application. … The methods [for handling them] are found in calculation procedures *suan shu* 算 術. These are propagated throughout the empire, and are taken as a pattern by elementary studies. Responsibility for such matters lies with the Grand Clerk, and they are handled by the Xi He [official].

The Xi He was of course Liu Xin himself. The material by Liu Xin is, we are told, divided into five sections, each dealing with one aspect of a quantitative approach to the cosmos:

一曰備數, 二曰和聲, 三曰審度, 四曰嘉量, 五曰權衡.

First: making the numbers complete. Second: harmonizing the sounds. Third: examining the measures. Fourth: making good the volumes. Fifth: weighting the balance. (*Han shu* 21a, 956)

The second of these sections is by far the longest, and sets out a detailed theory of the numerical basis and cosmic significance of ancient Chinese musical theory, culminating in a specification of how to derive the lengths of the standard pitch-pipes from that corresponding to the fundamental Yellow Bell note by 'lower' and 'upper' generation, that is, successive multiplications by $\frac{2}{3}$ and $\frac{4}{3}$ (*Han shu* 21a, 965). Space does not permit a detailed exposition here, but in addition to concepts of Yin–yang and the Five Phases, Liu Xin frequently returns to the numerical cosmology set out in the appendices of the *Book of Change*.[25] He uses

[24] This is a quotation from the *Book of Change*: see *Zhou yi* 周易 7, 29b in *Shi san jing zhu shu*, vol. 1, 157a.

[25] A forthcoming study by Michael Loewe will address aspects of this broader topic (Loewe, private communication). Meanwhile, see Kawahara Hideki 川原秀城 (1989). "San tong li yu Liu Xin de shi jie guan 三統曆與劉歆的世界觀 'The Triple Concordance system and the world-view of Liu Xin'" in *Chūgoku kodai kagakushi ron* 中国古代科學史論 *'Articles on ancient Chinese science'*, Kyoto, Kyoto University Research Institute for Humanistic Studies: 121–38.

this to derive values for all the basic constants used in the Triple Concordance system. It is clear that his overall project is to show that all possible visible and perceptible regularities of the world are not just capable of being described in general terms by the cosmology he favours, but that this can also be done quantitatively. This is an extremely ambitious project, which has in it something of the spirit of the search for a 'grand universal theory of everything', such as is still sought for by modern cosmologists.[26]

4.4 The structure of the Triple Concordance system

The Triple Concordance system[27] is set out in the chapter of the *Han shu* immediately following the one in which the material just discussed occurs. It is the first complete example of a long succession of similar documents giving complete specifications of astronomical systems as they were used by successive dynasties until the end of imperial China. We cannot say positively that it is the first document of its kind ever composed. Indeed, the summary of the imperial library catalogue composed by Liu Xin and his father lists a number of documents with the word *li* 曆 '[astronomical] system' in their titles (see section 3.4.1). But we have no indication of their structure or contents, and nobody in succeeding centuries quotes from or refers to them as texts. Going further back, to the time of the Grand Inception reform of 104 BCE that laid down the luni-solar constants used by Liu Xin, there is no reference to anybody writing or submitting a lengthy document embodying the new system—nor is there any title in the Lius' catalogue that is suggestive of such a document.

The possibility is therefore open that Liu Xin was the actual creator of the distinctive documentary format taken by all later *li*. Since in any case the Triple Concordance system is the first astronomical system of which we have a completely documented description, it is worthwhile to discuss its structure in some detail. The system description is divided into the following main sections, of which the titles are original to Liu Xin:

26 More historically, we may make a comparison with the Pythagorean project to found a view of the cosmos on number (W. K. C. Guthrie (1978) *A history of Greek philosophy. Volume 1, The earlier Presocratics and the Pythagoreans*. Cambridge, Cambridge University Press, 212–306) and Kepler's claim that the orbits of the planets were structured in accordance with a nested series of the regular solids; see Evans (1998), 428–30.

27 The reader who would like more detail of Liu Xin's system than the outline given here will find it in the full translation of the Triple Concordance system, with commentary and a number of worked examples, given in Cullen (2017), chapter 2.

Concordance Constants, tong mu 統母: this lists the basic constants relating to lunar and solar phenomena.

Sequence Constants, ji mu 紀母: this lists the constants relating to the cycles of appearance of the five visible planets.

The Five Pacers, wu bu 五步: for each planet, this section gives details of the daily motion of the planet against the background of the stars for each section of its cycle of apparent movement seen from the earth.

Concordance Workings, tong shu 統術: this gives step-by-step instructions for calculating the cyclical days on which fall the beginnings of lunar months, including intercalary months, and the principal events of the solar cycle.

Sequence Workings, ji shu 紀術: this gives instructions for calculating the times of occurrence of the principal events in the cycles of the planets.

There are also a number of other tabulations, giving such information on such matters as the widths of the 28 lodges, and the sexagenary days and associated historical events of major calendrical cycles of past centuries. All these are fully translated and discussed in Cullen (2017), chapter 2.

4.4.1 Concordance constants, *tong mu* 統母[28]

Here we are given a list of twenty-one constants that will be used in making calculations of solar and lunar cycles, the basic data required for producing a calendar. In addition, the list includes data that make possible the prediction of lunar eclipses. Each constant is introduced by giving it a descriptive name, followed by an integer value, after which there is a brief explanation of what may be called the 'derivation' of the given number.

Seven of the constants are independently defined, while all the rest are derived from those constants in various ways. These seven basic constants are certainly intended to represent observed astronomical reality, and they do so quite well. For instance, as we shall see in the discussion of the Grand Inception constants in 4.4.2, Lunation Factor [2,392] divided by Day Factor [81] gives a good approximation to the actual length of a mean synodic month in days. Liu Xin, however, sets himself the additional task of showing how such numbers may be derived from 'cosmological' considerations, using numbers taken from the numerology of the *Book of Change*, and in this he certainly succeeds.

[28] The term *mu* 母 (literally 'mother') used here has a more general significance than when it is used as a name for the denominator of a fraction, as opposed to the numerator, *zi* 子 (literally 'child').

In Box 4.1 is the complete list of constants in the order in which the numbers appear in the text; the seven basic cosmologically derived constants on which the rest depend are marked with asterisks. I have added the explanatory text in italics; Liu Xin's explanations of the derivation of the numbers will be discussed in sections 4.4.2–4.4.6.

Box 4.1: Triple Concordance solar/lunar constants

1 *Day Factor *ri fa* 日法 81
Multiplying the days in a lunation by this factor produces a whole number.

2 *Intercalation Factor *run fa* 閏法19
The number of years in a Rule Cycle, or more precisely the number of solar cycles precisely equal to Rule Months lunations.

3 Concordance Factor *tong fa* 統法 1,539
The number of years in a Concordance Cycle, after which system origin solar and lunar conditions repeat at midnight; multiplying the days in a solar cycle, or the du in a complete circuit by this factor produces a whole number, Circuits of Heaven. 1,539 is 19 × 81, the denominator of the fraction that results if one calculates the number of days in a solar cycle from the fact that there are a whole number of lunations (each with a fractional part with a denominator of 81) in 19 solar cycles.

4 Origin Factor *yuan fa* 元法 4,617
The number of years in an Origin Cycle, after which all system origin solar and lunar conditions and cyclical day numbers repeat at midnight.

5 *Coincidence Number *hui shu* 會數 47
An intermediate step in calculating the next constant, and in calculating Coincidence Months (see number 14).

6 *Rule Months *zhang yue* 章月 235
The number of months in a Rule Cycle of years, or the number of lunations precisely equal to Intercalation Factor [19] solar cycles.

7 *Lunation Factor *yue fa* 月法 2,392
The number of ($1/81$) day in one lunation – the number of days in one lunation, 29 $43/81$, 'scaled up' by Day Factor [81] to produce a whole number.

8 Compatibility *tong fa* 通法 Factor 598
Lunation Factor [2,392] divided by 4; used in calculating the next quantity.

9 Medial [Qi] Factor *zhong fa* 中法 140,530
The number of ($1/4,617$) days from one Medial Qi to the next, or the number of days from one Medial Qi to the next, 30 $2,020/4,617$ 'scaled up'

continued

Box 4.1: *Continued*

by Origin Factor [4,617] to produce a whole number.

10 Circuits of Heaven *zhou tian* 周天 562,120
The number of ($\frac{1}{1{,}539}$) days in a solar cycle, or the number of ($\frac{1}{1{,}539}$) du *in a circuit of the heavens – the number of days in a solar cycle, or* du *in a circuit, 365 $\frac{385}{1{,}539}$, 'scaled up' by Concordance Factor [1,539] to give an integer.*

11 *Year Medial [Qi] *sui zhong* 歲中12
The number of Medial Qi in one solar cycle.

12 Lunar Circuits *yue zhou* 月周 254
The number of circuits of the heavens performed by the moon in Rule Months [235] lunations, i.e. during Rule Factor [19] solar cycles.

13 *New and Full Moons Coincidence *shuo wang zhi hui* 朔望之會 135
The number of lunations over which a cycle of 23 lunar eclipses repeats precisely.

14 Coincidence Months *hui yue* 會月 6,345
The previous quantity 'scaled up' by multiplying by Coincidence Number [47] to produce a whole number of solar cycles.

15 Concordance Months *tong yue* 統月 19,035
The number of months in Concordance Factor [1,539] years, or the number of lunations precisely equal to Concordance Factor [1,539] solar cycles.

16 Origin Months *yuan yue* 元月 57,105
The number of months in Origin Factor [4,617] years, or the number of lunations precisely equal to Origin Factor [4,617] solar cycles.

17 Rule Medial [Qi] *zhang shong* 章中 228
The number of Medial Qi in Intercalation Factor [19] solar cycles.

18 Concordance Medial [Qi] *tong zhong* 統中18,468
The number of Medial Qi in Concordance Factor [1,539] solar cycles.

19 Origin Medial [Qi] *yuan zhong* 元中 55,404
The number of Medial Qi in Origin Factor [4,617] solar cycles.

20 Reckoning Surplus *ce yu* 策餘 8,080
The excess of one solar cycle over 360 days, scaled up by Concordance Factor [1,539] to give a whole number.

21 Circuits Culmination *zhou zhi* 周至 57
This quantity, three times Intercalation Factor [19], seems to serve little purpose in calculation. It is referred to only once in the text, when it is noted that after this period of years, the sexagenary day of winter solstice and the first month conjunction is either one less than the starting point, or the same.

One feature of all these constants stands out at once: as already mentioned, all of them are integers. This is not because the mathematical techniques of the time found simple fractions or 'mixed' numbers with an integral and fractional part difficult to represent. It would have been perfectly possible to state that the number of days in a month were 二十九日八十一分日之四十三, literally '29 days [and] 43 of the 81st parts of a day', or in modern symbolism 29 $^{43}/_{81}$ days. Why then state the quantity by giving two figures, the Day Divisor [81] and Lunation Factor [2,392]? It seems that Liu Xin prefers to carry out calculations using whole numbers only, which was simpler and less liable to error when using counting rods to represent rod numerals. This was achieved by simply multiplying up the quantity that included a fraction by the denominator of the fraction, so 'scaling it up' to a whole number. One no longer calculates with days, but with ($^{1}/_{81}$) day. In his third century CE commentary on the anonymous mathematical text known as the *Jiu zhang suan shu* 九章算術 'Mathematical procedures in nine sections',[29] Liu Hui refers to something like this process when he says, in relation to divisions that leave a remainder:

> 若有分, 以乘其實而長之. 則亦滿法. 乃為全耳.
>
> If there are parts, then multiply the dividend and scale it up [literally 'extend it']. Thus [the parts] will fill the divisor. Then it becomes a whole [number]. (Guo Shuchun 郭書春 2004: 15, commentary)

Similarly, one does not represent the days in a solar cycle or the number of *du* in a complete circuit by the mixed number 365 $^{385}/_{1,539}$, but by Circuits of Heaven [562,120], which represents that number 'scaled up' by the Concordance Factor [1,539].

Other numbers in the constants list are just there to make calculations simpler: the items number 15 to 20 are clearly of this type, and enable the calculator to save a little effort.

4.4.2 Deriving the constants of an astronomical system

But how are the basic constants derived? The first such factor is the Day Factor [81], for which we are told:

[29] Translated in Karine Chemla and Shuchun Guo (2004) *Les neuf chapitres: le classique mathématique de la Chine ancienne et ses commentaires*, Paris, Dunod, and Kangshen Shen, J. N. Crossley and Anthony W. C. Lun (1999) *The nine chapters on the mathematical art: companion and commentary*, Oxford, Oxford University Press. This book was probably completed around the time of Wang Mang.

元始黃鐘初九自乘, 一龠之數, 得日法.

At the Origin Initiation, the First 9 of the Yellow Bell multiplies itself, [giving 81] which is the number of one *yue*; one obtains the Day Factor. (*Han shu* 21b, 991)

The Yellow Bell pitchpipe has a length of 9 *cun*; the reference to 'first 9' recalls the *Book of Change*, in which this expression refers to a yang (unbroken) line in the lowest position of a hexagram—a yin line would have corresponded to 6.[30] A *yue* is a measure of volume equivalent to 200 millet grains, which will supposedly exactly fill the pitchpipe.

The Intercalation Factor [19] also has a simple cosmological derivation:

合天地終數, 得閏法.

Joining the ultimate numbers of Heaven and Earth, one obtains the Intercalation Factor. (*Han shu* 21b, 991)

According to the *Book of Change*, the numbers of Heaven and Earth are respectively the odd and even numbers in the sequence 1 to 10.[31] The 'ultimate' two numbers are thus 9 and 10, whose sum, Liu Xin says, is 19. Nineteen years, a Rule *zhang* 章, is the shortest period consisting of a whole number of solar cycles and lunations, and defines the cycle at which intercalation patterns repeat: see section 1.3.

Not all cosmological derivations are so simple. Perhaps the most elaborate is that of Lunation Factor [2,392]. This is not found in the specification of the Triple Concordance system itself, but in the part of the preceding chapter where Liu Xin brings together cosmological derivations of all the constants on which the Triple Concordance is to be based, only some of which are repeated fully in the present chapter:

是故元始有象一也, 春秋二也, 三統三也, 四時四也, 合而為十, 成五體. 以五乘十, 大衍之數也, 而道據其一, 其餘四十九, 所當用也, 故蓍以為數. 以象兩兩之, 又以象三三之, 又以象四四之, 又歸奇象閏十九, 及所據一加之, 因以再扐兩之, 是為月法之實. 如日法得一. 則一月之日數也, 而三辰之會交矣. 是以能生吉凶.

So the Beginning at the Origin has its counterpart in 1. Spring and Autumn are 2. The Three Concordances are 3. The Four Seasons are 4. Together these make 10, which makes up the Five Forms. Multiplying 10 by 5, that is the number of the

[30] See Richard Wilhelm and Cary F. Baynes (1967) *The I ching: or, Book of changes*. Princeton, N.J., Princeton University Press, 721–3.

[31] *Zhou yi* 7, 26b in *Shi san jing zhu shu*, vol. 1, 155b.

Great Expansion, but the Dao occupies one in it, so [we discard that, and] the remainder is 49, which is what is should be used [in continuing the divination]. So the yarrow stalks are used for the number [49], then one doubles it to correspond to two, one trebles it to correspond to three and one quadruples it to correspond to four. Then one adds on the odd 19 which corresponds to the intercalation, and adds the 'occupied 1' [for the Dao] to it, and goes on to double it for the two graspings [of the stalks when they are divided into two piles during divination]. This makes the product for the Lunation Factor [$(49 \times 2 \times 3 \times 4 + 19 + 1) \times 2 = 2{,}392$]. Count one for each accommodation of the Day Factor [81], then this is the number of days in a lunation, and the Three Markers meet and correspond. Thus good and evil fortune are able to be produced. (*Han shu* 21a, 983)

A modern reader can only admire Liu Xin's determination to get the result he needs. This he certainly succeeds in doing, since the value for the length of a mean lunation implied by his figures is:

$$2{,}392/81 \text{ days} = 29.53086 \text{ days}$$

A modern value to the same precision is 29.53059 days, which is only 24 seconds less. It would take nearly three centuries before this discrepancy built up to a whole day.

To take a further example:

參天九，兩地十，是為會數. 參天數二十五，兩地數三十，是為朔望之會. 以會數乘之, 則周於朔旦冬至, 是為會月. 九會而復元, 黃鐘初九之數也.

Treble the Heavenly 9, double the Earthly 10, and this makes Coincidence Number [$2 \times 9 + 2 \times 10 = 47$]. Treble the [total] Heavenly numbers 25, double the [total] Earthly numbers 30, and this makes New and Full Moons Coincidence [$3 \times 25 + 2 \times 30 = 135$]. Multiply it by Coincidence Number [47], then you cycle back to conjunction at the winter solstice, and this makes Coincidence Months [$135 \times 47 = 6{,}345$]. This is the number for the First Nine of the Yellow Bell. (*Han shu* 21a, 983)

Here the numbers calculated all relate to the prediction of lunar eclipses. Coincidence Months [6,345] is the main cycle during which patterns of lunar eclipses recur. It is equal to Coincidence Number [47] times New and Full Moons Coincidence [135], which is the number of lunations during which a pattern of 23 eclipses repeats itself. It is also precisely equal to 513 solar cycles; in calendrical terms the cycle is thus 6,345 months, or 513 years. Coincidence Months is not, however, a whole number of days. Three periods of this length are equal to a Concordance of 1,539 years, which is an integral number of days.

4.4.3 Sequence constants, *ji mu* 紀母

After detailing the constants needed for calculations relating to the sun and moon, the Triple Concordance specification passes straight on to the constants for the planets. These are given in the order Jupiter, Venus, Saturn, Mars and Mercury. As we shall see, the constants for the first three are generated differently from those for the last two, which may in part explain why the planets are presented in this order.

As the names just listed imply, in ancient western Eurasia the planets were given names that associated them with deities. In China, each planet might be referred to by one of two possible names, one of which linked the planet with one of the Five Phases *wu xing* 五行 which were seen as fundamental to the basic processes of the cosmos,[32] while the other was more descriptive:[33]

> 五星之合於五行, 水合於辰星, 火合於熒惑, 金合於太白, 木合於歲星, 土合於填星.
>
> The correspondence between the Five Stars and the Five Phases: Water corresponds to the Chronogram Star [Mercury], Fire corresponds to Dazzling Deluder [Mars], Metal corresponds to Great White [Venus], Wood corresponds to the Year Star, and Earth corresponds to the Garrison Star [Saturn]. (*Han shu* 21a, 985)

The aim of the constants given is to make it possible to calculate the date when each planet will make an 'Appearance', that is, will appear near the eastern horizon at dawn just before the sun rises. In the Triple Concordance system, a planet is counted as making such an appearance as soon as it has moved half a 'Jupiter station' to the west of the sun. A 'Jupiter station' is one-twelfth of a complete circuit of the heavens, so the elongation of the planet from the sun at that time is 15 degrees in modern terms. This criterion is not a good predictor of when human observers will begin to report planet visibility. Apart from variations of visual acuity between observers, and the fact that some planets (such as Jupiter and Venus) are always relatively bright and easy to see in comparison with a faint planet like Mercury, visibility is also affected by the angle of inclination

[32] These are five phases of cosmic activity named as Water, Fire, Wood, Metal and Earth: see chapter 3, section 3.2.1, on these. As noted there, 'Earth' in this scheme renders *tu* 土, meaning 'earth' in the sense of 'soil', not 'Earth' meaning the part of the cosmos on which we live, *di* 地, the cosmic correlate of Heaven *tian* 天.

[33] Mercury was said to appear once in each of the four seasons (which is roughly true), hence it is a time marker; Mars is a presage of conflict and confusion, Venus shines brilliantly, Jupiter moves through one 'station' (1/12 of a circuit of the heavens) per year, and Saturn is said to 'garrison' each of the 28 lodges for (on average) about a year as it circuits the heavens (see Cullen (2011a), 245).

to the horizon of the line from the sun to the planet in question (which will be close to the ecliptic).[34] Later astronomical systems no longer use the 'appearance' as their major reference point, but begin calculations from the moment of (unobservable) conjunction between the planet and the sun, although they still predict appearances by the same method.

As an example, Box 4.2 gives a summary list of the constants given for Jupiter in the order given by Liu Xin. Two of these constants, the Year Number and Appearance Number, marked in bold, are the basis of the whole list. All the other constants are simply calculated from these two, and serve as aids to shorten calculation. Of the two basic constants, Year Number is explicitly derived from cosmological considerations (see later in this section), while Appearance Number seems to have been adjusted to produce a fit with observation when used in combination with Year Number, without any explicit reference to cosmology.

Box 4.2: Constants for Jupiter given by Liu Xin, with added explanatory comments in italics.

1 **Year Number *sui shu* 歲數 1,728**

The number of years in which Jupiter makes Appearance Number [1,583] Appearances.

2 [Jupiter] Appearance Medial [Qi] Parts *xian zhong fen* 見中分 20,736

The number of Medial Qi in Year Number years, 12 times Year Number – or, the number of Medial Qi in one year, scaled up by Year Number.

3 [Jupiter] Accumulated Medial [Qi] *ji zhong* 積中 13; Medial [Qi] Remainder *zhong yu* 中餘 157

This is the number of Medial Qi in one Appearance, obtained by dividing Appearance Medial [Qi] Parts by Appearance Number.

4 **[Jupiter] Appearance Medial [Qi] Factor *xian zhong fa* 見中法 1,583. This is the Appearance Number *xian shu* 見數.**

The number of Appearances in Year Number years.

5 [Jupiter] Appearance Intercalation Parts *xian run fen* 見閏分 12,096

In 19 years, seven intercalary months accumulate. Thus, by multiplying Year Number by seven, we obtain the number of intercalary months in Appearance Number of appearances, at a scale of Intercalation factor [19].

continued

[34] See M. Robinson (2009) 'Ardua et Astra: On the Calculation of the Dates of the Rising and Setting of Stars.' *Classical Philology* 104: 354–75.

Box 4.2: *Continued*

6 [Jupiter] Accumulated Lunations *ji yue* 積月 13; Lunation Remainder *ji yu* 月餘 15,079

These numbers give the number of lunations in one Appearance. We know that in 19 solar cycles there are 235 lunations. We multiply Year Number [1,728] by $^{235}/_{19}$ and divide the result by Appearance Number [1,583] to find the number of lunations required, 13 $^{15,079}/_{30,077}$.

[Jupiter] Appearance Lunation Factor *xian yue fa* 見月法 30,077

This is the denominator of the calculation above, 19 × Appearance Number [1,583].

7 [Jupiter] Appearance Medial [Qi] Day Factor *xian zhong ri fa* 見中日法 7,308,711.

This is Appearance Number [1,583] × Origin Factor [4,617]. It is the factor required when calculating the number of days into an uncompleted Medial Qi when the Appearance occurs.

8 [Jupiter] Appearance Lunation Day Factor *xian yue ri fa* 見月日法 2,436,237

This is Appearance Number [1,583] × Concordance Factor [1,539]. It is the factor required when calculating the number of days into an uncompleted lunation when the Appearance occurs.

The other four planets are each given a similar listing. Those for the inner planets Venus and Mercury are somewhat more complicated, since those planets' cycles are divided into dawn and dusk appearances, reflecting the fact that they stay relatively close to the sun, appearing first to the west of it at dawn, and then to the east of it at dusk. In the Triple Concordance system, it is assumed that the times of visibility of both planets to the west of the sun (around dawn) and to the east of the sun (around dusk) are in the ratio 9:7. This is (in geocentric terms) a result of the sun's steady motion eastwards. The complete cycles for the inner planets relative to the sun (their synodic cycles in modern terms) thus contain two appearances, and are designated 'Returns' *fu* 復 rather than 'Appearances' *xian* 見 as for the other planets. When the listing for all five planets is complete, Liu Xin gives us a series of short comments on how the quantities relate to each other, parts of which have been recapitulated above.

Table 4.1 lists the key constants for each planet, with the resulting calculated values of synodic period—the time taken for the planet to return to the same position relative to the sun. For comparison, modern values of the mean synodic periods are also given; the correspondence with the values implied by the

Table 4.1 ***Year numbers, Appearance/Return numbers and synodic periods for the five planets***

	Year Number (Y)	Appearance or Return Number (A)	Triple Concordance synodic period (Y/A), years	Triple Concordance synodic period, calculated from (Y/A), days, using Triple Concordance year length	Modern value of mean synodic period, days[1]
Jupiter	1,728	1,583	1.092	398.7	398.9
Venus	3,456	2,161	1.599	584.1	583.9
Saturn	4,320	4,175	1.035	377.9	378.1
Mars .	13,824	6,469	2.137	780.5	779.9
Mercury	9,216	29,041	0.3173	115.9	115.9

[1] These modern values for mean synodic periods are taken from the NASA fact sheets on each planet available at http://nssdc.gsfc.nasa.gov/planetary/planetfact.html.

Triple Concordance is quite close. Note, however, that that length of a planet's synodic cycle (which depends on the earth's elliptical orbit as well as on that of the planet) is not constant. For Jupiter the period may vary by up to two days, but for Mars the variation may be by as much as five days. Using a fixed value of synodic period will therefore introduce some inaccuracy—though that will be small compared with other factors already discussed that have a much greater effect on the date of first visibility.

The first three planets in the list have similar explanations for Year Number. Jupiter:

> 木金相乘為十二, 是為歲星小周. 小周乘巛策, 為千七百二十八, 是為歲星歲數.
>
> Wood [3] and Metal [4] are multiplied to make 12. This is the Lesser Circuits for Jupiter. Lesser Circuits [12] multiplies the Chthonic Reckoning [144] to make 1,728. This is the Year number for Jupiter. (*Han shu* 21b, 992)

Venus:

> 金火相乘為八, 又以火乘之為十六而小復. 小復乘乾策, 為三千四百五十六, 是為太白歲數.
>
> Metal [4] and Fire [2] are multiplied to make 8. Further multiply it by Fire to make 16, which is Lesser Returns. Let the Lesser Returns multiply the Uranic Reckoning [216] to make 3,456. This is the Year Number for Venus. (*Han shu* 21b, 993)

Saturn:

> 土木相乘而合經緯為三十, 是為鎮星小周. 小周乘巛策, 為四千三百二十, 是為鎮星歲數.
>
> Earth [2] and Wood [3] are multiplied together and added, so as warp and weft they make 30 [i.e. $(2+3) \times (2\times3) = 30$], and this is the Lesser Circuits for Saturn. Lesser Circuits [30] multiply the Chthonic Reckoning [144] to make 4,320. This is the Year Number for Saturn. (*Han shu* 21b, 994)

In each case, the first phase is the one corresponding to the planet, and the second is the one that 'overcomes' it in the so-called 'conquest' order (see chapter 2).

There are two elements that enter into these calculations—the numbers representing which one of the Five Phases corresponds to each planet, and the so-called Uranic and Chthonic Reckonings *qian ce* 乾策 and *kun ce* 坤策. The first are commonplace enough, and were set out in the preceding chapter of the *Han shu* (21a, 985): see 3.2.1. The Uranic and Chthonic reckonings have their most basic origins in the ancient process of divination using yarrow stalks, as set out in the *Book of Change*. In manipulating the yarrow stalks, one builds up a hexagram line by line, beginning with the lowest line, by a process involving random division of 49 stalks into two piles, and counting off by fours. When an unbroken Yang line is generated, a total of 36 stalks will be left in the piles, and when a Yin line is generated, there will be 24. At one extreme, we might generate six yang lines, which gives us the hexagram *Qian* 乾 'Uranic':

䷀

The total of all the remainders will thus be $6 \times 36 = 216$, the 'Uranic Reckoning'. At the other extreme, if we generate six yin lines, we obtain the hexagram *Kun* 坤 'Chthonic':

䷁

The total of all the remainders will then be $6 \times 24 = 144$, the 'Chthonic Reckoning'.[35]

[35] See page *Zhou Yi* 7, 22a in the edition of *Shi san jing zhu shu*, vol. 1, 153. See also the explanation of the procedure in Wilhelm and Baynes (1967), 721–3. Liu Xin explains the two Reckonings as a multiplication of cosmic factors in the previous chapter (*Han shu* 21a, 985) without direct reference to manipulation of the yarrow stalks.

Table 4.2 ***How planetary constants are calculated***

Planet	Planet's phase	Number of phase	Antiphase	Number of antiphase	Process	Result: 'Lesser Circuits'	Uranic or Chthonic Reckoning	Product: Year Number
Jupiter	Wood	3	Metal	4	Multiply	12	144	1,728
Venus	Metal	4	Fire	2	Multiply, then multiply again by Fire [2]	16	216	3,456
Saturn	Earth	2	Wood	3	Multiply the sum [5] by the product [6]	30	144	4,320

Following (Teboul, Michel 1983: 10–12), we may tabulate the calculations specified above (Table 4.2).

Teboul attempts to find plausible reasons for the way each of these calculations is performed, but it is not clear that we really do understand why Liu Xin chose these procedures and could have chosen no other. For the remaining two planets, the specifications are somewhat different.

Mars:

> 火經特成, 故二歲而過初, 三十二過初為六十四歲而小周. 小周乘乾策, 則太陽大周, 為萬三千八百二十四歲, 是為熒惑歲數.
>
> Fire goes through a special completion, so in two years it goes past the start, and in going past the start 32 times it makes 64 years, so that is Lesser Circuits. Lesser Circuits multiplies the Uranic Reckoning (216), then this is the Greater Circuits of the Great Yang, making 13,824, and this is the Year Number for Mars. (*Han shu* 21b, 995)

Mercury:

> 水經特成, 故一歲而及初, 六十四及初而小復. 小復乘巛策, 則太陰大周, 為九千二百一十六歲, 是為辰星歲數.
>
> Water goes through a special completion, so in one year it reaches the start, and in reaching the start 64 times that is Lesser Returns. Lesser Returns multiplies the Chthonic Reckoning (144), then this is the Greater Circuits of the Great Yin, making 9,216, and this is the Year Number for Mercury. (*Han shu* 21b, 996)

The key to understanding what is happening here is given (as Teboul remarks) in a note by Liu Xin at the beginning of the section of explanation that follows the listings of planetary constants:

> 合太陰太陽之歲數而中分之, 各萬一千五百二十. 陽施其氣, 陰成其物.
>
> Join the Year Numbers of the Great Yin and Great Yang and divide them in the middle: each [part] is 11,520. Yang spreads forth the *qi*, Yin completes the creatures. (*Han shu* 21b, 997)

We have already been told that the Great Yin is linked with Mercury (Water) and the Great Yang with Mars (Fire). But what is this figure of 11,520?

Suppose that we consider all 64 hexagrams. If they were all composed of Yang lines, the total of the remainders generated in the divination process would be:

> $64 \times 6 \times 36 = 64 \times 216 = 13{,}824$, which is of course the Year Number for Mars.

If all 64 hexagrams were composed of Yin lines, the total of the remainders would be:

> $64 \times 6 \times 24 = 64 \times 144 = 9{,}216$, the year number for Mercury.

The total of these numbers is 23,040. But in fact, only half the lines in the 64 hexagrams are Yang, and half Yin, so that the total of the remainders will be half of 23,040, i.e. 11,520. As we were told 'Join the Year Numbers of the Great Yin and Great Yang and divide them in the middle: each [part] is 11,520'.

In modern terms:

> (Mercury Year Number [9,216] + Mars Year Number [13,824])/2 = 11,520

This number is given in the *Book of Change* immediately after the Uranic and Chthonic reckonings:

> 乾之策二百一十有六, 坤之策百四十有四, 凡三百有六十當期之日二篇之策萬有一千五百二十當萬物之數也.
>
> The Uranic Reckoning is 216, and the Chthonic Reckoning is 144, being in all 360, which corresponds to the days in the period.[36] The reckonings of the two sections [make] 11,520, which corresponds to the number of the myriad creatures. (*Zhou yi* 7, 22a–22b)

[36] The 'period' *qi* 期 here may be intended as an approximation to the length of the solar cycle, and the commentators take it in that sense, perhaps influenced by the use of the term in the *Book of Documents* where it clearly refers to the solar cycle, although the value there given is 366 days: see chapter 1 note 16. It may, however, simply refer to the period approximating to the solar cycle within which the sexagenary number of the first day of the count will repeat, since $60 \times 60 = 360$.

Evidently 11,520 is the number that represents all phenomena represented in the hexagrams—which is in effect, everything that matters in the cosmos. As the *Book of Change* puts it a few lines later *Tian di zhi neng shi bi yi* 天地之能事畢矣 'All possible phenomena in heaven and earth are complete' (*Zhou yi* 7, 23a). We shall see later (chapter 6, section 6.5.1) that Zhang Heng in the second century CE gives this as the total number of stars in the heavens, on the basis that there must one star to represent each *wu* 物 'creature'.

The origin of the two Year Numbers for Mars and Mercury in the *Book of Change* is thus clear enough, although the links with actual astronomical behaviour gestured at in the text are quite approximate. Mars does make a circuit of the heavens in about two years, and Mercury circuits the sky close to the sun, so that it will return to the same position relative to the stars in about one year. Quite why the resultant number for Lesser Circuits or Returns has to be 64 is not clear. In any case, it is fairly clear that dealing with actual astronomical circumstances is a task fulfilled by the Appearance/Return numbers, which are bound by no cosmological constraints.

Now all the year numbers are known, we may look back at the final section of Liu Xin's discussion of planetary constants in the previous chapter. After deriving the Uranic and Chthonic Reckonings, he performs further multiplications using a variety of cosmologically significant factors, until he arrives at 138,240, and tells us:

> 然後大成. 五星會終
>
> ... only then [we reach] the great completion, so that the Five Planets meet in conclusion (*Han shu* 21a, 985)

This figure is indeed the lowest common multiple of the Year Numbers of the five planets. To reach concordance with the cycles of the sun and moon, Liu Xin multiplies by Rule Years [19] to obtain 2,626,560, and more multiplying finally produces 23,639,040, which is a multiple of the Origin Factor [4,617], after which all data repeat their initial conditions—planetary appearances, winter solstice, conjunction of sun and moon at midnight, and sexagenary day number. At this point we are said to reached the *tai ji shang yuan* 太極上元 'supreme ultimate of the High Origin'. This number is, as Liu Xin points out, a multiple of the 'number of the myriad creatures' 11,520, and so, in words which slightly modify a sentence from the concluding passage of the section of the *Book of Change* cited above, on which he is drawing:

> 天下之能事畢矣
>
> All possible phenomena under heaven are complete. (*Han shu* 21a, 986)

4.4.4 The Five Pacers, *wu bu* 五步

Liu Xin now turns to matters where observation greatly outweighs cosmology. We have the numbers required to predict when each planet will make an 'appearance' as defined in the Triple Concordance system, but how do the planets get from one appearance to the next? Liu Xin answers this question by giving what may be called a 'motion template' for each planet, dividing up its movement between appearances into a number of discrete phases, during each of which the direction and speed of the motion of the planet is treated as constant. Let us consider the example of Jupiter as an illustration—see Box 4.3, in which material from *Han shu* 21b, 998 is translated (for the full context, with explanatory calculations, see Cullen 2017, 71–80).

Box 4.3: Triple Concordance phases of Jupiter

木, Wood [Jupiter]

晨始見, 去日半次.

1 It is first seen at dawn, distant half a station from the Sun.

順, 日行十一分度二, 百二十一日.

2 It moves direct, going $2/11$ *du* per day for 121 days.

始留, 二十五日而旋.

3 It begins to be stationary, and turns after 25 days.

逆, 日行七分度一, 八十四日.

4 It moves retrograde, going $1/7$ *du* per day, for 84 days.

復留, 二十四日三分而旋.

5 It is stationary once more, and turns after 24 days and 3 parts.

復順, 日行十一分度二, 百一十一日有百八十二萬八千三百六十二分而伏.

6 It moves direct once more, moving $2/11$ *du* per day, for 111 days and 1,828,362 parts, before it becomes invisible.

凡見三百六十五日有百八十二萬八千三百六十五分,

7 In all it is visible for 365 days and 1,828,365 parts.

除逆, 定行星三十度百六十六萬一千二百八十六分.

continued

Box 4.3: *Continued*

8 Casting out retrograde motion, its true travelling [relative to] the stars is 30 *du* and 1,661,286 parts.

凡見一歲, 行一次而後伏.

9 In all it is visible for a year, and becomes invisible after travelling through one station.

日行不盈十一分度一.

10 Its daily motion does not exceed $\frac{1}{11}$ *du*.

伏三十三日三百三十三萬四千七百三十七分,

11 It is invisible for 33 days and 3,334,737 parts.

行星三度百六十七萬三千四百五十一分.

12 It travels the stars 3 *du* and 1,673,451 parts.

一見, 三百九十八日五百一十六萬三千一百二分, 行星三十三度三百三十三萬四千七百三十七分.

13 One Appearance [is] 398 days and 5,163,102 parts; it travels the stars 33 *du* 3,334,737 parts.

通其率, 故曰日行千七百二十八分度之百四十五.

14 Making the rates compatible, thus we may say that in one day it travels $\frac{145}{1,728}$ *du*.

The data given in Box 4.3 are a striking mixture of low and high precision. This appears to be the result of using low precision observation data, which then have to be fitted into a scheme based on predetermined cycles derived from cosmological considerations, which are stated to high precision. Thus in (2) and (4) we are in effect told that the planet moves 22 and 12 *du* in the numbers of days given. But overall the motion of the planet from one appearance to the next is given as 33 *du* 3,334,737 parts, and in some phases—such as the period of invisibility given in (11)—the timing of a phase also involves parts specified to seven figures. These parts, as is clear from later specifications, are given using the Appearance Medial [Qi] Day Factor [7,308,711]. The reason why this factor is used for all parts in the phase listing is explained in Box 4.4.

Box 4.4: Liu Xin's choice of factor for phase calculations

Appearance Medial [Qi] Day Factor

If we want to calculate the number of days from one Jupiter Appearance to the next, we proceed as follows:

There are Appearance Number [1,583] Appearances in Year Number [1,728] years, and the number of days in one year (i.e. one solar cycle) is Circuits of Heaven [562,120] divided by Concordance Factor [1,539].

So the length of one Appearance (in modern terms, a synodic cycle) is:

$$(562{,}120/1{,}539) \times 1{,}728/1{,}583 \text{ days}$$
$$= 398 + 1{,}721{,}034/(1{,}583 \times 1{,}539) \text{ days}$$
$$= 398\ 1{,}721{,}034/2{,}436{,}237 \text{ days, or}$$
$$398.7064313 \text{ days to the same precision.}$$

The denominator in this fraction is [Jupiter] Appearance Lunation Day Factor, which is [Jupiter] Appearance Number [1,583] multiplied by Concordance Factor [1,539].

However, if we set out to find how many days of a given Medial Qi are uncompleted at the time that an Appearance takes place, we shall find ourselves with parts in the result which need a factor three times larger than this to produce a whole number – and this factor is [Jupiter] Appearance Number [1,583] multiplied by Origin Factor [4,617]:

$$3 \times 2{,}436{,}237 = 7{,}308{,}711$$

This factor, [Jupiter] Appearance Medial [Qi] Day Factor is the largest factor that occurs in calculations about the motion of Jupiter. For simplicity, therefore, Liu Xin has adopted this factor as the scale for all parts referred to in the listing of motion during each phase of Jupiter's synodic cycle.

Similar lists of motion during phases are given for each of the other four planets. We now have all the information given to find the position of a given planet on any given date. Essentially the process falls into three stages, each of which Liu Xin specifies in detail:

1. Find the date when the Appearance or Return immediately before the given date occurs. In essence, this is a simple matter of establishing how many years have elapsed since High Origin, when all planetary

cycles begin, and finding when the last complete cycle of Appearance or Return ended and the current cycle began by casting out complete cycles. In modern terms, we would make this calculation in terms of the simple day-count of the Julian Day system, but Liu Xin finds which day of which month of which civil year marks the cycle starting point.

2. Since we can calculate the position of the sun amongst the stars at this moment (see later in this section), and since we know that the planet is half a Jupiter station to the west of the sun at the start of its cycle, we can find the position of the planet amongst the stars at the start of the current Appearance or Return cycle.
3. Using the moment of the start of the cycle as our point of departure, we simply count off days to find out how far our given date is located from the start of the current cycle of the planet's phases. We can then simply add up the total motion of the planet so far through the cycle of phases, and add this amount to the position of the planet at the start of the cycle to find its position at the given date.

But how good a prediction of planetary positions actually results from this process? It is difficult to decide what would count as a fair test—as we shall see shortly, Liu Xin apparently thought it was important that his system should yield good retrodictions of planetary positions in the remote past. However, it seems unlikely that he would have wanted his system to produce predicted positions in his own day that were obviously very wrong. Let us try a test with a bright planet, Venus, around the time when Liu Xin constructed his system, around 10 CE. If we run the calculations and compare them with modern computations of planetary positions, we can produce a graph (Figure 4.1) of the distance between Venus and the sun—the 'elongation' of the planet. The vertical axis shows the elongation in *du*, and the horizontal axis shows the number of days elapsed in the current Return cycle for the planet, beginning on 13 November of that year.

On the whole, the actual behaviour of the planet (and modern calculations model this to within naked-eye accuracy) is quite well followed in detail by the predictions of the Triple Concordance. If we look at the predicted dates of first appearance of all five visible planets to the west of the sun at dawn—when they should have been 'half a station', 15°, from the sun—we may compare the results with modern calculations as in Table 4.3.

As indicated above, the principle underlying these calculations is very simple. All we need to do is, in effect, to count off complete Appearances (synodic

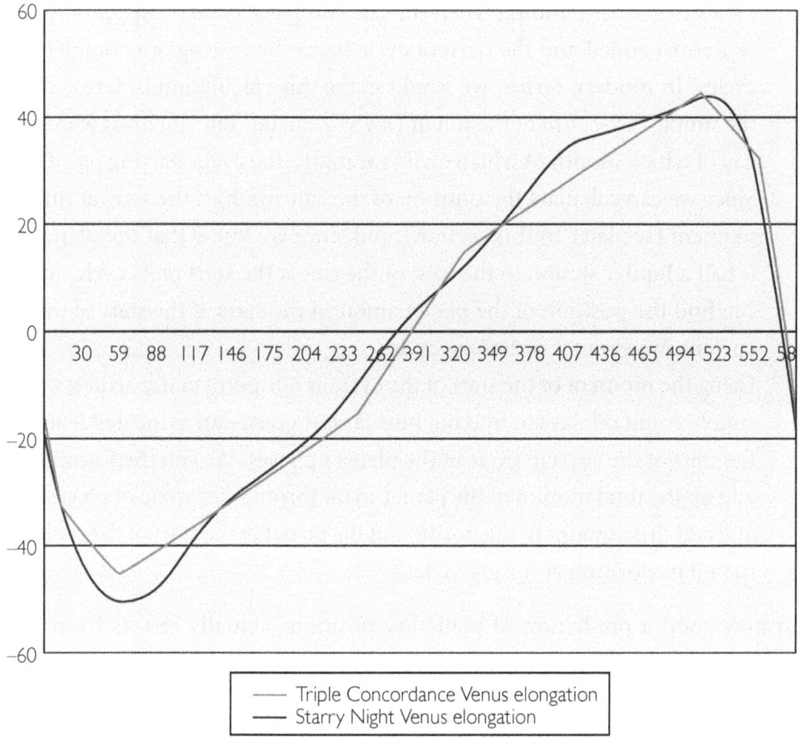

Figure 4.1 Elongation of Venus from the sun in *du*, for days counted from 13 November 10 CE, comparing values predicted by Triple Concordance and modern calculation using Starry Night Pro™.

Table 4.3 ***Triple Concordance predictions for appearances of planets in 10 CE***

	Triple Concordance predicted date of appearance in 10 CE	Modern calculation of angular distance from sun on predicted date	PLSV prediction of first visibility at dawn
Jupiter	5 August	12°	2 August
Venus	13 November	19°	5 November
Saturn	1 November	18°	24 October
Mars	21 June	15°	29 June
Mercury	11 September	18°	5 September

cycles) of the planet since the instant of High Origin on 2 December 143,232 BCE, a day which began at Chang'an at JD –50,593,729.80. If we add 131,221 synodic cycles of Jupiter as found in Box 4.4 to that JD, we find ourselves at JD 1,724,926.82, on 5 August 10 CE, which is therefore predicted to be the date of an Appearance of Jupiter in that year. Liu Xin goes through a process that begins by finding that 131,221 complete Appearance cycles of Jupiter have been completed some time during the year in question, and that $^{615}/_{1,728}$ of the next Appearance has been completed by the end of the year. From this, he finds on which day of which lunar month the Appearance began: a worked example for Jupiter is given in Box 4.9. A spreadsheet version of the Triple Concordance system confirms that the simplified calculation above gives the same result as Liu Xin's complete calculation.

In the tabulation, the angular distance from the sun was measured for the instant of sunrise on the predicted dates, using Starry Night Pro™. The last column shows a modern prediction of when the planet should first be visible at dawn at Chang'an (which was Wang Mang's capital), using Noel Swerdlow's programme 'Planetary, Solar and Lunar Visibility' (PLSV).[37] It is noteworthy that while in every case the Triple Concordance prediction of the date when the planet will first be seen in the east at dawn is not far from the modern prediction, it is in general a few days later—in some cases by over a week. The effect of such a delayed prediction would have been to ensure an increased chance that the planet would indeed have been seen in the east if an observer were to look for it on the day predicted by the Triple Concordance.

4.4.5 Concordance workings, *tong shu* 統術

4.4.5.1 INITIAL CONDITIONS, ORIGINS AND CONCORDANCES

Now all the data on which all calculations depend has been set out, the next major section of the Triple Concordance system specification takes us back to the core purpose of an astronomical system—to generate the data for the imperial calendar to be issued in any given year. We have already seen the basic principles on which the necessary calculations must be based:

[37] The method used by this programme depends upon what is called the *arcus visionis* ('arc of vision'), the vertical distance of the sun below the horizon at which a planet can be seen close to the horizon. The *arcus visionis* is an empirical parameter, different for each planet, that depends upon the magnitude of the planet and its difference of azimuth from the sun, both of which are variable. This method is based on that used by Ptolemy: see *Almagest* XIII.7, in Toomer (1998), 636–45.

1. A starting point when all important cycles are at their initial point—typically midnight on a *jiazi*.1 day that is the first day of the first month of the 'celestial' count (which is three months before the first month of the 'Xia' count commonly used for the civil year), at which time conjunction of the sun and moon occurred at the same instant as winter solstice. In addition, Liu Xin specified that the five visible planets were also simultaneously at their starting positions, which for him meant 'half a station' to the west of the sun, so that they all made their first dawn appearance together. As already mentioned (see section 4.2), for the Triple Concordance this moment was specified as being the winter solstice immediately preceding the Julian Year 143,231 BCE.
2. From that origin point, one counts off complete cycles governing the repetition of solar, lunar and planetary conditions up to any date of interest, until one is left with an incomplete part of a cycle, and one then finds out how far the given date is through the cycle—thus yielding the position of the sun, the phases of the moon and the positions of the planets, along with many other data.

4.4.5.2 PREDICTING MONTHS AND SOLAR CYCLES

Let us look at the way the text achieves this aim. We may pass lightly over the initial stage of calculation, which deals with a very obvious preparation for what follows. We are told to take the total number of years between the High Origin and the year we are interested in, and cast out the number of whole Origin Cycles of 4,617 years—which makes sense, since all solar and lunar positions, together with cyclical day numbers, repeat at the start of every Origin Cycle. Then we find out which of the three Concordance Cycles of 1,539 years each contains our present date. This is helpful, because solar and lunar conditions repeat at the start of each Concordance Cycle, just as at the start of an Origin Cycle. The only difference is the cyclical day of the first day of each Concordance Cycle is 20 days earlier than in the preceding Concordance. So we have:

First day of first Concordance Cycle in an Origin Cycle: *jiazi*.1
First day of second Concordance Cycle in an Origin Cycle: *jiachen*.41
First day of third Concordance Cycle in an Origin Cycle: *jiashen*.21
First day of first Concordance Cycle in next Origin Cycle: *jiazi*.1 once more.

Box 4.5: Finding the number of intercalary months

推天正,以章月乘(人)[入]統歲數,盈章歲得一,名曰積月,不盈者名曰閏餘 . 閏餘十二以上,歲有閏.求地正,加積月一;求人正,加二.

To predict the Celestial Standard Month: Multiply Rule Months [235] by the number of years of entry into the Concordance, and obtain 1 for each filling of Rule Years [19]. This is called Accumulated Months. What does not fill is called Intercalation Surplus. If Intercalation Surplus is 12 or above, the year has an intercalation. To seek the Terrestrial Standard Month, add one to Accumulated Months. To seek the Anthropic Standard Month, add 2. (*Han shu* 21b, 1001; Cullen 2017, 90–1)

The 'celestial standard month' is the one that normally contains the moment of winter solstice; this month is the first of the 'celestial' month count which is used for astro-calendrical calculations. The civil year (from 104 BCE onwards) begins two months later, with the first month in the usual Xia count (the 'anthropic standard month'). The 'celestial standard month' is thus the 11th Xia month preceding the civil year in question.

Suppose we take the example of the civil year that began in 102 BCE. Then the 'number of years of entry into the Concordance' is two, since the most recent Concordance Cycle began with the winter solstice preceding 104 BCE.

$$235 \times 2 = 470, \text{ and } {}^{470}\!/\!{}_{19} = 24, \text{ remainder } 14$$

So Accumulated Months is 24, which is the time interval in months between the start of the concordance and the start of the current celestial year, which begins with the 11th Xia month preceding 102 BCE. There have in fact been two complete civil years since the Concordance Cycle began, and the fact that the number of months so far is $24 = 2 \times 12$ shows us that both of those years have been normal years of 12 months each, with no intercalary month. Since there are seven intercalations in 19 years, each passing year contributes another $^{7}/_{19}$ of a year to the surplus. 14 is the Intercalation Surplus for this year, meaning that $^{14}/_{19}$ of an intercalary month has accumulated by the start of this celestial year. Since

$$^{14}/_{19} + {}^{7}/_{19} = {}^{21}/_{19}$$

continued

Box 4.5: *Continued*

an intercalary month will build up during the year ahead. By the beginning of the next year the Accumulated Months will be 12 + 12 + 13 = 37 rather than 36. The Intercalation Surplus will also be useful in later calculations. The Intercalation Surplus can also be explained in connection with the sequence of Medial Qi, which begins each year with the Winter Solstice. At the start of a Rule, winter solstice coincides with the conjunction beginning the first celestial month. This condition recurs at the beginning of the next Rule, but each successive winter solstice is $\frac{7}{19}$ of a lunation later than its predecessor with respect to its first celestial month conjunction, and it is the total of this lag that is represented by Intercalation Surplus. 24 months is, as stated, the number of months between the start of the current Concordance and the start of the first month of this celestial year.

The three Concordance Cycles in each Origin Cycle are named in sequence *tian* 天 (heavenly), *di* 地 (earthly) and *ren* 人 (human)—we might use the more formal terms 'celestial', 'terrestrial' and 'anthropic'.

Now the main task begins: finding the days on which the months of the year begin. We cannot do this by simply assuming that all preceding years in the current Concordance Cycle had 12 months, since some of them may have had 13—meaning that the first month of the current year may be one or more months later than expected. The method for finding how many intercalary months there have been so far has been is set out in Box 4.5 (see Cullen 2017, 90–1). As shown in the box, there were no intercalary months between the start of the current Concordance Cycle and the start of our example year of 102 BCE—but there will be an intercalary month during the corresponding civil year. Our next task to is to calculate on what day (that is, what day in the sexagenary cycle) the first month of the year will begin. Essentially we do this by calculating how many whole days have elapsed since the beginning of the current Concordance Cycle in which that month begins. Since we know the sexagenary day number on which that Concordance Cycle begins—calculated as above—we can find the number of (celestial) New Year's day, from which the day numbers of all subsequent beginnings of months can be found. The details of the calculation are given in Box 4.6 (see Cullen 2017, 91–2).

Box 4.6: Finding the sexagenary day of the conjunction at the start of the first Celestial month

推正月朔，以月法乘積月，盈日法得一，名曰積日，不盈者名曰小餘．小餘三十八以上，其月大．積日盈六十，除之，不盈者名曰大餘．數從統首日起，算外，則朔日也．求其次月，加大餘二十九，小餘四十三．小餘盈日法得一，從大餘，數除如法．求弦，加大餘七，小餘三十一．求望，倍弦．

To predict the conjunction of the [Celestial] Standard Month: Multiply Accumulated Months by Lunation Factor [2,392]. Count one for each filling of the Day Factor [81], and the name of this is Accumulated Days. What does not fill is called the Lesser Remainder. If the Lesser Remainder is 38 or above, the month is long. If Accumulated Days fills 60, cast it out. What does not fill is called the Greater Remainder. Count starting from the Concordance Head, and outside the count is the day of conjunction. To seek the next month, add to the Greater Remainder 29, and to the Lesser Remainder 43. Count one for each time the Lesser Remainder fills the Day Factor [81], and let it go with the Greater Remainder. Then count and cast out according to the method. To seek the first quarter, add to the Greater Remainder seven, and to the Lesser Remainder 31. To seek full moon, double [the amounts] for the first quarter. (*Han shu* 21b, 1,001; Cullen 2017, 91–2)

Here the aim is to predict the cyclical day on which falls the conjunction of the first day of the first month of the celestial count. We have already found that for 102 BCE Accumulated Months is 24. Now:

$$24 \times 2{,}392 = 57{,}408$$

So Accumulated Days must be $^{57{,}408}/_{81}$ = 708 with a remainder 60.

Turning first to the Accumulated Days, 708, this is the number of days from the start of the current concordance to the start of the first day of the first month of the present celestial year.

$$708 = 11 \times 60 + 48$$

continued

Box 4.6: *Continued*

So 11 complete 60-day cycles and 48 odd days have elapsed since the start of the current concordance, and thus 48 is the Greater Remainder. Following the instructions, then if the first day is number 1 and we are to count 48, the day 'outside the count'—the day of conjunction—is number 49. Now the first day of the current Concordance Cycle was *jiazi*.1. So the conjunction for the first month of this celestial year falls on day *renzi*.49. The Lesser Remainder, 60, is the fraction of a day by which the (mean) conjunction starting this month falls after midnight, at a scale of Day Factor [81]—i.e. the conjunction falls $^{60}/_{81}$ day after midnight.

Since the mean interval between conjunctions will be 2,392/81 days = 29 + $^{43}/_{81}$ days, it is clear that 29 and 43 are the amounts we must add to the Greater and Lesser Remainders to get to the next conjunction.

Since 38 + 43 = 81, if the Lesser Remainder at the start of this month is 43 or more, then a whole extra day will have to be counted before we get to the next conjunction, which will be 30 days away rather than 29, giving a 'long month' as stated. At the start of this month the Lesser Remainder is 60, so the month is long. The figures for the first quarter, and the full moon, follow from the fact that:

$$2{,}392/4 = 598 \text{ and } 598/81 = 7 \text{ remainder } 31.$$

It is assumed that the lunar phases are equally spaced.

To sum up the results of the calculation—the first day of the Concordance Cycle was *jiazi*.1, and since then 708 whole days have elapsed, which is 11 whole sixty-day cycles plus 48. Therefore the month we are targeting begins 48 days later than *jiazi*.1, i.e. on *renzi*.49. We also know that the conjunction beginning this month falls at $^{60}/_{81}$ day after midnight. All we have to do to find when subsequent conjunctions fall—and hence on what days subsequent months begin, is to add 29 $^{43}/_{81}$ days, counting an extra day each time the fractional parts add up to 81. If we do this twice, we shall find the day on which the first Xia month begins, the start of the civil year.

So much for the lunar cycle, which underlies the sequence of months. What about the solar cycle, for which the main mark-point is the moment of winter solstice? We proceed along the same lines as we did to find the conjunction of the first celestial month: see Box 4.7 (see Cullen 2017, 96).

Box 4.7: Finding the day of winter solstice

推冬至, 以(算)[策]餘乘(人)[入]統歲數, 盈統法得一, 名曰大餘, 不盈者名曰小餘. 除數如法, 則所求冬至日也.

To predict winter solstice: By Reckoning Surplus [8,080] multiply the number of years into the Concordance, obtaining 1 for each filling of the Concordance Factor [1,539]. [This] is called the Greater Remainder. What does not fill is called the Lesser Remainder. Cast out and count off according to the method, then that is the winter solstice day of the year sought. (*Han shu* 21b, 1,001; Cullen 2017, 96)

It would have been possible for us to find the total number of days elapsed between the start of the current Concordance Cycle (when winter solstice fell at midnight beginning a *jiazi*.1 day), and cast out multiples of 60.

But to avoid unnecessarily large numbers, this method exploits the fact that every solar cycle from one winter solstice to the next is 360 days plus a few extra days and a fractional part of a day. Since the 360 days do not change the sexagenary day number, we need only count the 'few extra days and a fractional part of a day' – and that is Reckoning Surplus [8,080], at a scale of Concordance Factor [1,539], equivalent to 5.25016 days.

Using our example year of 102 BCE, which is two years into the current Concordance Cycle, we calculate:

$$8{,}080 \times 2 = 16{,}160$$

$^{16{,}160}/_{1{,}539} = 10$, remainder 770, so 10 and 770 are the Greater and Lesser Remainders respectively.

Hence the sexagenary day number of the winter solstice is $1 + 10 = 11$, a *jiaxu*.11 day. The solstice will occur $^{770}/_{1{,}539}$ of a day after midnight on that day.

The next section of the text informs us that we may calculate the days on which fall the eight main subdivisions of the solar cycle, the Eight Nodes *ba jie* 八節, by adding 45 to the Greater Remainder, and 1,010 to the Lesser Remainder. This of course reflects the fact that the solar cycle has Circuits of Heaven [521,210] days at a scale of Concordance Factor [1,539], so that one eight of a cycle is

$$(1/8) \times 562{,}120/1{,}539 \text{ days} = 70{,}265/1{,}539 \text{ days},$$

which is 45 days, remainder 1,010.

By subdividing each node into 3, we can find when each of the 24 *qi* occurs.

Now we have two key dates—the first day of the first (celestial month), when conjunction of sun and moon occur, and the day of winter solstice. What is more, we know the precise instants of the conjunction and solstice, by means of the two 'Lesser Remainders' which give the fractions of a day elapsed between midnight and the events in question. For 102 BCE there are $^{60}/_{81}$ day and $^{770}/_{1,539}$ day respectively. It is then a simple matter to calculate the date and time of day of the next conjunction by (in modern terms) counting forward by the length of one lunation, 29 $^{43}/_{81}$ days, and to locate the next winter solstice by counting forward by the length of the solar cycle, 365 $^{385}/_{1,539}$ days. The instants of the 24 *qi* can be found by successively counting forward by $^{1}/_{24}$ of the solar cycle length.

4.4.5.3 MONTH NUMBERS AND INTERCALATIONS

In numbering the months, we must remember that the first month we locate by the methods set out in the previous section will be the first 'celestial' month, whereas the first 'anthropic' month, i.e. the first month according to the Xia count which begins the civil year, will not begin until the third celestial month. It is important to recall that references to months by their numbers in historical dates of the imperial period will always use the Xia count—so that, for instance, before the 104 BCE reform the Western Han dynasty saw no problem in saying that the civil year began in the tenth month.

Month numbering will pause when we encounter an intercalary month. Thus a month designated as *san yue* 三月 'third month' may be followed by a month designated *run san yue* 閏三月 'intercalary third month', and the month after that will be labelled *si yue* 四月 'fourth month'. How do we know when intercalary months will fall? The basic principle is that a month will be designated as intercalary if it does not contain a 'medial *qi*' *zhong qi* 中氣, i.e. one of the odd-numbered *qi* if winter solstice is number one. Each month is associated with a particular medial *qi*, which will always fall somewhere within it if intercalation is carried out as it should be. Thus the first celestial month (the 11th Xia month) should contain winter solstice. The third celestial month (the first Xia month, which nowadays marks 'Chinese New Year') should contain the fifth medial *qi*, *yu shui* 雨水 'rain waters'. There is, however, a slight difference between the method set out in the Triple Concordance system and actual practice as revealed by records of dates from the period when the system was in use. The Triple Concordance system asks whether the medial *qi* in question falls within the lunation corresponding to the given month (which may not begin until some way into the first day of the month, and may not end until some way into the first day of the next month), and decides on that basis whether the month is intercalary. Actual practice appears to have been simpler: one simply asked whether the relevant medial *qi* fell within the whole days of that month.

Applying this to 102 BCE, we find by the normal process of counting that the sixth Xia month began on a day corresponding to 27 June, with subsequent months beginning on 26 July and 25 August. Now the medial *qi* corresponding to the sixth month is number 15, 'Great Heat', and it falls on the last day of that month, 25 July. However, the next medial *qi*, number 17, 'Enduring Heat', does not fall until 25 August. This is therefore designated as the first day of the seventh month, and the month beginning on 26 July, which contains no medial *qi*, is thus designated 'intercalary sixth month'.

4.4.5.4 LUNAR ECLIPSE PREDICTION

The last major topic to be dealt with in the 'Concordance workings' section is the prediction of lunar eclipses. Lunar eclipses are fairly simple phenomena in physical terms, but since the basic facts are less well known than the phases of the moon and the seasons of the year, it may be helpful to give a little background detail. The moon shines in the sky because it is illuminated by light from the sun. When the moon is 'full', it is more or less opposite the sun in relation to the earth, so that when we look at it from the side of the earth that faces away from the sun (i.e. the side where it is night), we see it fully illuminated. However, the earth casts a long cone of shadow on the side away from the sun, and if the moon passes through that shadow, its light will be dimmed. Since the sun is an extended object, the earth's shadow will have a central zone, the 'umbra' within which no light from any part of the sun reaches the moon directly (though some red-tinged light may reach it by refraction through earth's atmosphere), and an outer zone, the 'penumbra' in which the moon is partially illuminated, since light from some parts of the sun is not blocked by the earth: see Figure 4.2.

Looking at an example of what might be seen in the sky, we may consider the eclipse of the moon that was visible from Chang'an around 23:00 local time on 8 January 104 BCE.

The central circle in Figure 4.3 is the umbra, and the outer circle is the penumbra. Clearly this is only a partial eclipse, although there will be significant dimming of a large portion of the moon's disc. At most full moons, the moon misses the shadow cone of the earth altogether, and so no lunar eclipse is seen. If on the other hand the moon is completely within the umbra, the eclipse will be total.

How does the Triple Concordance predict lunar eclipses? Like all its predictions, the procedure is based on the idea that at the system origin all elements were in their initial states, from which appropriate cycles may be counted off

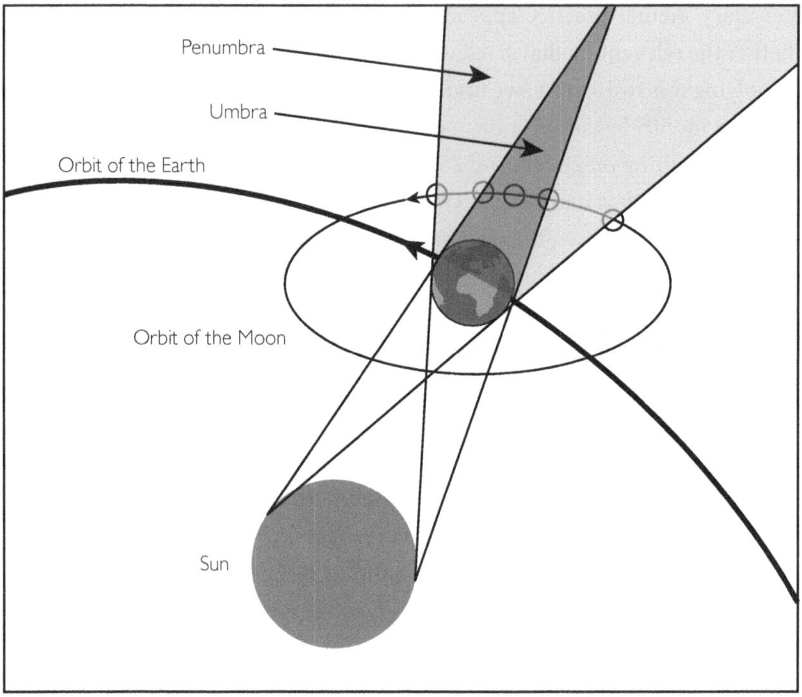

Figure 4.2 Earth's shadow causing lunar eclipse (not to scale).

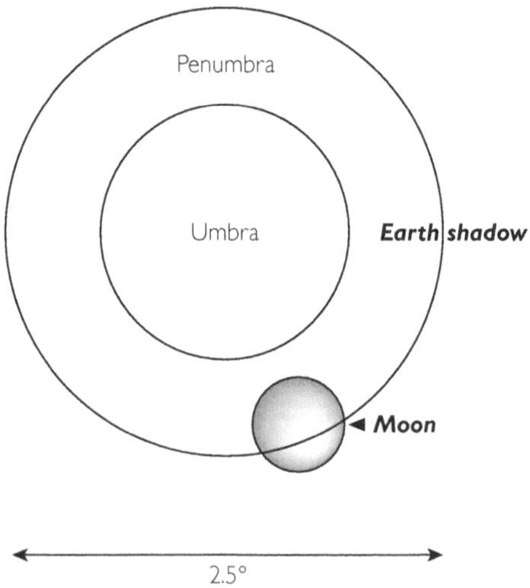

Figure 4.3 Lunar eclipse of 8 January 104 BCE.

to predict all future states. The Triple Conjunction therefore takes the position that there was a lunar eclipse near system origin. This cannot have occurred at the actual moment of origin, when the sun and moon were in conjunction, but would have fallen at the first full moon thereafter. The eclipse shown in Figure 4.3 did in fact fall on the day of the first full moon after the winter solstice of late 105 BCE, which is an Origin for the Triple Concordance, with all solar and lunar conditions at their initial points. As explained in Box 4.8, the Triple Concordance then proceeds by its usual method of counting off cycles, and in this case the basic cycle used is 135 lunations long.

However, while this cycle can be used to say that pairs of lunar eclipses follow one another at 135-lunation intervals, there are also other eclipses between each such pair—in fact the Triple Concordance predicts 23 lunar eclipses roughly equally spaced within each 135-month cycle. Again, the explanation will be found in Box 4.8.

Box 4.8: Calculating lunar eclipses

Oppositions and nodes

If the plane of the moon's orbit round the earth coincided with the plane of the earth's orbit round the sun, then the moon would always be on the ecliptic, which is simply the projection of the earth's orbit out onto the imaginary celestial sphere. At every full moon, the moon would be precisely opposite the sun (as seen from the centre of the earth), and would pass through the centre of the earth's shadow, producing a total lunar eclipse. In fact, however, the moon's orbit is inclined to the plane of the ecliptic, and can be up to 6° away from it (for the first explicit discussion of this, by Liu Hong, see chapter 8). Since the diameter of the earth's shadow is only about 2.5°, most full moons will not take place near the shadow zone, and there will be no eclipse. There can only be a lunar eclipse if full moon falls close enough to the instant when the moon crosses the ecliptic, which it will do twice in every circuit of the heavens. The two points where it does so are called nodes. Suppose there has just been a full moon near a node, producing a total lunar eclipse. How long will it be to the next such eclipse at a full moon? Clearly the time elapsed must be a whole number of lunar phase cycles, that is, a whole number of synodic months. We know that:

$$1 \text{ mean synodic month} = 29.53059 \text{ days}$$

continued

Box 4.8: *Continued*

The time for the moon to move from one node to the other node and back again (a 'draconitic month') is not the same period:

1 mean draconitic month = 27.21227 days.

The time taken from one node to the next is half this.

The tritos cycle

To get another full moon near a node, and hence another eclipse, we need a period that is a whole number of synodic months, and a whole number of (half) synodic months. We get a quite good match like this:

135 synodic months = 146.5 draconitic months. This is the so-called 'tritos' cycle. Since 135 × 29.53059 days = 3,986.63 days, and 146.5 × 27.21222 days = 3,986.59 days, the equivalence is close. The number of complete circuits of the two nodes is greater than the number of phase cycles by 146.5−135 = 11.5, which implies that the number of extra complete cycles from one node to the next is 11.5 × 2 = 23. There is thus a good chance of there being another eclipse 135 lunations after a given eclipse. But in fact there will be others, since we do not require exact coincidence of full moon and passage through a node to get an eclipse of some sort. The Triple Concordance attempts to predict these eclipses by, in effect, finding how many full lunations it will take for the moon to make about an extra half-cycle of the nodes. Clearly this will be $^{135}/_{23}$ lunations = 5.87 lunations. This is a little short of six lunations, so we might expect another eclipse after that period, that is, at the seventh full moon in the series. The procedure laid down is as follows:

> 推月食,置會餘歲積月,以二十三乘之,盈百三十五,除之.不盈者,加二十三得一月,盈百三十五,數所得,起其正,算外,則食月也.加時,在望日衝辰.
>
> To predict lunar eclipses: Set out the Accumulated Months in Coincidence Remainder Years. Multiply by 23, and cast out what fills 135. As for what does not fill, count one month for each addition of 23, until it fills 135. With the number you obtain, count off from the [celestial] first [month]. Outside the count, then that is the lunar eclipse. At its time of occurrence, it is at the double-hour [mark] directly opposite to the sun. (*Han shu* 21b, 1,002; Cullen 2017, 99–100)

The calculation is based on an underlying cycle of 135 lunations, during each repetition of which a pattern of 23 eclipses is assumed to repeat. We could

continued

Box 4.8: *Continued*

simply count off lunations since the last Origin, but the text, as usual, follows a procedure that reduces the size of the numbers used. It exploits the fact that in one 'Coincidence' cycle of 513 solar cycles, which is exactly 513 × $^{235}/_{19}$ = 6,345 lunations, there will be 47 cycles of 135 lunations, and hence 47 cycles of 23 eclipses. So we can cast out complete Coincidence cycles from the years since the most recent Origin before starting to calculate. What is more, since 1,539 = 513 × 3, we can start calculating from the most recent Concordance, thereby reducing numbers even further. Finally, we take the remaining months, and in effect multiply by $^{23}/_{135}$, which is the fraction of an eclipse that builds up in one lunation. The whole number result of that is the number of eclipses that have been completed by the start of the year in which we now are. The fractional part will be the number of ($^{1}/_{135}$) of an eclipse left over. Every month adds 23 to that, so all we have to do is to add 23 each month until the total exceeds 135. Then a lunar eclipse is predicted for the next full moon.

How accurate are such predictions? Firstly, we must take account of the fact that there is no guarantee that a lunar eclipse will take place while the moon is visible above the horizon. So even if the predictions were perfect, one would still expect to miss seeing a certain number. However, since an eclipse may take up to about 6 hours from the time that the moon touches the outside of the penumbra to when it emerges on the other side, and given that at opposition the moon is above the horizon all night, it is always possible that some part of an eclipse may be visible above the horizon, even if the moment of maximum eclipse takes place out of sight of the observer, before moonrise or after moonset. Secondly, no prediction based on simple cycles can take account of the complex variations in the speed of the moon, and in the position of its orbit. Having said that, it is interesting to look at the first 23 eclipse predictions made by the Triple Concordance system after the Grand Inception reform, given in Table 4.4.

Interestingly, there are only three occasions on which nothing at all can potentially be seen of the eclipse near the predicted date. In about half the cases, the eclipse would not have required much attention to notice it if the sky was clear. So far as this sample of predictions goes, it seems that we have to class the Triple Concordance method as at least a moderate success.

Table 4.4 ***First series of lunar eclipses from 104 BCE onwards***

Year BCE	Month	Day	Eclipse as seen from Chang'an	Local time of maximum eclipse	JD
104	January	8	Visible, partial penumbral.	23:15	1,683,445.169
104	July	4	Visible, penumbral, shortly before moonset	03:45	1,683,621.357
104	December	29	Visible, almost total umbral	05:00	1,683,799.409
103	June	24	Visible, total umbral, on evening of June 23	20:00	1,683,976.034
103	December	18	Visible, total umbral	04:48	1,684,153.401
102	June	13	Invisible, total umbral	12:57	1,684,330.74
102	December	7	Visible, total penumbral	05:21	1,684,507.424
101	June	2	Visible, partial penumbral	03:26	1,684,685.344
101	October	27	Visible, partial penumbral	00:08	1,684,832.206
100	April	22	Visible, partial umbral	20:52	1,685,010.07
100	October	17	Visible, half penumbral on horizon on Oct 16	17:33	1,685,186.932
99	April	12	Visible, total umbral on evening of April 11	21:40	1,685,364.103
99	October	6	Visible as total penumbral near horizon before moonset at 05:07; total umbral below horizon	06:52	1,685,541.49
98	April	1	Visible, half umbral	01:36	1,685,718.267
98	September	25	Visible, total penumbral	18:34	1,685,895.974
97	March	21	Very partial, not visible	12:34	1,686,072.724
97	August	15	Visible, partial penumbral before moonset	03:46	1,686,220.357
96	February	8	Visible, Total	19:48	1,686,398.025
96	August	5	Visible, partial umbral before moonset on August 4	04:38	1,686,574.402
95	January	29	Invisible, total	10:14	1,686,752.627
95	July	25	Invisible, total on July 24	12:39	1,686,928.727
94	January	18	Visible, Penumbral	18:41	1,687,106.979
94	July	14	Penumbral	02:44	1,687,283.314
94	December	9	Visible, penumbral but very partial	23:08	1,687,431.164

4.4.6 Sequence workings, *ji shu* 紀術

In this section of the text, we are told how to calculate the phases and motions of the planets. Box 4.9 follows the example of Jupiter in 10 CE, when it is predicted that the planet will have an 'Appearance' on 5 August of that year in the Julian calendar. This is supposedly the date when the planet will be seen for the first time on the eastern horizon just before the sun rises and the daylight conceals it from view. On each day thereafter it should rise earlier and earlier, and be visible for longer periods further back into the night. In the Triple Concordance system, all the planets are supposed to have their appearances when they are half a Jupiter station (in modern terms 15°) from the sun. Modern calculations place Jupiter 12° from the sun at dawn on that date, so the distance of Jupiter from the sun will not be 15° till three days later, when the sun will have moved by the extra amount required. Given the simplicity of the underlying calculations, this is not too inaccurate a result.

Of course, as already noted, the distance of the planet from the sun is at best only a crude indicator of whether the planet will be visible or not. In reality, the visibility of a planet depends not only on its distance from the sun, but also on its position relative to the horizon, its brightness, the observer's visual acuity, and climatic conditions. As we saw earlier in the 'Five Pacers' section (section 4.4.4), once we have the date of an Appearance, we can calculate the position of the planet at any subsequent date with reference to the system of the 28 lodges. Applying the methods prescribed, we may find, for instance, that on 30 December 10 CE Jupiter should be near the eastern end of the lodge 'Spread' *Zhang* 張, and just commencing retrograde motion, which should continue until 22 March 11 CE, when the planet should be near the western end of the lodge. Modern calculations show a retrogradation beginning on 30 November, and continuing until the last days of March. While the first of these two dates is not well predicted, the retrograde arc does in fact fall within 'Spread' more or less as it should.

Once planetary phenomena have been dealt with, all the principal procedures of the Triple Concordance have been specified. Remaining sections tell us how to calculate the Jupiter cycle, tabulate the 28 lodges, and tabulate the sexagenary days when Rules, Concordances and Origins commence.

Box 4.9: Finding the date of Appearance of a planet

推五星見復,置太極上元以來,盡所求年,乘大統見復數,盈歲數得一,則定見復數也.不盈者名曰見復餘.見復餘盈其見復數,一以上見在往年,倍一以上,又在前往年,不盈者在今年也.

To predict the Appearances or Returns of the Five Stars:

Set out the years from the High Origin of the Great Ultimate that exhaust the year sought, and multiply the Appearance or Return Numbers of the great concordance. Count 1 for each filling of the Year Number, then this is the Determined Appearance or Return Number. The remainder is called the Appearance or Return Remainder. When the Appearance or Return Remainder fills its [appropriate] Appearance or Return Number, if [it does so] 1 [time] or more then it is in the past year; if it is 2 [times] or more, then again it is in the year before the past year, but if it does not fill it, it is in the present year. (*Han shu* 21b, 1,002; Cullen 2017, 100)

The first aim is to find whether there is an Appearance in the current year (in this connection, this means the interval between the last winter solstice and the next, one solar cycle). We therefore attempt to find when the last Appearance before the end of this period took place. For example in 10 CE, we know that at the start of the period of interest (the last winter solstice) 143,240 years have elapsed since High Origin. So to the end of the year (i.e. the winter solstice of late 10 CE) we have 143,241 years. Taking Jupiter as an example, we know that in Year Number [1,728] years (in the sense of solar cycles) there are Appearance Number [1,583] of Jupiter Appearances. So in 143,241 years the number of complete appearances will be:

$$(143{,}241 \times 1{,}583)\ /1{,}728 = 226{,}750{,}503/1{,}728 = 131{,}221 \text{ remainder } 615$$

Thus so far there have been 131,221 complete Appearances (the Determined Appearance Number), and the last Appearance was 615 parts (the Appearance Remainder, at a scale of 1,728) before the end of the current year. The last part of the procedure deals with the fact that if the planet has a long enough period between Appearances, the last Appearance may actually have fallen before the winter solstice beginning the current year. Since one year is $^{1,583}/_{1,728}$ of an Appearance, this will clearly happen if the Appearance Remainder is greater

continued

Box 4.9: *Continued*

than Appearance Number [1,583]—but in this case the Appearance falls in the current year.

Now we proceed to find the day of the Appearance. Large numbers could not be avoided in the previous calculation, but in what follows the procedures cast out complete calendrical cycles in the usual way in an effort to keep the numbers small. We find as a result that the appearance falls on a *gengzi*.37 day, the seventh day of the ninth Celestial month (i.e. the 7th Xia month, Wang Mang's eighth civil month), which is the 11th day of the eighth medial qi of the year, at which time the sun is in the tenth *du* of the eighth Jupiter station. The day of Appearance corresponds to 5 August 10 CE.

Similar procedures apply for Mars and Saturn. For Venus and Mercury, the calculations are twofold, since the aim is to predict the behaviour of each planet during both its dawn and dusk visibility phases.

4.5 Testing the system? The *Canon of the Ages*, *Shi jing* 世經

Liu Xin makes a strong effort to justify the cosmological basis of the constants he uses in the Triple Concordance system. As we have seen, the basic solar and lunar constants of the system, as well as the system origin for sun and moon, are those that were put in place in 104 BCE, so we cannot reasonably ask Liu Xin how they came to be chosen in the first place. It does seem possible that the planetary parts of the system may be his own work, but we are not given any explicit idea what kinds of observations might have given him the data needed to construct them, or what process of testing, if any, led him to feel that the system could actually make useful predictions.

However, immediately after the end of the specification of the Triple Concordance systems in the *Han shu*, the editors give us several more pages of material, apparently by Liu Xin, under the heading *Shi jing* 世經, a title that may be rendered as 'Canon of the Ages' or 'Canon of the Generations'. This document goes systematically through a list of rulers from high antiquity onwards, and discusses astronomical and calendrical data from their reigns. Much of the discussion is derived from the *Chun qiu* 春秋 'Spring and Autumn [annals]', it seems likely that this is the document referred to as a 'Listing' (in the sense of a systematic setting out of calendrical data) by the editors in the preceding chapter:

至孝成世, 劉向總六曆, 列是非, 作五紀論. 向子歆究其微眇, 作三統曆及譜以說春秋, 推法密要, 故述焉.

In the time of Chengdi 成帝 [r. 33–7 BCE], Liu Xiang 劉向 [79–8 BCE] went through all six astronomical systems [known in his day], and set out where they were right and wrong, to make his 'Discussion of the Five Eras' *Wu ji lun* 五紀論. His son [Liu] Xin went into all the subtleties of such matters, and wrote his Triple Concordance system, together with his Listing for explaining the Spring and Autumn Annals. His methods of inference keep close to the essentials, so we have set them out. (*Han shu* 21a, 979)

We reviewed the question of the 'six systems' at the end of chapter 3. I have discussed the *Canon* at length in an article, and I shall therefore not go into much detail here.[38] A few examples may serve to illustrate the extent of Liu Xin's ambitions to match his system against demanding problems of chronology.

Liu Xin begins his account by reviewing the sequence of the rulers mentioned in accounts of the remotest past, without any attempt to discuss the state of the heavens in connection with their activities. His first application of celestial calculations to history come when he discusses two dynastic transitions: the conquest of Xia by Shang, and the Zhou conquest of Shang. The second of these conquests is recognized by historians in east and west as a real historical event. Nearly all western historians, and some Chinese historians, doubt whether there was ever anything that could have been called a unified Xia state that governed a large area in the Yellow River basin and was overthrown by a group that founded the Shang as a successor state.[39]

For the fall of Xia, which he places 141,480 years after the High Origin of the Triple Concordance system, i.e. in 1751 BCE, Liu Xin does no more than verify that Jupiter was in a position later said to have been favourable to the Shang. It is not clear how he establishes the date he uses.

If we turn to the Zhou conquest of Shang, we find a quite different situation. Liu Xin evidently knew many references to this event in historical sources, some of which are extant today and some not, in which the correlation of celestial events with the conquest itself and the period of preparation for the conquest are spelled out in detail. It seems clear, for instance, that Liu Xin had access to material equivalent to the following section of the *Guo yu* 國語 'Conversations

[38] Cullen (2004).

[39] See for instance the notes by Kwang-chih Chang on 'The question of the Xia dynasty' in Loewe and Shaughnessy (1999), 71–3, in which it is suggested that the Xia may have been one amongst many states co-existing in the Yellow River basin in the early Bronze Age.

from the States', a text which may have reached its present form by the end of the fourth century BCE:

> 昔武王伐殷, 歲在鶉火, 月在天駟, 日在析木之津, 辰在斗柄, 星在天黿
>
> Formerly when Wu Wang attacked Yin, the Year [star] (i.e. Jupiter) was at [the station] Quail Fire, the moon was at Celestial Team, the sun was at the ford of Ximu, the conjunction was at the handle of Dipper, and the star [taken by all the commentators as the 'water star' *shui xing* 水星, Mercury, in the light of the context] was at Celestial Tortoise. (*Guo yu* 國語, *Zhou yu* 周語, B, 3, 22b–23a)

We do not know exactly what Liu Xin's sources were, but he refers to all these data in the course of his discussion, with considerable extra detail, introducing each item as taken from 'the Tradition' *zhuan* 傳. Liu Xin sets a date for the conquest equivalent to 1122 BCE, and on that basis he is able to reconstruct the entire set of astronomical conditions with some exactitude, using his Triple Concordance system.[40] It is besides the point to ask whether Liu Xin's calculations were correct in modern terms, if only because all historians of early China nowadays agree that the actual Zhou conquest happened nearly a century later than the date Liu Xin gives. We cannot tell whether Liu Xin adjusted the conquest date to fit his astronomical system, or adjusted the system to match a date he had chosen for other reasons, or both. For Liu Xin and his contemporaries, these calculations were a very effective exhibition of the success of the Triple Concordance in fitting in with the records of one of ancient history's most significant events. No-one, so far as we can tell, had ever attempted anything like this before.

Moving onwards through history, Liu Xin eventually came to the challenge of matching the best-known sequence of datable ancient celestial events—that given by the Spring and Autumn annals. This work was reputed to have been edited by Confucius himself, on the basis of the annals of his home state of Lu. Here Liu Xin achieved a reasonable degree of success. Thus, for instance, calculations based on his system agreed with the well-known statement in the entry for the fifth year of Duke Xi of Lu (which fell mainly in 655 BCE) stating that the preceding winter solstice was observed on a *xinhai*.48 day, when there was also a conjunction of sun and moon.[41] However, when there are failures he does not go

[40] See Cullen (2004), 47–51.

[41] The day in question corresponds to 25 December 656 BCE in the proleptic Julian calendar. Modern calculations place the winter solstice at 21:00 local time on 27 December, and the conjunction at 20:00 on 26 December. It is unlikely that techniques in use in the seventh century BCE could have detected the discrepancy in the solstice date, and the only potentially observable consequence of the one-day conjunction error would have been a very faint (2%) illuminated crescent visible on the morning of 25 December.

out of his way to advertise them. A striking example comes from his treatment of solar eclipses—events which in his day were recognized as infallible indicators of a solar-lunar conjunction. There are several datable eclipses mentioned in the *Chun qiu*, of which we may consider three examples:

Duke Xiang 27th year, 546 BCE: 12th month, conjunction *yihai*.12
Duke Zhao 31st year, 511 BCE: 12th month, conjunction *xinhai*.48
Duke Ding fifth year, 505 BCE: third month, conjunction *xinhai*.48

All three of these correspond to verifiable eclipses on the cyclical days in question when checked against modern calculations, although the first would have fallen on the conjunction of the 11th celestial month. There are other examples of such one-month errors in eclipse records in the *Chun qiu*, probably the result of irregularities in the insertion of intercalary months, so there is no need to doubt the record. Liu Xin, however, has to rely on the conjunctions given by the Triple Concordance system, and there he has major problems. For the first eclipse, Liu Xin must deal with the fact that he cannot easily find a conjunction date that fits. According to the Triple Concordance, the conjunction of the 12th Celestial month of late 546 BCE fell on *guimao*.40, which is far out. The 11th Celestial month had conjunction on *jiaxu*.11 according to the Triple Concordance, one day wrong. But Liu Xin ignores that, and instead points to the fact that the 9th Celestial month does have a conjunction on *yihai*.12. For this month to have been recorded as the 12th month would have required no less than three intercalations to have been missed. To make matters a little better, Liu Xin draws on the *Zuo zhuan*, which records the *yihai*.12 eclipse as falling on the conjunction of the 11th month—a date that requires only two missed intercalations to make it consistent with the Triple Conjunction—bad, but not so unlikely as the *Chun qiu* record would have made it. For the other two eclipses, in 511 and 505 BCE, there is simply no correct Triple Concordance conjunction anywhere near the months given. Liu Xin deals with this difficult situation by ignoring it altogether, and makes no mention of either eclipse in the *Canon*.

Finally, we may consider another example where rather than ignoring eclipses that actually were observed, Liu Xin appears to have referred to an eclipse that could not have been observed. This event is mentioned in his discussion of the winter solstice preceding the regnal year that began in spring of 47 BCE, the second year of the Chuyuan reign period.

元帝初元二年十一月癸亥朔旦冬至, 殷曆以為甲子, 以為紀首. 是歲也, 十月日食, 非合辰之會, 不得為紀首.

> Yuandi: The second year of the Chuyuan reign period, the 11th month, [day] *guihai*.60 [25 December 48 BCE] was conjunction and winter solstice. The Yin system made this *jiazi*.1, and made this an Era Head. In this year, there was a solar eclipse in the tenth month, but this was not a Coincidence [Head] at conjunction [i.e. it was not the start of one of the 6,345 months periods used to predict lunar eclipses], and [thus] cannot be treated as an Era Head. (*Han shu* 21b, 1024).

The 11th month here refers, as was customary, to the civil month, which was also the first Celestial month preceding the start of the civil year in question in spring 47 BCE. This month, according to the Triple Concordance, did indeed begin on a *guihai*.60 day, and winter solstice was predicted for the same day. As indicated, the Yin system placed both events a day later, and had conjunction and winter solstice coinciding at midnight, so that the day was a Yin system Era Head. According to the Triple Concordance, the preceding two months were in fact intercalary tenth month, beginning on 25 November, and tenth month beginning on October 27, a *jiazi*.1 day. On neither of those days was there a solar eclipse.

It is odd to find Liu Xin making an incorrect statement about events in a period when he must have had direct access to official observation records. Even more odd is the fact that there actually was an eclipse at the conjunction beginning the 11th month—but it took place about three hours before sunrise at Chang'an, and would therefore not have been visible. It seems possible therefore that Liu Xin retrodicted this eclipse using a record of some later solar eclipse that actually was observed, and counted back using an eclipse cycle. If so, he made an error in the month, perhaps because of a slip in reckoning the total number of intercalations to be counted. Thus for instance, there was a large magnitude partial solar eclipse visible from Chang'an in the afternoon on the last day of the eighth civil month of the third year of the He ping 河平 reign period, cyclical day *yimao*.52, which corresponds to 23 October 26 BCE. This eclipse was recorded in *Han shu* 27 B.2, 1505. This eclipse was two 'coincidence cycles' of 135 lunations away from the eclipse of 25 December 48 BCE, at the start of the eleventh month.[42] Further, it is unlikely that Liu Xin could have thought it plausible that there might have been a solar eclipse at either of the conjunctions

[42] We have no explicit statement by Liu Xin or any of his contemporaries saying that the 'coincidence cycle' could be applied to solar as well as lunar eclipses. The earliest statement implying this is not found until the seventh century CE: see Nathan Sivin (1969) *Cosmos and computation in early Chinese mathematical astronomy*, Leiden, E. J. Brill, 46–8.

preceding 25 December, since the Triple Concordance predicted a lunar eclipse for the opposition of the eleventh month on 9 January, which would have suggested to him that there might have been a solar eclipse at either of the two conjunctions on either side of it, the first of which was 25 December, but excluded a solar eclipse at an earlier conjunction such as that of the tenth month.[43]

Despite the inconsistencies and loose ends in the *Canon of the Ages*, Liu Xin's attempt at systematic checking of the astronomical events recorded in major historical sources was a major departure. It was certainly successful enough to set the pattern for many subsequent attempts, particularly in relation to the solar eclipses of the *Spring and Autumn Annals*. I cannot trace any sign of detailed critical discussion of Liu Xin's work in the rest of the Han dynasty: as we shall see later (in chapter 7), a time did come when Han scholars not only debated publicly on such questions, but also left us detailed records of those debates. But in Liu Xin's time that was not yet the way things were done.

[43] This lunar eclipse did in fact take place, and was a partial umbral eclipse, with maximum when the moon was about 10° above the horizon from Chang'an.

CHAPTER 5

The measures and forms of heaven

We begin this chapter by listening to the first voice from ancient China that speaks to us directly and personally about the experience of observing the sky with instruments that produce numerical results. The speaker is Huan Tan, who lived through the time of Wang Mang and into the early Eastern Han. This leads into consideration of what it was he measured, and hence into an explanation of the nature and purpose of the system of 28 lodges, *xiu* 宿, into which the circuit of the sky was divided. Since we began with the personal, it is appropriate that this part of the chapter should contain an account of other personal experiences of observation, designed to re-enact and test what Huan Tan tells us he did. We then discuss how changing ideas about the heavens were linked with the introduction of new instruments and new ways of measurement. Finally we return to Huan Tan, as he tells us of his arguments with his friend Yang Xiong on the shape of the cosmos.

5.1 Prologue: modern measures for the stars

During the two centuries that followed the Grand Inception reform of 104 BCE, the way that the heavens were imagined and measured—even, perhaps, the way the heavens were perceived—changed radically. We shall follow that story through the eyes of two people whose lived around the middle of that period of change, and whose discussions on this topic have come down to us in enough detail to show us what issues were at stake.

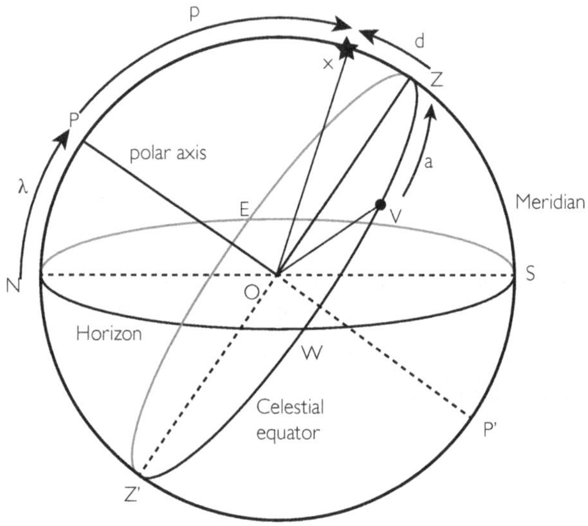

Figure 5.1 Locating a star on the celestial sphere in modern equatorial co-ordinates.

First, however, it will be helpful to set out the way that modern astronomers express the observed positions of celestial bodies in quantitative terms. The fundamental concept here is the celestial sphere, as shown in Figure 5.1.

Apart from the star X, the observer O is the only part of the diagram that has any physical reality. All the lines and circles on the diagram merely serve as ways to impose a simple mathematical structure on what an observer at a particular point on the earth's surface sees if she looks at the sky over a prolonged period. The appearance of the sky as she sees it can be modelled by assuming she is at the centre of a vast rotating sphere bearing the celestial bodies on its inner surface, a sphere so large that it is effectively infinite. Thus her local movements do not cause any change in the relative positions of the celestial bodies on the sphere. The diagram assumes that the observer is somewhere in the northern hemisphere.

Let us start with the observer's horizon. This is, in effect, the projection outwards into space of the horizontal plane through the position on earth where she is standing. It intersects the celestial sphere in the circle NESW. If the observer lies on her back with her head pointing north towards N (her 'north point') and her feet pointing south towards S (her 'south point'), then her 'east point' E is on her left and her 'west point' W is on her right. If she watches the heavens for a day and a night, it will appear that the celestial sphere rotates about an axis PP′, where P and P′ are the north and south celestial poles, so that a series of celestial bodies will rise over the eastern horizon on her left, moving steadily higher until they cross the line on the heavens that runs from N to S and passes overhead (her 'meridian line'), and then sink down until they set over the western horizon on her right. In reality, all this is simply the effect of the daily rotation of the spherical earth on its axis. Stars that are close enough to the north celestial pole P will not appear to rise and set, but simply circle round the pole in the course of a day and a night. These are the observer's 'circumpolar' stars. Conversely, there are stars close enough to P′ so that they will never rise above her horizon The angle PÔN, equivalent to the arc λ, is equal to the observer's latitude. So if the observer is at (say) the earth's geographical equator, where latitude is zero, the axis PP′ will be horizontal, whereas if she is at the north geographic pole (latitude 90°), the axis will be vertical.

We may locate a star such as X, shown here at the instant it crosses the meridian, by stating two angles relative to fixed points on the sphere. A fundamental reference here will be the celestial equator, a circle on the sphere lying equidistant from the two poles, in a plane perpendicular to the polar axis and passing through O, shown in the diagram as ZEZ′W; because the celestial equator is basic to our frame of reference, we are said to be locating the star using equatorial co-ordinates.

The distance of the star X from the equator is measured by the so-called *declination* of the star, the angle XÔZ corresponding to the arc d. The complement of this angle (i.e. the remainder when it is subtracted from 90°) is the angle XÔP corresponding to the arc p: this is known as the *north polar distance* (NPD) of the star, and is the quantity normally used in Chinese sources rather than declination. The declination of a star will lie between zero and 90°; it is counted positive north of the celestial equator and negative south of the celestial equator. North polar distance will, however, lie between zero and 180°.

As for measurements 'round the sphere', our reference point is V, the position of the sun at the vernal (spring) equinox, at which time it will rise due east at E and set due west at W. This point V lies on the celestial equator. In this diagram, if the equator cuts the meridian on which the star lies at Z, then the angle VÔZ, corresponding to the arc a, is what is called the *right ascension* (abbreviated RA) of the star. Although nowadays RA is nowadays mainly thought of in angular terms, this concept originated from the problem of finding the time taken for a given section of the sky to rise above the horizon. That time was its 'ascension'.

The simplest case considered in the ancient west was that known as *sphaera recta* 'the right sphere', when the two poles P and P′ lie on the observer's horizon. Celestial bodies will then rise and set at right angles to the horizon. In that case the rising times are called 'right ascensions', and like all ascensions these will naturally enough be expressed in units of time rather than angle—with 24 hours of RA equivalent to 360° of angle.[1]

On the scale of centuries, the stars on the celestial sphere do not move enough relative to one another for changes in their patterns to be detected by naked-eye observation. However, the right ascensions and declinations of stars do change with epoch, due to the phenomenon of 'axial precession'. This is caused by the effects of the gravitational forces of the moon and sun on the earth's equatorial bulge, which make the earth's axis (and hence the line PP′) precess like the axis of a spinning top, so that P and P′ trace out circles on the sphere with a radius of about 23½°, only returning to their original positions after about 26,000 years. As a result, both the right ascension and declination (and hence the north polar distance) of a star like X will both change slowly with time. For historians of astronomy, this phenomenon may make it possible to assign a date to ancient observations if they are capable of being understood in terms of modern co-ordinates, and we shall see an example later in this chapter of how this may be done.[2]

But let us now return to the Han dynasty, and in particular to a young gentleman whose job description makes him of interest to us in the context of this book.

[1] See for instance the tables of rising-times of each 10° section of the 12 signs of the zodiac given by Ptolemy in *Almagest* II.8, Toomer (1998), 99–104. After listing rising-times in *sphaera recta*, Ptolemy tabulates values for ten named places of different latitudes. The time units in which right ascension is measured are not ordinary clock hours, but are hours of 'sidereal time', a concept to which we shall return later in this chapter. 24 hours of sidereal time is the interval during which the stars return to the same position relative to the observer. This is about 4 minutes shorter than the mean solar day, which is the interval between two returns of the mean sun to noon. The difference is due to the sun's daily eastwards displacement of about one degree relative to the celestial sphere—in reality the effect of the earth's annual orbital motion round the sun.

[2] For an elementary discussion of the effects of precession, see Evans (1998), 245–7. Mathematical details of how to calculate the resultant changes in the co-ordinates of celestial bodies are given in W. M. Smart and Robin M. Green (1979 (reprint of 6th edition 1977)) *Textbook on spherical astronomy*. Cambridge; New York, Cambridge University Press, 226–231. Instead of performing such calculations manually, it is nowadays possible to use software such as Starry Night Pro™ to read off the co-ordinates to high precision for any historical epoch. The reader interested in the algorithms behind such calculations may consult, for instance, Jean Meeus (1991)*Astronomical Algorithms*. Richmond, Virginia.

5.2 Huan Tan: the Gentleman with the clepsydra

In earlier chapters, we have seen brief mentions of the use of the *du* as a measure of the circuit of the heavens, and of the system of the 28 lodges, *xiu* 宿 into which that circuit is divided. By what procedures, and using what instruments, were these lodges defined quantitatively? Further, what was the shape of the heavens thought to be, and how did that shape relate to the process of measurement? We can find answers to those questions if we look at some of the writings of Huan Tan 桓譚 (c. 43 BCE–28 CE).

Huan Tan was the son of the Director of Music *Tai yue ling* 太樂令 under Chengdi 成帝 (r. 32–7 BCE). He was eventually given the rank of Gentleman *lang* 郎 at court in recognition of his father's service, and remained at that grade during the reigns of Aidi 哀帝 (r. 6–1 BCE) and Pingdi 平帝 (r. 1–5 CE). He was apparently given a good musical training by his father—good enough so that in later life he was criticized for playing 'licentious music' at court—but he showed talent in several other areas:

> 善鼓琴. 博學多通, 徧習五經, 皆詁訓大義, 不為章句. 能文章, 尤好古學, 數從劉歆, 楊雄 . . .
>
> He excelled as a lute player, studied widely, and understood many subjects. He was well acquainted with the Five Classics, preferring to explain their broad meaning rather than making detailed commentaries. He had a talent for writing prose, and was particular devoted to the study of antiquity. In numbers, he followed Liu Xin and Yang Xiong . . . (*Hou Han shu* 28A, 955)

Like Liu Xin, Huan Tan went on to serve under Wang Mang. This did not prevent him being invited to take office when the Han was restored.[3] He can therefore serve as a bridge between the periods before and after what the Han came to see as an unfortunate interruption to their legitimate rule. Although his writing was admired for centuries after his death, it has only survived in the form of fragmentary quotations. A passage from one of these lost works, *Xin lun* 新論 'New Discussions', refers to matters that concern us here.[4] Here, Huan Tan

[3] For his biography, see *Hou Han shu*, 28a, 955–61.

[4] On this work, see Loewe (1993), pp. 158–60, and Timoteus Pokora (1975) *Hsin-lun (New treatise), and other writings by Huan T'an (43 B.C.–28 A.D.): an annotated translation with index.* Ann Arbor, Center for Chinese Studies, University of Michigan. Exceptionally, Huan Tan has biographies in both Loewe (2000), 164–165 and De Crespigny (2007), 338. The former is more detailed. The passage here quoted survives in a single source, an encyclopedia compiled in the early seventh century CE.

tells us that as a young man he had experience of work that involved careful and regular observations of the heavens both by day and by night at all seasons, and in all weathers:

> 余前為郎. 典漏刻. 燥, 濕, 寒, 溫, 輒異度. 故有昏明晝夜. 晝日参以晷景. 夜分参以星宿. 則得其正.
>
> Formerly, when I served as a Gentleman, I was in charge of the clepsydra. If [conditions varied between] dry and humid, cold or warm, then there were different *du* 度. So [in order to] have [timings for] dusk and dawn, daylight or night-time, I checked against the solar shadow in the daytime, and in the night portion [of the clepsydra run] I checked against the stellar lodges. Thus I got the correct [*du*]. (Quoted in *Bei Tang shu chao* 北堂書鈔 130, 12b. p. 208)

The mention of observation of the solar shadow shows that Huan Tan must have been using a gnomon in conjunction with his clepsydra. We do not know exactly what kind of clepsydra might have been in use at the Han court at Chang'an in the late first century BCE, but a number of specimens have been found from other locations. Figure 5.2 shows a device found in Inner Mongolia in 1977.[5]

This device bears an inscription as follows:

> 千章銅漏一重卅二斤河平二年四月造
>
> Bronze clepsydra [from] Qianzhang. Weight 32 *jin*. Manufactured in the second year of the Heping period, fourth month.

Such inscriptions are typical of objects produced in official workshops, and may well correspond to an entry in some written registry of equipment. The date of manufacture corresponds to 27 BCE, so this device may have still been in use in a provincial office while Huan Tan was attending to clepsydras at the capital. Like other clepsydras from this period, it is a simple outflow device with a frame on top, evidently designed to hold in place an indicator rod that would have been attached to a float inside the water vessel. Marks on the rod would indicate the passage of time as the water level fell, presumably as they passed by one of the horizontal members of the upper frame, which were pierced with rectangular slots to hold the rod vertical.[6]

[5] For a discussion of this device, see Chen Meidong 陈美东 (1989). 'Shi lun Xi Han lou hu de ruo gan wen ti 试论西汉漏壶的若干问题 (On some questions relating to Western Han clepsydras)' in *Zhong guo gu dai tian wen wen wu lun ji* 中国古代天文文物论集 *(Collected articles on astronomical artefacts from ancient China)*, Beijing, Wenwu Press: 137–44.

[6] The method of graduating these rods in accordance with seasonal changes in the lengths of day and night were discussed in detail a century after the time of Huan Tan: see chapter 6, section 6.4.

Figure 5.2 Western Han clepsydra with inscription dated to 27 BCE; height 479 mm, diameter 187 mm. (Chinese Academy of Social Sciences Institute of Archaeology 1980: 41, fig. 39).

Whatever kind of clepsydra Huan Tan may have used, his reference to different measurements being obtained under different climatic conditions make it plain that he was a painstaking observer.[7] There may have been other 'Gentlemen in charge of the clepsydra' whose interest was limited to looking in at the office in from time to time to make sure the waterclock was being regularly refilled by the clerk in charge, but Huan Tan was certainly not one of them. Let us look more closely at what he tells us.

He is clearly very concerned by the fact that measurements under different conditions give *yi du* 異度 'different *du*', and he claims to have found a way to make these measurements 'correct' *zheng* 正. Although the word *du* can be used

[7] See the discussion of the effects of temperature on clepsydra rates in J. Fermor and J. M. Steele (2000) 'The design of Babylonian waterclocks: Astronomical and experimental evidence.' *Centaurus* 42 (3): 210–22.

as a general term for 'measurement', in the context of observations of the 'stellar lodges' *xing xiu* 星宿, it seems very likely to be referring to the *du* measure we have met before as a measure of movement of the sun and other heavenly bodies round the heavens. What is more, we have already seen it stated in the *Han shu* that at the time of the Grand Inception reform in 104 BCE, those working on the new system used gnomons and clepsydras 'in order to find the extents of the 28 lodges in the four quarters'.[8] What Huan Tan was doing was therefore nothing new—except that he is the first named individual to give us a first-hand account of such observations.

5.2.1 Listing the lodges

The system of the lodges antedates the foundation of the empire by at least a few centuries: the names of all 28 lodges appear in an approximate circle on the lid of a lacquer box found in a tomb dated to 433 BCE.[9] There is, however, no indication of any measurements of lodges on this object. The earliest list of the lodges with measures in *du* is that found in chapter 3 of the *Huai nan zi* book around 139 BCE, and given in Table 5.1.[10] Each lodge is associated with an asterism, and the continuity of Chinese tradition means that there is no great difficulty in identifying these with the modern names of stars, just as there is no problem in identifying the stars referred to by Ptolemy and other ancient writers in the Greek language.[11] For convenience I have numbered the lodges in sequence, and listed in the fifth column of the table from the left the modern names of the stars at the westernmost end of each lodge, which marks the point where the sun enters that lodge in its annual west to east circuit of the heavens.[12]

[8] See *Han shu* 21a, 975, translated in chapter 3, section 3.3.2.

[9] See Wang Jianmin 王健民, Liang Zhu 梁柱 and Wang Shengli 王胜利 (1979) 'Zeng hou Yi mu chu tu de er shi ba xiu qing long bai hu tu xiang 曾侯乙墓出土的二十八宿青龙白虎图像.' *Wenwu* (7): 40–45.

[10] I have divided the lodges into the four directional groups given a few pages after the listing of lodge widths; see *Huai nan hong lie ji jie*, chapter 3, 122 and 127 and Major (1993), 127 and 138. The *du* measurements associated with each lodge differ slightly from those given by Liu Xin in his Triple Concordance system c. 10 CE; it seems likely that the values used by Huan Tan would have followed the latter. But the differences were small.

[11] See for instance *Almagest* VII.5–VIII.1 in Toomer (1998), 341–99. The lack of a continuous tradition makes it more difficult to identify the stars referred to in ancient cuneiform sources—see for instance the discussion of the MUL.APIN 'Plough Star' text (a composite text that reached its final form in the seventh century BCE) in Hunger and Pingree (1989), 137–46.

[12] These are taken from Sun Xiaochun and Jacob Kistemaker (1997) *The Chinese sky during the Han: constellating stars and society*. Leiden; New York, Brill, Table 3.1, column V. In three cases I have added index numbers to clarify which of two visually separable stars, sharing the same Bayer Greek letter, is intended.

Table 5.1 ***Names and widths of lodges in* Huai nan zi*, and modern names of leading stars***

	Name of lodge and number of *du* in *Huai nan zi*	Translated lodge name	Width of lodge, *du*	Modern identification of leading star of lodge
1	角十二	Horn	12	α Virginis
2	亢九	Gullet	9	κ Virginis
3	氐十五	Base	15	α Librae
4	房五	Chamber	5	π Scorpii
5	心五	Heart	5	σ Scorpii
6	尾十八	Tail	18	μ^1 Scorpii
7	箕十一四分一	Winnower	11 ¼	γ Sagittarii
8	斗二十六	Dipper	26	φ Sagittarii
9	牽牛八	Ox [Leader]	8	β Capricorni
10	須女十二	[Serving] Woman	12	ε Aquarii
11	虛十	Barrens	10	β Aquarii
12	危十七	Rooftop	17	α Aquarii
13	營室十六	[Lay out the] House	16	α Pegasi
14	東壁九	[Eastern] Wall	9	γ Pegasi
15	奎十六	Straddler	16	ζ Andromedae
16	婁十二	Harvester	12	β Harvester
17	胃十四	Stomach	14	35 Arietis
18	昴十一	Mane	11	17 Tauri
19	畢十六	Net	16	ε Tauri
20	觜巂二	[Turtle] Beak	2	φ^2 Orionis
21	參九	Triaster	9	δ Orionis
22	東井三十三	[Eastern] Well	33	μ Geminorum
23	輿鬼四	[Carriage-borne] Ghost	4	θ Cancri
24	柳十五	Willow	15	δ Hydrae
25	星七	Stars	7	α Hydrae
26	張、翼各十八*	Spread	18	υ^1 Hydrae
27		Wing	18	α Crateris
28	軫十七	Axletree	17	γ Corvi

*The text here reads 'Spread, Wing: 18 each'.

The total of the widths in *du* in the *Huai nan zi* list comes to 365 ¼ *du*, a complete circuit of the heavens, round which the sun was thought to move at a constant 1 *du* per day during an annual solar cycle of 365 ¼ days. But what is the significance of these lodge widths in modern terms? At the beginning of this chapter, we saw that the modern co-ordinate corresponding to distances 'round the sky' from west to east was right ascension (RA). Let us consider the first lodge in the list, Horn, from this point of view. This lodge begins with α Virginis, and ends with κ Virginis, the leading star of the next lodge, Gullet.

Firstly we find the RAs of these two stars in 139 BCE, the year when the *Huai nan zi* book was presented to the throne. These will not be the same as their modern values, because of the phenomenon of precession and the resultant shift of the celestial axis relative to the stars in the course of over two millennia. The values we require are:

α Virginis: 173.87°
κ Virginis 185.58°

The difference of RA is 11.7° (to three significant figures), equivalent to:

$$11.7° \times (365\ ¼\ du)/360° = 12\ du \text{ to the nearest } du$$

This is indeed the width of the lodge Horn as listed in *Huai nan zi*. Similar results may be found for other lodges—for seventeen lodges out of the series, the results calculated for 139 BCE are identical to the nearest *du* with those in the list given in Table 5.1, and for nine others the results differ by only one *du*. It therefore appears that the *Huai nan zi* lodge width values were obtained by some process equivalent to a modern determination of right ascension. What the process might have been we shall shortly discuss. By the time of Huan Tan a century or so later, the lodge widths listed in Table 5.1 would still have corresponded closely to right ascension differences.[13]

But if the lodge widths checked by Huan Tan were given in terms equivalent to differences of right ascension, how were they determined? Right ascension is nowadays mostly thought of as a measurement of angle on the celestial sphere. In the ancient west, such measurements would have been made by means of some kind of graduated protractor equipped with sights, enabling angles to be measured directly. Thus, in the first century BCE Geminus stated that the zodiac may be divided into twelve equal parts by the use of the instrument he calls a

[13] Because neighbouring stars change their RA by similar amounts due to precession, the difference in RA between two stars changes markedly less than the two individual RAs. As we shall see later in this chapter, the north polar distances of stars provide a more time-sensitive measure than lodge widths: see Table 5.6.

dioptra, meaning something like 'sighter', whose graduated scale and sights enable the angle between the sightlines to two heavenly bodies to be measured.[14] Two centuries later, Ptolemy made such measurements using an instrument consisting of a nest of graduated rings with sights—what we would now call an armillary sphere.[15] But Huan Tan mentions nothing of either kind.

Nevertheless, a method can be found to carry out the required task of finding differences in right ascension by using the apparatus available to Huan Tan, apparatus which yields measurements of time rather than angle. Given the origins of the concept of right ascension as a measure of time taken for the rising of a constellation, it should not be surprising to find that this is possible. To understand exactly how such measurements might have been carried out by Huan Tan, let us start by looking back at the earliest evidence we have of how the lodge system was used by those concerned with systematic observation of the heavens. This is found in a work produced only a few years before the beginning of the imperial age—the *Lü shi chun qiu* 呂氏春秋 'Spring and autumn [annals] of Mr Lü', a compendium of knowledge completed in the kingdom of Qin in 239 BCE by a group of scholars convened under the patronage of Lü Buwei 呂不韋 (c. 291–235 BCE), a statesman who did much to ensure Qin's later rise to imperial power.

5.3 What was the use of the lodge system?

The first systematic indication of what purpose the lodges might serve in practice comes from a document known as the *Yue ling* 月令 'Monthly ordinances', which lists the natural phenomena and rituals appropriate to each lunar month of the year—assuming, that is, that the relation of the months to the seasons has been preserved by the insertion of intercalary months when appropriate. The text was included in the *Lü shi chun qiu* 呂氏春秋, so even though we cannot be sure of the date when this listing originated, we know that the thinking behind it was of interest to a group of well-informed scholars around the middle of the third century BCE.[16]

[14] See *Geminos's introduction to the Phenomena: a translation and study of a Hellenistic survey of astronomy*, i.4, p. 114 and 38–42.

[15] See *Almagest* V.1 in Toomer (1998), 217–19.

[16] The scholars in question belonged to a group assembled in the state of Qin 秦 by the powerful minister Lü Buwei 呂不韋, and were commissioned to produce the compendium of knowledge that bears his surname. See John Knoblock and Jeffrey K. Riegel (2000) *The annals of Lü Buwei = [Lü shi chun qiu]: a complete translation and study*. Stanford, Stanford University Press for a full translation and study. There are also version of this material in *Huai nan zi* chapter 4, and in the *Li ji* 禮記 'Record of the Rites'. On the confused question of the date and origins of the latter text in the Han period, see Loewe (1993), 293–7.

As well as such matters as the expected behaviour of animals, birds and fishes, the seasonal growth and flowering of plants, the appropriate colours of ritual robes and implements, and the sacrifices to be made in the given month, the *Yue ling* gives information about the heavens by referring to the lodges in two ways:

1. We are told in which lodge the sun will be 'located' *zai* 在 in a given month.
2. We are told which two lodges will be 'centred' *zhong* 中 at dusk and at dawn respectively in that month.

Thus for example, for the first month of the year (i.e. the first in the Xia sequence), we read:

> 孟春之月: 日在營室, 昏參中, 旦尾中. [. . .] 是月也, 以立春.
>
> The month of early spring: the sun is in the lodge House, the lodge Triaster is centred at dusk, and the lodge Tail is centred at dawn. [. . .] In this month, [there falls the *qi*] Establishment of Spring . . . [17]

And for the fifth month:

> 仲夏之月: 日在東井, 昏亢中, 旦危中. [. . .] 是月也, 日長至.
>
> The month of mid summer: the Sun is in the lodge Well, the lodge Gullet is centred at dusk, and the lodge Rooftop is centred at dawn . . . In this month, the day has its extreme length . . . [18]

Leaving aside the question of the position of the sun for the moment, what could be meant by saying that a lodge is 'centred'? This term *zhong* has a well-known meaning with relation to celestial bodies, and in all our sources clearly refers to a body being in a position exactly to the south of the observer—that is, on the observer's meridian line. The earliest commentator on the *Lü shi chun qiu*, Gao You 高誘 (fl. 210 CE) notes after the dusk and dawn lodges named for the first month:

> 是月昏旦時皆中於南方.
>
> In this month, at dusk and dawn [respectively] they are both centred in the south. (*Lü shi chun qiu* 1, 1a, p.17, commentary).

For an example of what 'centring' means in practice, if we were to look south from Chang'an around 6 a.m. local time on 20 February 139 BCE, we would see the asterisms associated with a number of lodges in the southern sky: see

[17] Knoblock and Riegel (2000), 60, modified.
[18] Knoblock and Riegel (2000), 133–135, modified.

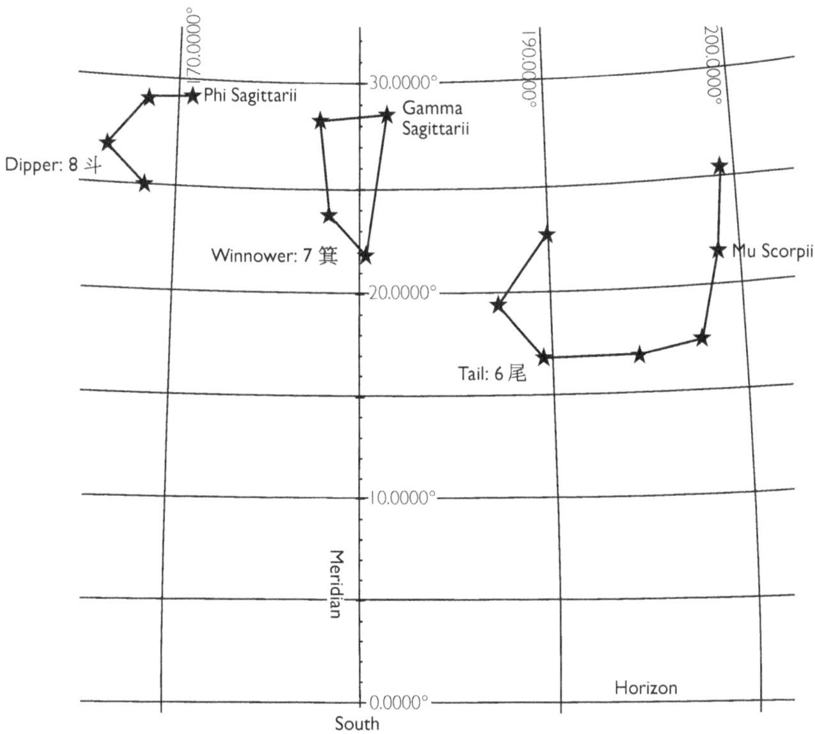

Figure 5.3 Lodges seen when looking south from Chang'an around 20 February 139 BCE, 06:00. The vertical line above the label 'South' is the observer's meridian, and the figures on it represent altitude above the horizon in degrees. The observer is directing his or her gaze upwards at a bit more than 25° to the horizon.

Figure 5.3. Winnower is currently 'centred': it is on the meridian about 25° above the horizon, and is moving from east to west with the overall diurnal movement of the heavens. In modern terms, it is said to be performing a 'meridian transit'. Tail, to the west, was centred about an hour ago, and Dipper, currently to the east, will be centred in about 40 minutes time.

In Figure 5.4, an observer such as Huan Tan sights on a star that has just become 'centred', using a gnomon. OS is a line through the observer's position to his 'south point' S on the horizon. Such a line can be easily found, either by finding the direction of the sun when the shadow is at its shortest, or by bisecting the angle between a morning and afternoon shadow of equal length.[19] The vertical line through S is part of the observer's meridian.

[19] *Huai nan zi* contains instructions for setting out an east-west line using the shadows cast by the rising sun and the setting sun; the *Zhou bi* 周髀 '*Gnomon of Zhou*', a text probably in final form by the early first century CE, uses these to define a north-south line directly. See *Huai nan hong lie ji jie*,

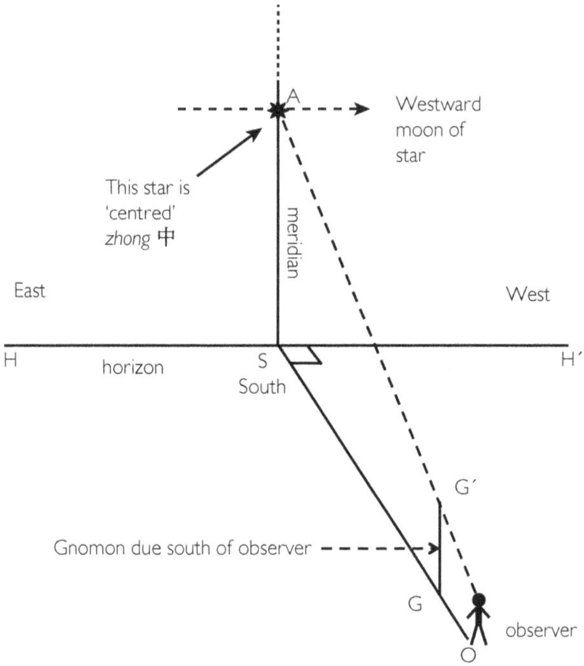

Figure 5.4 Observing a centred star.

The stars listed in the *Yue ling* are supposed to be observed as 'centred' at dusk and dawn. Because of the phenomenon of precession explained in section 5.1, the stars likely to be seen as 'centred' at a given time of day on a given date will change slowly through the centuries; this has suggested to some scholars that it should be possible to find a plausible date for the observations described in texts such as the *Yue ling*.[20] It is, however, extremely difficult to predict from first principles the day of the year when a given star will first become visible at dusk or dawn exactly on the meridian. The result will depend not only on the brightness of the star in question, the observer's visual acuity, and horizon conditions, but also on how close the star is to the horizon (which in turn depends on the observer's latitude and the star's declination), and the angle that the sun's path makes with the horizon on the day in question.[21] Then there is the question of how much certainty the

chapter 3, 128–9 and Christopher Cullen (1976) 'A Chinese Eratosthenes of the flat earth: a study of a fragment of cosmology in *Huai nan tzu*.' *Bulletin of the School of Oriental and African Studies* 39 (1): 106–27, Cullen (1996), 192.

[20] See for instance Needham and Wang Ling (1959), 247.

[21] On such questions, see Robinson (2009). What is more, the 'centrings' given refer to entire lunar months, which can shift by up to 30 days in relation to the seasons.

observer demands before being willing to say definitely that the star has indeed become visible. A more fundamental difficulty is that we do not know whether the listing of lodges given in the *Yue ling* is in any sense a record of observation, or whether it is the result of calculation based on some idealized scheme of the motion of the sun and the times of dawn and dusk. All this means that attempts to date the *Yue ling* list on the basis of astronomical calculation cannot produce results that combine precision with high levels of confidence.

On the other hand, this certainly does not mean that the *Yue ling* list of centred stars is in principle useless for its ostensible purpose—which is to provide a reference for whether a lunar month is falling in the appropriate position in the solar cycle.[22] The heavens turn relative to the observer at the rate of about one degree (close to 1 *du*) every four minutes, and the sun moves close to one degree eastwards relative to the heavens every day. Thus, if we ignore the shift in the timing of dusk or dawn from day to day, if a star first becomes visible after dusk on a given date just at the moment when it crosses the meridian, then for every day after that date, the star will pass the meridian four minutes earlier. It will therefore already be to the west of the sightline when it is first seen. Similarly, before the date of first visibility on the meridian, it will first be seen to the east of the meridian.

I have in fact had occasion to test this in practice, as an initial step in a series of experiments conducted on the balcony of my college room at Clare Hall, Cambridge,[23] in 1979 from 7–19 June, which will be further described in section 5.4.1. Around 7 June the bright star Arcturus (α Bootis) was well to the east of a meridian sightline (defined by two simple pole gnomons 1 metre apart) at its first dusk visibility. Each evening, this star had moved observably closer to the sight-line by the time it first became visible. On 11 June, it was still just to the east of the sight-line at first visibility, and despite some clouds on 12 June, it had obviously passed the sight-line at dusk on that date. Thus, without the need for any measurement, a list of centred stars could produce a sequence of checks of progress through the solar cycle that can be reproduced to within a day or two every year. The obvious utility of such observations lies in the fact that if the appropriate stars for the given month were not yet centred at dusk or dawn during that month, that would show that an intercalary month needed to be inserted to allow the calendar to get back in step with the heavens. Thus the calendar could be managed quite accurately without calculation, and with only the simplest of observations.

[22] Compare Confucius's reference to the behaviour of Antares as a check on the proper placement of a lunar month—see chapter 1, section 1.2.

[23] Latitude 52.20° N, longitude 0.10° E.

5.4 How could Huan Tan have found the 'correct *du*' of lodges?

The interval in days between (say) dusk centrings of the leading stars of successive lodges will give some indication of the width of the lodges in *du*, since the change in centred stars is ultimately due to the daily motion of the sun through the lodge system, assumed to be at a rate of one *du* per day. The total of the days counted until the same star is once more centred at dusk will of course be a whole number of days close to the length of the solar cycle, 365 ¼ days, which is equal to the number of *du* in a complete circuit. But since stars vary in brightness and height above the horizon at meridian transit, and solar and horizon conditions change throughout the year, the indication thus given of the relative widths of lodges in terms of right ascension can only be very rough at best. The *du* values referred to by Huan Tan must have been found by some other procedure.

The key to this problem is to look more closely at the two things Huan Tan tells us he did with his clepsydra:

(a) 'I checked it against the solar shadow in the daytime'
(b) 'In the night portion [of the clepsydra run] I checked it against the stellar lodges.'

The first procedure must surely have been a process of calibrating the clepsydra by running it from one moment of noon (as marked by the shadow cast by a gnomon), when the sun will be 'centred' due south, to the next. In modern terms that interval is 24 hours,[24] but in terms of the divisions of clepsydras in early imperial times it would be 100 *ke* 刻 'marks'.[25]

[24] We ignore here the fact that the interval from one noon to the next is not constant, a fact unknown in China in the period we are discussing. 24 hours is in fact the length of the *mean* solar day.

[25] If Huan Tan had used a single vessel clepsydra of the type shown in Figure 5.2, then even if it had been large enough (and had a narrow enough outlet bore) to have run for 24 hours on a single filling, it would certainly have run much slower near the end of the run than at the start, simply because of falling pressure at the outlet as the water level went down. Since this fact would have been very obvious to an observer concerned with the rate of the clepsydra (such as Huan Tan) I provisionally follow the view expressed in the discussion of Babylonian waterclocks in Fermor and Steele (2000), 213, who hold that water running out of the clepsydra is likely to have been returned to the reservoir in periodic 'top-ups' to maintain a fairly constant flow rate. Later in Chinese history, this was done by a system of automatic top-up vessels, each one feeding into the one below it so as to keep its level as close to constant as possible. On these 'polyvascular' devices, first well attested centuries after the end of the Han, see Needham and Wang Ling (1959), 522–6. An alternative might have been to use a single clepsydra that emptied in a short time, and refill it repeatedly. Any timing of a long period would have been included several full runs, plus one fractional run, and since any lack of uniformity of rate would only apply to the fractional final run, the effect on the accuracy of the overall timing would be much reduced.

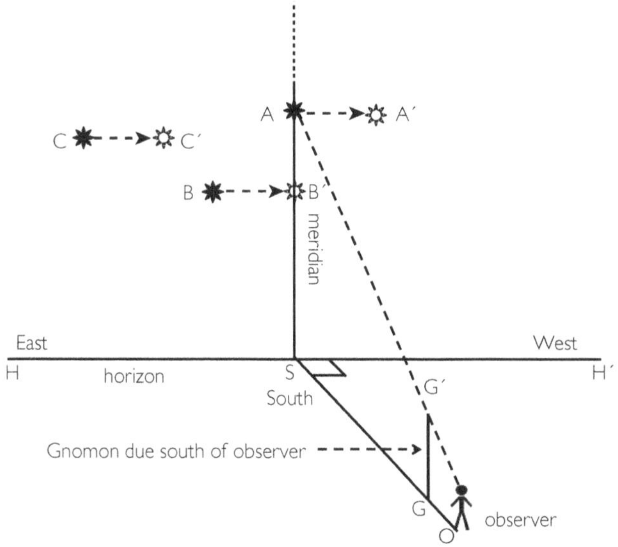

Figure 5.5 Observing a sequence of centred stars.

The second procedure would have involved Huan Tan observing the lodges at night, and *timing* them in some way with his clepsydra. That would imply a sequence of observations of centrings (or meridian transits) as shown in Figure 5.5.

Suppose A is the westernmost star of a lodge, the first of that lodge to be 'centred' on the meridian; B is the westernmost star of the next lodge (which will be centred when A has moved off to the west), and C is the westernmost star of the lodge after that. If Huan Tan is running his clepsydra while watching the stars cross his sight-line, there are a number of observations he can make. For a start, he might treat try to treat star A as if it was the noon sun, and run his clepsydra for 100 *ke* (the interval from one noon to the next) from the moment when star A is 'centred' so as to find the same time of night on the next night. But when that time comes, he will then notice that star A has already moved to the west of the meridian—because the eastwards motion of the sun relative to the stars at the rate of one *du* a day means that in 100 *ke* the heavens make a little more than a whole revolution. After a few more days, it will be star B that is centred at that time when A used to be centred, and so on. The number of days between the centring of two successive stars at the same time of night corresponds to the motion of the sun by the same number of *du*.

There is, however, a quicker way to measure lodge widths, and it works as follows. In 100 *ke* (24 hours in modern terms), the daily 1 *du* motion of the sun means that the sky turns by 1 *du* more than a whole circuit of 365 ¼ *du* so as to bring the sun back to the meridian. (We are of course considering the mean sun, as did Han specialists in astronomical systems.)

So the movement of the heavens in 1 *ke* is:

$$(365\ \tfrac{1}{4}\ du + 1\ du)/100 = 366\ \tfrac{1}{4}/100\ du$$

All we have to do is to measure the number of *ke* between the centrings of (say) star A and B, and from that we can find the number of *du* in the lodge beginning with A. Thus, if we find (for instance) that a given lodge takes 3 *ke* from the time that its leading star is centred to the time that the leading star of the next lodge is centred, the width of the lodge is:

$$3 \times 366\ \tfrac{1}{4}/100\ du = 11\ du, \text{ to the nearest } du.$$

It seems likely that a procedure of this kind was followed by Huan Tan to check the calibration of his clepsydra—since he already knew the results he ought to get for each lodge.

As explained earlier in this chapter (note 246), the rotation of the stars relative to the observer completes a circuit in about 4 minutes less than the 24 hours it takes for the sun to return to the meridian. While the interval from one noon to the next is a solar day,[26] the interval from one 'centring' of a star on the meridian to the next is known as a *sidereal* day. A sidereal day is itself divided into 24 hours of sidereal time, and corresponds to a 360° rotation of the heavens.[27] If we measure the interval of sidereal time between the instants when two successive stars are centred, i.e. perform meridian transits, that quantity is the difference in right ascension of the two stars. It is commonly measured in hours and minutes of sidereal time, but may also be expressed in degrees, with 24 hours of sidereal time equivalent to 360° of right ascension. Since the spring equinox is the reference point for right ascension, the absolute value of the right ascension of a star is the interval of sidereal time between the transit of the position of the mean sun at the spring equinox and the transit of the star.

[26] Since this period is not constant throughout the year, the original definition of the hour as used on common clocks was $1/24$ of a *mean* solar day. But since variation in day-length was unknown in early imperial China, the distinction need not be made here.

[27] Older observatories frequently had clocks that ran in sidereal time for convenient reference.

5.4.1 Reconstructing Huan Tan's measurements

We do not have access to the results of any measurements made by Huan Tan. Under such circumstances it seems worthwhile to attempt an experimental reconstruction of the procedures he seems likely to have followed, using the instruments available to him—the gnomon and the clepsydra—and to see the kind of results that can be obtained by such means. Now, as Hasok Chang has pointed out, all attempts at such reconstruction demand careful consideration of what is really being reconstructed (or perhaps simply constructed).[28] In Chang's terms, I was aiming to make at least an approximation to a 'historical recreation' of Huan Tan's experiments—that is, I was attempting to make something that resembled his apparatus, and to use it in a way that may have resembled his actual observational procedures. By contrast, the use of a modern transit circle with a telescope and electronic timing apparatus would have been more like what Chang calls a 'physical replication'.

As mentioned in section 5.3, my experiments were carried out on a college balcony in Cambridge in summer 1979.[29] My clepsydra was a 20-litre water vessel (in polythene rather than bronze) with a narrow glass tube as an outlet, and instead of a float and rod I used a measuring cylinder to receive the water flowing out of the main vessel. I emptied the cylinder back into the vessel each time it filled, thus maintaining a fairly constant flow rate, thus anticipating the pattern later hypothesized by Fermor and Steele for ancient Babylon (see footnote 270). I set up a 2 m gnomon on the balcony, with a 0.5 m back-sight gnomon 1 metre away on a north-south line, giving a clear sight-line to the southern horizon. I calibrated the clepsydra by running it from one noon to the next; I found that it was easy to judge the moment when the shadow of the main gnomon fell on the north-south line to within 1 minute, or 0.07% day. The total clepsydra run from noon on 11 June to noon on 12 June was 25,515 ml.

[28] See for instance Hasok Chang (2011) 'How Historical Experiments Can Improve Scientific Knowledge and Science Education: The Cases of Boiling Water and Electrochemistry.' *Science & Education* 20 (3–4): 317–341.

[29] Cullen (1982a), 'Early Chinese measurements of right ascension before the armillary sphere', First International Conference on the History of Chinese Science, Louvain, Belgium An updated version of the content of that paper was presented in Christopher Cullen (2014), 14 January. 'Étoiles et Saisons: peut-on reconstituer l'observation *du* ciel dans la Chine ancienne?'. Seminaire: Histoire des sciences, des techniques et de la médecine en Asie orientale, EHESS, Paris. I am grateful to the President and Fellows of Clare Hall, Cambridge, for the Stipendiary Research Fellowship that I held while carrying out these experiments, and to the British Academy for a grant to purchase apparatus.

Following the pattern of thought set out in section 5.4, but using modern units for convenience of calculation, we may say that this corresponds to a rotation of the heavens by:

$$360° + 1° = 361°$$

So we may convert clepsydra run to RA difference in degrees by multiplying by:

$$361°/25{,}515 \text{ ml}$$

During the night of 11/12 June, I used my waterclock to measure the times between the 'centrings' on my north-south line of a sequence of 12 easily visible stars, and also of the centre of the moon, which was about two days past full. It was easy to judge the moment of 'centring' to within half a minute of time, corresponding to ⅛° of RA. The results are given in Table 5.2, to which modern calculations of the right ascensions of the bodies concerned have been added for comparison, with the differences between the two sets of values being given in the final column on the right.[30]

The root mean square error of less than ⅓° between the differences in right ascensions of the stars found by the waterclock and modern calculations compares favourably with the discrepancies in the *Huai nan zi* data noted in Table 5.1. It appears that the limit on accuracy is set by the timing rather than the sighting procedure. Conversely, there is no reason why Huan Tan might not have used a list of known widths of lodges to check the calibration of his clepsydra—which is just what he says that he did. There was certainly no need for him to have used any kind of graduated circle to obtain lodge widths.

5.5 Locating the sun among the lodges, and the shape of the heavens

For each lunar month the *Yue ling* told us which lodges will be 'centred' at dawn and dusk if the month is properly situated in relation to the solar cycle—but it also told us in which lodge the sun was located (*zai* 在) in that month. What could that mean in practical terms?

[30] These have been read off directly from the values for 00:00 hours on 12 June 1979 given by Starry Night Pro™.

Table 5.2 ***Finding RA differences by clepsydra timing***

Celestial body performing meridian transit (*zhong* 中 'centring')	Modern calculated RA/hours	Modern calculated RA/ degrees	Increment RA /degree	Total clepsydra run/ml	Increment in run/ml	RA difference:, calculated from run increment × 361°/25,515 ml	RA diff. error/degree
Noon sun 11 June 1979 (observed using gnomon)				0			
α Bootis	14.2452	213.678		9,675			
			7.3425		530	7.50	−0.156
ε Bootis	14.7347	221.0205		10,205			
			7.953		547	7.74	+0.214
β Librae	15.2649	228.9735		10,752			
			4.479		318	4.50	−0.020
α Corona Borealis	15.5635	233.4525		11,070			
			2.3595		180	2.55	−0.187
α Serpentis	15.7208	235.812		11,250			
			5.247		390	5.52	−0.271
β^1 Scorpii	16.0706	241.059		11,640			
			5.976		427	6.04	−0.065
α Scorpii	16.469	247.035		12,067			
			1.9695		133	1.88	+0.088
ζ Ophiuchi	16.6003	249.0045		12,200			

(continued)

Table 5.2 *Continued*

Celestial body performing meridian transit (*zhong* 中 'centring')	Modern calculated RA/hours	Modern calculated RA/ degrees	Increment RA /degree	Total clepsydra run/ml	Increment in run/ml	RA difference:, calculated from run increment × 361°/25,515 ml	RA diff. error/degree
			8.2935		590	8.35	−0.054
η Ophiuchi	17.1532	257.298		12,790			
			6.195		425	6.01	+0.182
α Ophiuchi	17.5662	263.493		13,215			
			2.1195		120	1.70	+0.422
β Ophiuchi	17.7075	265.6125		13,335			
			16.4655		1,202	17.01	−0.541
Moon centre	18.8052	282.078		14,537			
			15.366		1,033	14.62	+0.751
α Aquilae	19.8296	297.444		15,570			
			142.6935		9,945	140.71	
Noon sun 12 June 1979 (observed using gnomon)	5.3425	80.1375		25,515		Root Mean Square error/degree	0.324

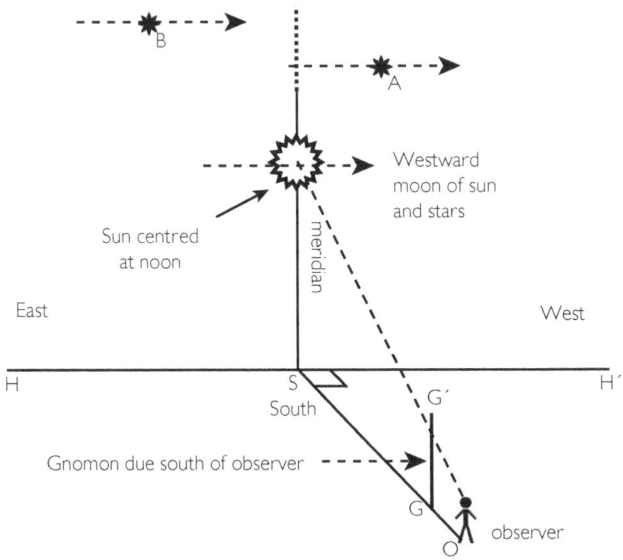

Figure 5.6 Observing the sun in a lodge—a hypothetical procedure.

Suppose, for the moment, that we can observe the stars at the same time as the sun. Then, if in Figure 5.6 star A is the leading star of a given lodge, and if B is the leading star of the next lodge, the sun may be said to be in the given lodge if its moment of centring (i.e. noon) falls between the centrings of A and B. Since, however, we cannot observe the stars when they are near the sun, we shall have to do calculations in order to find which lodge will centre around noon. But such calculations are not difficult: for instance, in Table 5.2 we can see:

Between the centring of α Aquilae at night, and of the sun at noon on 12 June, the volume of water running from the clepsydra was:

$$25{,}515 \text{ ml} - 15{,}570 \text{ ml} = 9{,}945 \text{ ml}$$

thus the implied difference of right ascension between that star and the sun was:

$$9{,}945 \text{ ml} \times 361°/25{,}515 \text{ ml} = 140.71°,$$

whereas the calculated value is 142.69°, corresponding to about 8 minutes out of some 9 hours and 20 minutes, an error of only 1.4%. So Huan Tan could not only easily check the widths of the lodges with his clepsydra, but also check the location of the sun in relation to the stars at any time he wished.

Anybody performing measurements of this kind on a regular basis would inevitably have a general picture of the starry heavens as turning from east to west daily as one looks south, as well as becoming aware that the sun moves slowly across the background of the stars from west to east. But what kind of physical picture of the heavens and the earth might they have in mind as they made their observations? We have a device from the early Han dynasty that shows us what that picture may have been.

The object shown in Figure 5.7 was found in the tomb of the Marquis of Ruyin 汝阴 (closed in 165 BCE).[31] It consists of a pair of lacquered wooden plates, the lower plate being square, with sides measuring 13.5 cm, and the upper plate being a disc of diameter 9.5 cm, arranged so as to turn on a small pin passing through the centres of both plates. Such objects, known as *shi* 栻 ('[cosmic] models') were widely used for hemerological divination.[32] The disc also bears round its circumference in anti-clockwise order what is essentially the same sequence of twenty-eight lodge names as listed in *Huai nan zi*. The lodges are repeated in the outer border of the square earth-plate, divided into the four groups of seven corresponding to north, south, east and west listed in *Huai nan zi* (see Table 5.1). I have marked the position of the name of the lodge Horn on the diagram for both sequences. Within the border of the square plate is another band, this time bearing the standard set of twelve cyclical characters, apparently playing their role as direction-markers. The four main directional signs are indicated by the labels in roman script and modern-style characters on the diagram. The disc is marked with the emblem of the Dipper at its centre—but the figure of the Dipper is reversed, as it would be if seen from outside the heavens.

The whole object embodies an ancient view of the cosmos summed up in the words of a poet of the fourth century BCE, Song Yu 宋玉:

[31] On the tomb in general, see Anhui cultural relics working group 安徽省文物工作队 (1978) '阜阳双古堆西汉汝阴侯墓发掘简报 Fu yang shuang gu dui Xi Han Ru Yin hou mu fa jue jian bao (An outline report on excavations at the Western Han tomb of the marquis of Ru Yin at Shuang Gu Dui, Fu Yang).' *Wen wu* (8): 12–30 & 98–9. On the object discussed here, see particularly Yin Difei 殷涤非 (1978) 'Xi Han Ru Yin hou mu chu tu de zhan pan he tian wen yi qi 西汉汝阴侯墓出土的占盘和天文仪器 (The divining board and the astronomical instrument excavated from the Western Han tomb of the Marquis of Ru Yin).' *Kao gu* 5: 338–43. The photographs given in this article are not clear enough to justify reproduction here; the drawings shown were prepared by the archaeologists who first studied this and other objects in the tomb, including the 'lodge dial' discussed later in this section.

[32] For a discussion of these devices from the point of view of astronomy, see Cullen (1981).

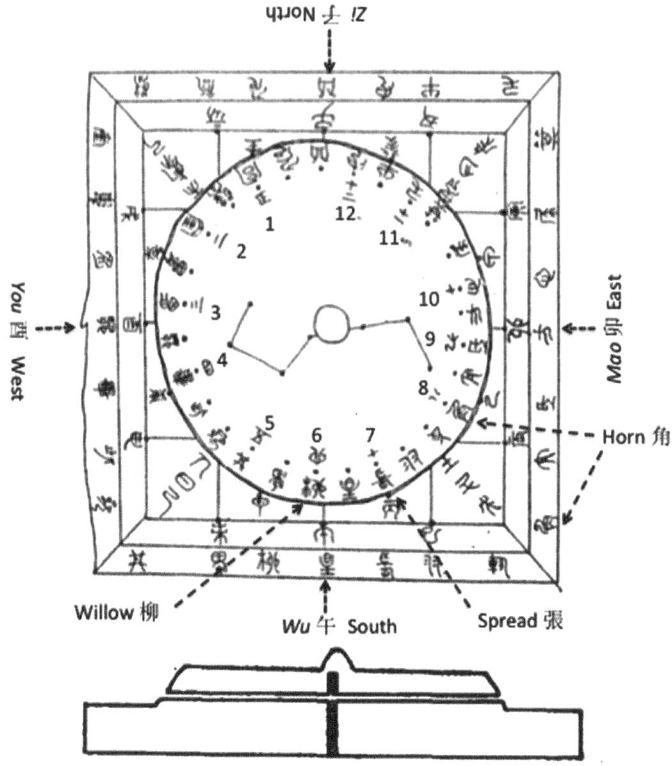

Figure 5.7 Cosmic model *shi* 栻 from the tomb of the Marquis of Ruyin (165 BCE). Drawing of original, with cross-section below (Yin Difei 殷滌非 1978: figs 3 & 4).

> 方地為輿, 圓天為蓋
>
> The square earth is my chariot, and the round heaven is my canopy (*Bei tang shu chao* 149, 3b, comm.)

If we imagine an observer somewhere underneath the heaven-disc, on a line running south from its centre, and looking upwards towards the south, then if the disc rotates clockwise as we look down on it, he or she will see the lodges moving from left to right (i.e. from east to west) exactly as they are seen in the sky—as well as seeing the Dipper the right way round on turning to gaze northwards.

Twelve of the lodge names on the heaven-disc are accompanied by Chinese numerals with *zheng* 正 'standard [month]' starting the sequence. I have added Arabic numbers next to these. By listing these numbered lodges in comparison with the 'solar location' lodges given in the *Yue ling* for the months bearing those numbers, as in Table 5.3, we may see that the information is largely identical.

Table 5.3 ***Comparison of lodges given in the* Lü shi chun qiu *and on the* shi *device***

	Data from the *Yue ling*, as listed in *Lü shi chun qiu*			Lodges on heaven-disc of *shi* device: asterisk * shows lodge is different from that found in the *Yue ling*.
Month order in *Yue ling* or number on disc	Lodge centred at dusk	Lodge centred at dawn	Lodge where sun is located	
1	Triaster	Tail	House	House
2	*Hu* 弧 'Bow' = Ghost[1]	*Jian xing* 建星 'Establishment Star' = Dipper	Straddler	Straddler
3	Star	Ox	Stomach	Stomach
4	Wings	Woman	Net	Net
5	Gullet	Rooftop	Well	Well
6	Heart	Straddler	Willow	Willow
7	Dipper	Net	Wings	Spread*
8	Ox	Beak	Horn	Horn
9	Barrens	Willow	Chamber	Base*
10	Rooftop	Stars	Tail	Heart*
11	Wall	Axletree	Dipper	Dipper
12	Harvester	Base	Woman	Woman

[1] 'Bow' and 'Establishment Star' are identified in the earliest commentaries as asterisms close to the better known Ghost and Dipper, and functionally equivalent to them in this context.

We can see that the lodges on the heaven-disc of the *shi* match with those given for the monthly locations of the sun in the *Lü shi chun qiu* in all but three cases—the seventh, ninth and tenth months, when the lodge on the *shi* is one step anti-clockwise from the one listed in the *Lü shi chun qiu*. Given that the position of the sun is not directly observable, and that we have no idea when, where, or by whom the two sets of data were determined, it is no surprise to find some discrepancies between the sources as we do here.

Looking back at Figure 5.7, we can see that the heaven-plate in the position in which it is drawn shows the situation near noon in the days of the sixth month, when the sun is in the lodge Willow, which thus passes due south (marked by the seventh cyclical sign, *wu* 午 at the bottom of the diagram) at noon for some part of this month. During the course of the next 24 hours, we would see the heaven-plate performing a little over a clockwise circuit (as seen from above) so that Willow will be back near the noon position on the next day. If we look ahead a month, we should find that another lodge, Spread, occupies the noon position—since the sun has shifted into that lodge.

We do, however, need to be cautious about treating this device as a straightforward analogue computer, in which rotation of the disc directly models movement seen in the sky.[33] One obvious problem is that the lodges are shown as evenly spaced round the circumference of the disc, whereas in fact their widths vary from 2 *du* for Beak to 33 *du* for Well in the *Huai nan zi* list. But since the sun is supposed to move steadily through the lodges at the rate of 1 *du* per day (and hence at 29 or 30 *du* per month), the numbered lodges in which it is found each month cannot be evenly spaced round the disc. And indeed they are not—thus for instance, as we move round the disc from the lodge marked for the first month to that marked for the second, then to that for the third, and for the fourth month, we shift by two lodges each time. But from the fourth to the fifth month, we shift by three lodges, as we also do from the seventh to eighth, tenth to eleventh, and from the twelfth back to the first month. This device shows the *sequence* of the lodges, as it does of other time and direction markers—but so far as the lodges are concerned, their depiction does not appear to be intended to show their *extent in space*.

However, another object found in the same tomb as the *shi* cosmic model shows that the extent of the lodges could be physically modelled in due proportion if required.[34] This object, shown in Figure 5.8, consists of two circular discs of lacquered wood, the smaller of which (23 cm in diameter) fits on top of the larger (25.6 cm diameter), the two discs being arranged on a common axis passing through their centres, so that one disc could be rotated with respect to the other. The upper disc has small marks in the form of indentations near its rim at equal intervals, and despite damage it may be deduced that there were originally 365 of these. The larger lower disc has the names of the lodges round its rim, with numbers under each name. Where it is possible to count the marks and read the numbers, the number of marks on the upper disc between each lodge

[33] In case this term is unfamiliar, note that the great majority of computers in the early 21st century are *digital*, in that they basically do nothing but perform calculations on numbers, expressed by the binary digits 0 and 1. The continuously varying physical quantities found in the world we inhabit are thus represented in a discontinuous way, since the precision with which numbers are represented is necessarily limited. *Analogue* computers (nowadays much less common than they once were) function quite differently, using one continuously varying physical quantity (such as the electric current in a circuit) to represent another (such as the speed of a vehicle).

[34] Wang Jianmin 王健民 and Liu Jinyi 刘金沂 (1989). 'Xi Han Ruyin hou mu chu tu yuan pan shang er shi ba xiu gu ju du de yan jiu 西汉汝阴侯墓出土圆盘二十八宿古距度的研究 (Researches on the old du widths of the 28 lodges on the basis of the disc excavated from the Western Han tomb of the Marquis of Ruyin)' in *Zhongguo gudai tianwen wenwu lunji*, Beijing, Wenwu publishing house: 59–68. As the title of the article implies, the widths of the lodges marked on this disc differ slightly from those found in later sources, and may represent an older listing.

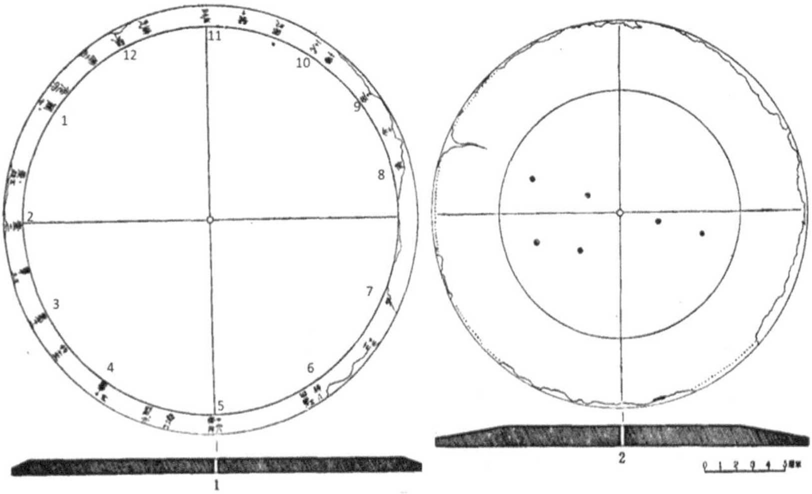

Figure 5.8 'Lodge Dial' from tomb of Marquis of Ruyin (165 BCE), with addition of numbers for months as marked on the *shi* from the same tomb (Anhui cultural relics working group 安徽省文物工作队 1978: 19).

name and the next name in an anti-clockwise sense corresponds to the number marked by the lodge name on the lower disc. Since, as we have seen, there are 365 ¼ *du* in a circuit of the heavens, this device obviously shows the extent of each lodge in *du*, neglecting the fraction of ¼ *du*. Some Western researchers have given this device the self explanatory title of 'lodge dial'; the ancient name is unknown.

Both the numbers marked for each lodge, and the different extents of the lodges round the circumference of the lower disc make it plain that the lodges do indeed differ greatly in extent. But if we mark month numbers against the corresponding lodges numbered on the *shi* heaven-disc, as in Figure 5.8, the effect of the uneven extent of the lodges is to make the displacement from one month to the next much more regular. Indeed, if it were not for the large size of the gaps between points 6 and 7, and 7 and 8, there would be three monthly intervals in each quadrant. We must recall, however, that we cannot expect that the sun will actually fall at the start of a given numbered lodge at the start of the corresponding month—at best, we can only hope that it will fall somewhere within that lodge, with the precise position shifting from year to year as the months shift relative to the solar cycle.

Given the precise depiction of the extent of each lodge in terms of rotation of the heaven-disc, we could use this device as an analogue computer for a number of purposes.[35] Thus, given the position of the sun relative to the lodge system on a given day, we could find the position on any subsequent day by simply counting anti-clockwise by the requisite number of *du*. If we know the position of the sun on that day, then we can not only say that the sun's lodge will be 'centred' on the meridian at noon that day, but also predict which lodge will be centred at any other time of day—thus, for instance, the lodge that will be centred at midnight will be diametrically opposite the noon lodge.[36]

5.6 From model to cosmos: the *Zhou bi* 周髀

5.6.1 The size and shape of the *gai tian* 蓋天 cosmos

I have used the term '[cosmic] model' as a rendering for the name of the *shi* device shown in Figure 5.7. What, however, would the full-scale cosmos itself be like if this device was a model of it? Interestingly, we have a text of the Han period which describes just such a cosmos: this is the *Zhou bi* 'Gnomon of Zhou'.

I have translated this text and discussed a range of questions arising from its content in a book-length study.[37] In my view, the *Zhou bi* is composed of a collection of related material from different sources rather than being a single book written by an author with an overall plan in mind. The first section of the text, which contains a dialogue between the Duke of Zhou and a worthy of the Shang dynasty, suggests that the collection may have been completed in the time of Wang Mang (r. 9–23 CE) who is known to have seen himself as following the Duke's example. If that is the case this dialogue may have been the last element to be added to the collection.[38] The *Zhou bi* is certainly not referred to in any text dating from before the Eastern Han; in fact we have to wait until the time of Cai Yong 蔡邕 in about 180 CE before finding an explicit reference to this work.[39]

[35] Some have suggested that it might have been possible for such a device to have been set in the plane of the celestial equator and used for measurement. In my view, it does not seem well adapted to such a purpose and was more likely to have been a calculating device of some kind—see Cullen (1981).

[36] Given that this device is constructed with a precision of no less than one *du*, we can neglect the effect of the sun's daily displacement of one *du* in this context.

[37] Cullen (1996).

[38] Cullen (1996), 148–56.

[39] See Cullen (1996), 38–9.

Parts of the text describe a cosmos which is in effect a *shi* enlarged to a vast scale, with the heaven-disc 80,000 *li* above the flat square earth that lies below it. This view of the cosmos became known as the *gai tian* 蓋天 'chariot-umbrella heaven', since, as indicated in the words by Song Yu quoted earlier, heaven turned about a vertical axis above the earth that lay beneath it, just as the large umbrella-like cover *gai* of a Han period chariot was held above the square body of the vehicle by a vertical pole.

The *Zhou bi* describes in considerable detail the layout and dimensions of the *gai tian* cosmos. It justifies these dimensions through the use of a simple rule for turning gnomon shadow-lengths into distances of celestial bodies from the observer: every inch, *cun* 寸, of shadow cast by a gnomon eight feet, *chi* 尺 (= 80 *cun*), tall illuminated by a celestial body on the heavenly disc (assuming the body to be luminous) corresponds to 1,000 *li* 里 of distance from the observer to the point on the flat earth directly below the celestial body in question. So (for instance) if we move 1,000 *li* southwards, the noon solar shadow will shorten by one inch.[40] Figure 5.9 shows an example, given in the *Zhou bi*, of how this rule may be applied.[41] Applying the rule, a shadow 6 *chi* (= 60 *cun*) long implies a distance of 60,000 *li* to the sub-solar point. Pythagoras' theorem shows that the top of the gnomon is 10 *chi* from the tip of the shadow. Thus, by simple proportion, the sun is 100,000 *li* distant from the observer.

If in every case a one-*cun* change in the length of the shadow cast by an 80-*cun* gnomon always represents 1,000 *li* of distance to the subsolar point, then again simple proportion implies that the sun must always be 80,000 *li* above the flat earth, and that the heavens themselves, on which the sun is located, must be flat like the earth and parallel to it. Note too that the 'shadow rule' is not only applicable to a body like the sun, whose emitted light causes a gnomon to cast a visible shadow. If we replace the sun in the diagram by (say) the position of the north celestial pole, then we may generate a 'pseudo-shadow' by tying a string to the top of the gnomon, and holding the other end on the ground so that the string aligns with the sightline from the gnomon to the celestial pole. The length of the 'pseudo-shadow', that is, the distance from the end of the string on the ground to the base of the gnomon, will yield the distance from the gnomon to the point below the pole by application of the 'inch for a thousand *li*' rule.

[40] The values of the units used here were approximately as follows in Han times: 1 *chi* = 23.1 cm (making an 8 *chi* gnomon about 1 m 85 cm high), 1 *cun* = 2.31 mm, 1 *li* = 0.415 km: see Twitchett, Loewe and Fairbank (1986), xxxviii. The 'inch for a thousand *li*' rule is in fact far from reality. Actual observations in the Yellow River region would have shown a change of one *cun* of summer solstice shadow length for a north-south displacement under 200 *li*. By the middle of the first millennium CE, the error of the traditional figure had been recognized: see Cullen (1996), 113–14.

[41] Cullen (1996), 78–80.

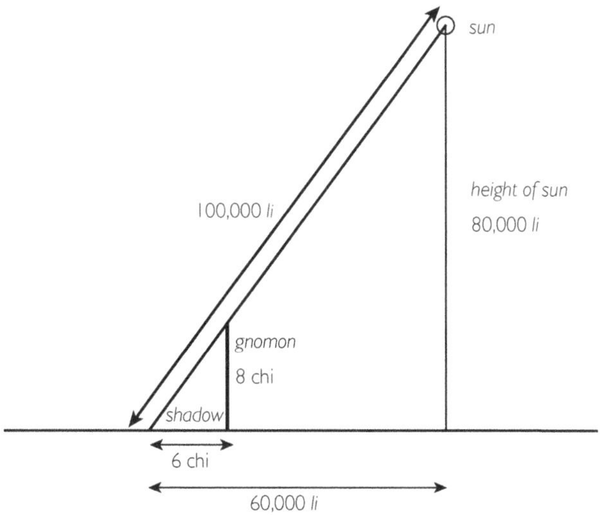

Figure 5.9 Applying the shadow rule of the *Zhou bi*.

The *Zhou bi* cites three crucial measurements on which its picture of heaven and earth is based—the shadows cast by the noon sun at the winter and summer solstices, and the 'pseudo-shadow' from sighting on the north celestial pole. All these measurements are said to have been made at 'Zhou 周', which is presumably meant to indicate the site of one of the ancient capitals of the dynasty of that name. We may tabulate the shadow lengths given (expressed in *cun*) and convert to distances as in Table 5.4.

From the data in the first column, we may derive those in the second column using the 'shadow rule'. Taking the example of the summer solstice, since the sub-solar point is 16,000 *li* to the south of the observer, and the sub-polar point (directly below the north celestial pole) is 103,000 *li* to the observer's north, then the distance from the sub-solar point to the sub-polar point must be:

$$16{,}000\ li + 103{,}000\ li = 119{,}000\ li$$

Table 5.4 ***Basic observations of distances using gnomon***

	Shadow of gnomon/*cun*	Distance from gnomon/*li*	Distance of sun from sub-polar point/*li*
North celestial pole	103	103,000	–
Summer solstice noon sun	16	16,000	119,000
Winter solstice noon sun	135	135,000	238,000

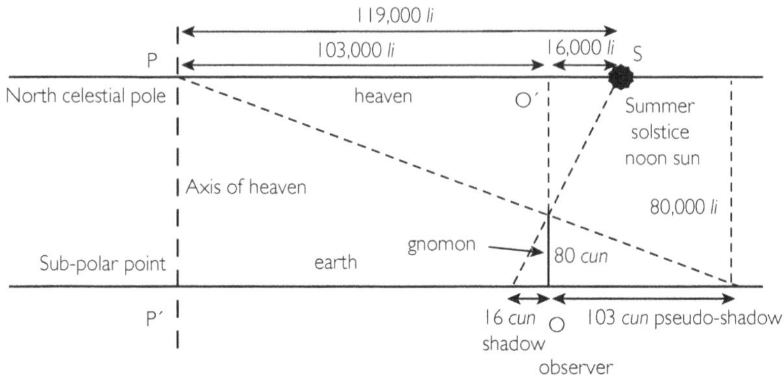

Figure 5.10 Finding distance of summer solstice sun from the pole.

As shown in Figure 5.10, this is also the distance from the summer solstice sun to the north celestial pole. Similarly, the distance from the sub-solar point at winter solstice to the sub-polar point must be:

$$135{,}000\ li + 103{,}000\ li = 238{,}000\ li$$

which is also the distance from the winter solstice sun to the north celestial pole.

The lengths of the shadows cited here are values that might well have been obtained by an observer in latitude 35° N, which is reasonable enough for the ancient Chinese sites in the Yellow River basin. But some adjustment has apparently taken place to produce the neat doubling of the pole to sun's distance from the summer to winter solstices. We are also told that the noon shadow at the equinoxes is 17 *chi* 8.5 *cun*—which is clearly not an observed value, and is simply the mean of the two solsticial values, implying a distance of 178,500 *li* from the north celestial pole.

We may draw a schematic plan view of the position of the pole, the observer, and the daily paths of the sun at the summer and winter solstices, as in Figure 5.11.

The *Zhou bi* subdivides the intervals between the daily paths of the sun at the solstices, by defining five intermediate daily paths, equally spaced between the two solsticial extremes. Together with the solsticial paths, these make up the seven *heng* 衡 (perhaps 'levels'), and represent the daily paths of the sun at the 'nodal *qi*', that is, the odd-numbered of the 24 *qi* counting the winter solstice as number 1 and the summer solstice as number 13; see Figure 5.12.

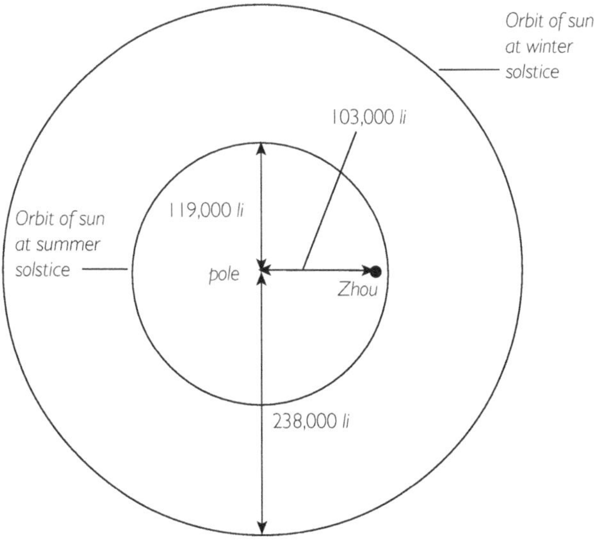

Figure 5.11 Plan view of *gai tian* cosmos showing daily paths of sun at solstices.

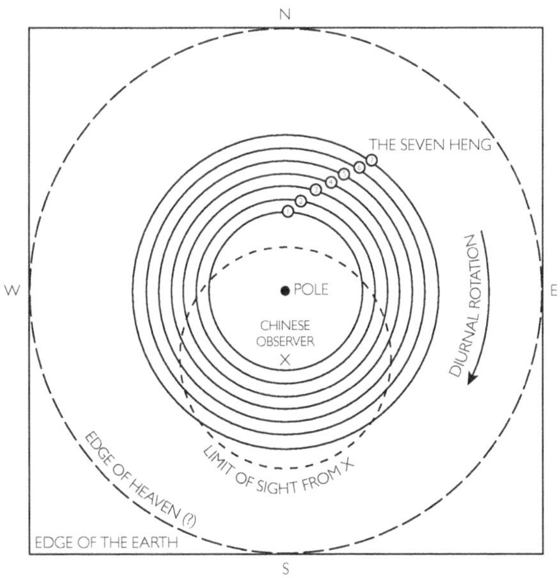

Figure 5.12 The seven *heng* and the 167,000 *li* range of sight.

One final peculiarity of the *Zhou bi* must be noted here. If the sun and other heavenly bodies are always 80,000 *li* above the earth, why are they not always visible? To deal with this obvious difficulty, the *Zhou bi* introduces a large *ad hoc* assumption, which is that beyond a certain distance, 167,000 *li*, all objects become invisible. The fact that heavenly bodies appear to rise and set is thus explained as an optical illusion caused by their moving within that distance from the observer, or moving beyond it. Thus, in Figure 5.12, any celestial body inside the dotted circle centred on the observer will be visible, but all others are invisible. An object that comes within the circle appears to rise over the horizon, but an object that leaves it appears to set.

5.6.2 The strange case of the north polar distances

As we have seen, the *Zhou bi* provides a self-consistent system for calculating distances of the sun from the north celestial pole, and the resultant doubling of the summer solstice distance at the winter solstice is satisfyingly neat. These distances are of course linear distances in *li*, not angular north polar distances in degrees (or rather *du*) as they would be measured on the celestial sphere. The *Zhou bi* describes no instrument for making such angular measurements directly, and without the use of trigonometry (which is wholly absent from this text) there is no way to calculate them from gnomon observations. It is therefore striking that this is exactly what the *Zhou bi* does in fact purport to do in one section of its text. The results in Table 5.5 are given for the lodges at the winter solstice, equinoxes and summer solstice.

The first thing to say about the strange mixture of angular and linear measures in Table 5.5 is that the *du* values given are reasonable approximations to values for north polar distances of the sun at the given seasons accepted in the Eastern Han. Thus for instance in a memorial of 92 CE we are told by Jia Kui 賈逵 that the relevant values are 115 *du*, 91 *du* and 67 *du* (*Hou Han shu*, *zhi* 23,029). A careful examination of the text reveals that the calculations leading to the results above are based on the use of the following equivalence (1 li = 300 bu 步):

1 *du* = 1,954 *li*, 247 *bu* and 933⁄1,461 *bu*

Table 5.5 ***Angular north polar distances in the*** **Zhou bi**

Season	Lodges	Distance of sun from north celestial pole.
Winter solstice	Ox	115 *du*, 1,695 *li*, 21 *bu* and 819⁄1,461 *bu*
Equinoxes	Harvester, Horn	91 *du*, 6,10 *li*, 264 *bu* and 1,296⁄1,461 *bu*
Summer solstice	Well	66 *du*, 1,481 *li*, 155 *bu* and 1,245⁄1,461 *bu*

This figure is obtained by dividing up the circumference of the innermost *heng* into 365 ¼ intervals—each one of which represents one *du* with reference to that circumference. This circumference is, be it noted, found from the radius of the innermost *heng*, 119,000 *li*, on the basis that $\pi = 3$, the value used throughout the *Zhou bi*. The circumference is thus:

$$119{,}000\ li \times 2 \times 3 = 714{,}000\ li.$$

We then use this equivalence to convert the linear distances of the sun from the north celestial pole already given, noting that as already stated each *li* contains 300 *bu* 步 'double paces'. In the case of the middle (equinoctial) *heng*, the logic of the calculation may be set out as follows.

Since the radius of the middle *heng* is the mean of the innermost and outermost *heng*, and the outermost *heng* is twice the radius of the innermost *heng*, thus the radius of the middle *heng* is:

$$\begin{aligned} R_M &= (\text{radius of innermost } heng) \times (1+2)/2 \\ &= (3/2) \times (\text{radius of innermost } heng). \end{aligned}$$

Now in linear terms, 1 *du* round the circumference of the innermost *heng* is:

$$\begin{aligned} d &= [(\text{radius of innermost } heng) \times 2 \times 3]/(365\ 1/4) \\ &= 6 \times (\text{radius of innermost } heng)/(365\ 1/4) \end{aligned}$$

So the angular distance of the sun at the equinoxes will be:

$$\begin{aligned} R_M/d &= [(3/2) \times (\text{radius of innermost } heng)]/ \\ &\quad [6 \times (\text{radius of innermost } heng)/(365\ 1/4)]\ du \\ &= (365\ 1/4\ du) \times (3/2)/6 = (365\ 1/4\ du)/4 \end{aligned}$$

This is a quarter of a revolution, 91 5⁄16 *du* or 90°, precisely the right result for the north polar distance of the sun at the equinoxes. As can be seen, however, this result is nothing to do with observation, but is the inevitable result of the use of the approximation $\pi = 3$, and the decision to make the equinoctial *heng* lie exactly halfway between the two solsticial *heng*.

It is less easy for the *Zhou bi* to produce realistic results for the north polar distances of the sun at the solstices in angular terms. Simply using the linear distances calculated from the shadow rule (which are ⅔ and ⅓ those for the equinoxes) will produce the following figures:

$$\text{Summer solstice: } (2/3) \times (365\ 1/4\ du)/4 = 60\ 7/8\ du$$

$$\text{Winter solstice: } (4/3) \times (365\ 1/4\ du)/4 = 121\ 6/8\ du$$

These are obviously wrong. The *Zhou bi* solves the problem by introducing a contrived correction factor, the 11,500 *li* radius of a circle round the pole described by the orbit of an idealized circumpolar star called the *xuan ji* 璿璣. For no evident reason apart from getting the right result, this is added to the summer solstice radius and subtracted from the winter solstice radius, thus producing the results given in Table 5.5, which are close to those actually observed.

This outline of the methods used by the *Zhou bi* to obtain, by whatever means necessary, realistic values for the north polar distances of the sun at the solstices and equinoxes, is a summary of the account I have given at full length elsewhere.[42] It is, I trust, detailed enough to show that whoever wrote this material must have known these values already, and was determined to show that they could be obtained by using the basic procedures of the *Zhou bi*, suitably adapted. The fact is, however, that the only way to obtain such values without the use of trigonometry (which was unknown in Han times in East Asia) is to measure them directly using a device capable of measuring angles on the heavens. Clearly such devices must have been in use by the time the material from the *Zhou bi* discussed here was written, even though the *Zhou bi* chooses to make no mention of them.

But what were these devices, and when did they appear? And what were the implications of such devices for the view of the cosmos taken by those who used them for observation? These are the questions to which we shall now turn.

5.7 Changing measurements, instruments, and pictures of the cosmos

As mentioned in the last section, we do have documents known to date from the Eastern Han that give the north polar distance of the sun at the solstices and equinoxes. But there are no such texts that tell us the north polar distances

[42] Cullen (1996), 119–27.

of stars, such as those that mark the starting points of the 28 lodges (the 'lodge determinatives' as they are sometimes called). However, a compendium of divinatory and astronomical information compiled in the eighth century CE, the *Kai yuan zhan jing* 開元占經 'Divination manual of the Kaiyuan reign period' does include such information.

The material given in this text is ascribed to Shi Shi 石氏 'Mr Shi', i.e. Shi Shen 石申, a somewhat shadowy figure of the pre-imperial age.[43] Thus, for instance, the entries for the lodges Base, Chamber and Heart include the following:[44]

> 石氏曰: 氐四星, 十五度. (古十七度) 距西南星先至. 去極九十四度
>
> Mr Shi says: Base, 4 stars, 15 *du* (anciently 17 *du*).[45] [Measure] distance from the south-western star that arrives first [at the meridian]. 94 *du* from the pole.
>
> 石氏曰: 「房四星, 鉤鈐二星, 五度. (古七度) 距房西南第二星. 去極一百八度
>
> Mr Shi says: Chamber, 4 stars [with] Hooked Lance, 2 stars, 5 *du* (anciently 7 *du*). [Measure] distance from the second star in the south-west. 108 *du* from the pole.
>
> 石氏曰: 「心三星, 五度. (古十二度) 距前第二星. 去極一百八度半
>
> Mr Shi says: Heart, 3 stars, 5 *du* (anciently 12 *du*). [Measure] distance from the second star from the front. 108 ½ *du* from the pole.

And the same kinds of data are given for other lodge determinatives, as well as for other stars. Now, north polar distance (often abbreviated to 'NPD') is much more sensitive than lodge width to a change in the date of observation: the phenomenon of precession means that once in every 26,000 years any given star will find itself aligned with the winter solstice once, and with the summer solstice once, thus causing an overall variation of about 48° in its NPD.

It is therefore interesting to tabulate the values of north polar distance according to modern calculations at a number of sample dates for the three stars considered, as in Table 5.6.

[43] In *Shi ji* 27, 1343 he is listed as one of those who have *chuan tian shu* 傳天數 'passed on the celestial reckonings' (or we might say 'the heavenly numbers'), and is said to have been from the state of Wei 魏.

[44] For these entries, see *Kai yuan zhan jing* 開元占經 *(Divinatory manual of the Kai yuan reign period).* (c. 725 CE). Qutan Xida 瞿曇悉達, Taipei, Si ku quan shu MS edn. c. 1782, photorepr. Commercial Press 1983–1986, 60, 4a–6a.

[45] For the significance of these 'ancient' lodge widths, which are given in smaller characters in the text, see footnote 279.

Table 5.6 ***North polar distances of stars at different epochs***

Reference star used (Sun & Kistemaker, Table 3.1 column V)	North polar distance given in *Kai yuan zhan jing* / *du*	North polar distance in 200 BCE/*du*	North polar distance in 100 BCE/*du*	North polar distance in 200 CE/*du*	North polar distance in 700 CE/*du*
α Librae	94.0	96.6	97.2	98.8	101.5
π Scorpii	108.0	108.9	109.4	110.8	113.1
σ Scorpii	108.5	109.2	109.7	111.0	113.1

In 700 CE, around the time that the *Kai yuan zhan jing* was compiled, the values given in the text were markedly different from what observation at the time would have shown. But as the table shows, the further back in time we go, the better the match—although the actual values of north polar distance remain larger than those given in the *Kai yuan zhan jing*, which suggests a systematic error of some kind, leading to north polar distances being underestimated. A sample of three stars proves very little—although it certainly does suggest that the figures in the *Kai yuan zhan jing* are more likely to have come from early imperial China than from the eighth century, when that work was compiled. A mathematically sophisticated analysis of the entire data set sampled here was given by Sun Xiaochun and Jacob Kistemaker (1997: 53–67), together with a summary and evaluation of previous work on this problem. Their conclusion is a striking one: the north polar distances set out in the *Kai yuan zhan jing* fit an epoch of observation of (78 ± 20) BCE, but with a systematic underestimate of north polar distance at that date of 0.86°, about 0.9 *du*.

Now simple gnomon sightings cannot produce estimates of the north polar distances of stars unless basic trigonometry is available to process the observational data by turning length measurements into angle. But there is no sign at all of such techniques in China until the eighth century CE.[46] The only alternative is to conclude that somebody in the first half of the first century BCE had access to some kind of simple angle measuring device. This might, at its simplest, have consisted of a graduated disc such as the one shown in Figure 5.8, held in a vertical north-south plane (the observer's meridian plane) and provided with sights of some kind—which, in the Chinese tradition, would have consisted of a

[46] Christopher Cullen (1982b) 'An eighth century Chinese table of tangents.' *Chinese Science* 5: 1–33.

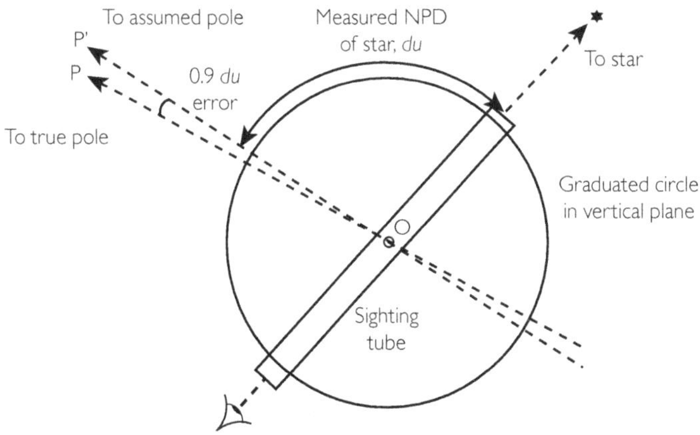

Figure 5:13 Finding NPD of star using graduated circle in a meridian plane.

sighting tube rather than the 'gun-sights' favoured in Europe. An arrangement like the one shown in Figure 5.13 would have been enough.[47]

The sighting tube is pivoted at O. OP is the actual polar axis, but if the circle is set up so as to produce the kind of observations found in the *Kai yuan zhan jing*, then there is an error of about 0.9 *du* in the setting of the disc, so that the assumed axis is OP′ and NPDs will be systematically underestimated. In this conjectural reconstruction, the graduations are shown as marked on a disc; the resulting instrument would thus resemble the device thought to have been the *dioptra* of Geminus c. 70 BCE (see footnote 14). Had a graduated ring been used instead of a disc, the device might have been similar to a simple armillary instrument of the type used by Ptolemy in about 135 CE. Ptolemy describes two versions of such a device in the *Almagest*, and claims to have used it to find the altitude of the sun at the solstices.[48]

Some significant methodological changes seem to be taking place here, even if the historical record does not yet permit us to see them as they are happening. The only thing we can be fairly sure of, if Sun and Kistemaker (and their

[47] During the Western Han, there was no bright star near the north celestial pole: the nearest was the second magnitude star β Ursae Minoris, which was about 8° (sixteen moon-widths) from the pole. The polar position could not therefore have been fixed by a single observation. It would have been possible to locate the pole by taking the point between the extreme eastern and western positions of the pole star—as in fact proposed in the *Zhou bi* (Cullen (1996), 191). It is not surprising to find an error in location of the pole of the order of magnitude found by Sun and Kistemaker.

[48] *Almagest* 1, 12, in Toomer (1998),61–3. It seems possible that Ptolemy did not in fact derive the data he used from actual observation with such an instrument. See Alexander Jones (2002) 'Eratosthenes, Hipparchus, and the obliquity of the ecliptic' *Journal for the History of Astronomy* 33 (1): 15–19 and note 336.

predecessors) are correct, is that something that we would today think of as an angle measuring device of some kind was being applied to the heavens in the Western Han. But what might the use of such a device say about the heavens themselves?

We have already seen that a particular physical picture of the universe, the *gai tian* view of a flat and disc-like heaven rotating above the flat earth (see Figure 5.10), is closely connected with observational procedures based on the use of clepsydras and gnomons—the latter used for shadow measurements as well as for observing meridian transits of celestial bodies. But what kind of simple view of heaven would be inspired by the measuring circle shown in Figure 5.13? Surely it would be a shape that matched the graduated disc used to map heaven into *du*, as in Fig. 5.14?

If we imagine that the observer and the graduated disc are concentric with a much larger heaven of circular vertical cross-section, then there will always be a simple correspondence between the measurements between positions on heaven and between positions on the disc. Thus in the diagram, the quantities marked NPD and npd measured round the circumferences of the heavens and the measuring disc are in simple proportion. Extending the circular image of heaven to planes other than the vertical, it is evident that the kind of heaven that fits in most easily with measurements with a graduated disc will be a heaven that is a sphere, with the observer at its centre. And this is precisely the idea that we shall find explicitly stated in sources of the Eastern Han period,

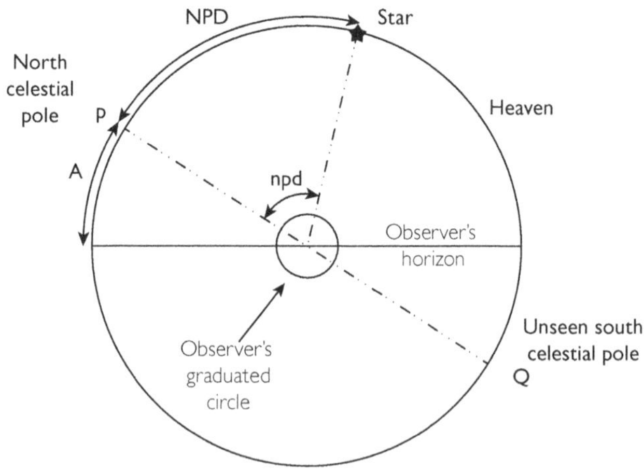

Figure 5.14 Vertical section through heaven and graduated disc.

where this view is given the name *hun tian* 渾天 'spherical heaven' in contrast with the *gai tian* 蓋天 'umbrella heaven' view described earlier. The first full description of the *hun tian* view was given by Zhang Heng 張衡 (78–139 CE), and is discussed in section 6.5.1. It is interesting to note that Zhang Heng seems to have set his polar axis too high by an amount similar to the error found by Sun and Kistemaker.

5.8 Huan Tan criticizes the *gai tian*, and Yang Xiong adopts the *hun tian*

One of Huan Tan's youthful acquaintances was the poet and thinker Yang Xiong 楊雄 (53 BCE–18 CE), another associate of Liu Xin who managed to survive in office some way into Wang Mang's reign. In another fragment of Huan Tan's writing, we hear of an interesting conversation with his older friend on the subject of the shape of the heavens. First, we hear of Yang Xiong's own interest in this topic:

> 通人揚子雲, 因衆儒之說天, 以爲蓋, 常左旋, 日月星辰, 隨而東西. 乃圖畫形體行度, 参以四時曆數昏明晝夜, 欲爲世人立紀律, 以垂法後嗣.
>
> Yang Ziyun [= Yang Xiong], that perceptive man, based himself on the discussions of heaven by many *Ru* scholars and took it to be [like] an umbrella (*gai*). It always turns to the left [i.e. as seen from outside the heaven disc looking towards the pole], carrying the sun, moon and stellar mark-points with it so they move from east to west. So he drew a diagram of its shape and the *du* of its movement, checking it against the calendrical data for the four seasons, for dusk and dawn and for day and night, intending that it should serve as a rule for his contemporaries, and be left as a pattern for posterity. (*Tai ping yu lan* 2, 6b–7a, in (Li Fang 李昉 et al. 983 CE, 1960 reprint of Song edition))

In Yang Xiong, we evidently have the earliest example of a named historical person who is said to have held the *gai tian* 蓋天 view of the heavens derived from gnomon and clepsydra observations, as set out earlier in this chapter. His 'diagram' might have resembled either of the objects shown in Figure 5.7 and Figure 5.8. But Huan Tan had two points to make on that. His first point is as follows:

> 余難之曰:「春秋分晝夜欲等平, 旦日出於卯, 正東方; 暮日入於酉, 正西方. 今以天下之占視之; 此乃人之卯酉, 非天卯酉. 當北斗極, 北斗極天樞, [. . .] 彼晝夜刻漏之數, 何從等平？」子雲無以解也.

> I objected: at the spring and autumn equinoxes, day and night must be equal. Now at dawn the sun rises in the direction *mao*, exactly due east, and at dusk it sets in the direction *you*, exactly due west. But we are looking at things from a position in the [Han] empire: this east-west line [through] people [in that position] is not the east-west line of heaven. That passes through the pivot of the northern Dipper, which is the pivot of heaven. [. . .] So how could the numbers of *ke* for the day and night clepsydras be equal? Yang Xiong had no explanation. (Same reference)

Figure 5.15 shows the situation that Huan Tan appears to have had in mind.P is the celestial pole at the centre of the disk of heaven, and the daily circuit of the sun at the season in question is the circle NESW. The movement of the sun ESW takes up half of a day-night cycle. However, if an observer at O sees the sun rising due east (at E′) and setting due west (at W′)—as is the experience of any observer at the equinoxes—then clearly the daytime portion of the sun's path is E′SW′, which is significantly less that the night-time portion W′NE′. So, contrary to observation, if the sun rises and sets east-west at the equinoxes, day and

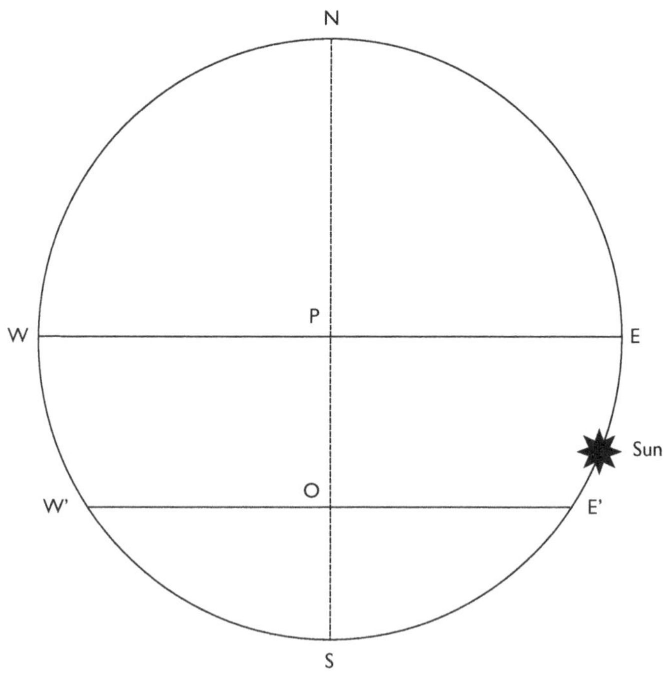

Figure 5.15 Huan Tan's discussion of risings and settings according to the gai tian (plan view).

night cannot be equal. Huan Tan goes on to tell us about a further objection, this time based on the experience of the two men when waiting to make a report on the veranda of a hall of audience in the palace. Such an audience would normally have taken place not long after dawn, and Huan Tan notes that at first the sun was low in the sky and shone on their backs, which must have been a comfort during what might have been a long cold wait for the summons to enter the hall. But as time passed, the sun no longer warmed them—because it had risen too high in the sky to shine under the eaves of the overhanging roof of the hall. If the sun was just moving horizontally as the disc of heaven rotated, why should that happen? Was not this observation in better accord with the *hun tian* 渾天 notion that the heavens were a sphere—so that the sun's markedly increasing altitude would be a simple consequence of it rising up over the horizon with the turning heavenly sphere? Yang Xiong, we are told, then agreed that his views must be mistaken.[49]

Yang Xiong's views on the heavens (presumably after his discussions with Huan Tan) are stated in a passage of his *Fa yan* 法言 'Exemplary words', a work modelled on the *Analects* of Confucius in which Yang Xiong comments on various issues through responses to an un-named questioner.

> 或問「渾天」. 曰: 「落下閎營之, 鮮於妄人度之, 耿中丞象之, 幾乎！幾乎！莫之能違也.」請問『蓋天』.」曰: 「蓋哉！蓋哉！應難未幾也.」

> Someone asked about the *hun tian*. [Yang Xiong] said: 'Luoxia Hong devised it, Xianyu Wangren calculated it, and the Palace Assistant Geng made an image of it. So subtle, so subtle! No one can contradict it! [The questioner] asked if he might enquire about the *gai tian*. [Yang Xiong] said: 'The *gai*! The *gai*! In answering objections it is not so subtle. (Fa yan, 10.3; translation modified from Bullock, Jeffrey S., 2011: 141.)

We have already met Luoxia Hong as a collaborator in the Grand Inception reforms of 104 BCE; Xianyu Wangren took part in the observation programme launched in 78 BCE to settle the controversy associated with the views of Zhang Shouwang 張壽王 (for both men, see 3.4). The 'Geng' who held the rank of 'palace assistant' was almost certainly Geng Shouchang 耿壽昌, palace assistant

[49] In other sources, we hear of as many as eight objections to the *gai tian* attributed to Yang Xiong, though no textual source said to be by him is cited: see *Song shu* 23, 679, which simply refers to his having made eight objections, while *Sui shu* 19, 506 gives details of each of the eight. The second objection in that list amounts to Huan Tan's point about day-length at the equinoxes. While it is always possible that Yang Xiong really is the source of these objections, it is at least equally likely that they are by a later hand.

to the Director of Agriculture, who is said to have measured the motions of the sun and moon with a tu yi 圖儀 'diagram/plotting instrument' in 52 BCE (*Hou Han shu, zhi* 2, 3029)—see 6.3.2. We cannot say whether Yang Xiong's views of the origin of the *hun tian* (whether here referring to a measuring instrument or a view of the shape of heaven, or both) are correct; no other early source refers to these men in that connection. But at least this was the opinion of the first person to discuss the topic, and, what is more, of a person who was a well-informed scholar at the end of the century in which the events referred to had occurred.[50] In addition, the suggestion that the *hun tian* originated in the first century BCE is fully consistent with Sun and Kistemaker's conclusion that direct measurements of north polar distance were made during that period.

[50] In another fragment of the *Xin lun*, which exists in a number of different versions, Yang Xiong is said to have met an old palace workman who had made a *hun tian*. Other versions say he actually met Luoxia Hong, which is simply impossible in dating terms. See Pokora (1975), 114–15.

CHAPTER 6

Restoration and re-creation in the Eastern Han

This chapter moves onto different ground from any of the preceding ones. We have already seen a short example of argument related to the heavens in Huan Tan's efforts to persuade Yang Xiong to renounce the gai tian view of the shape of the cosmos. But now for the first time, we find ourselves in a world where the written sources contain extensive persuasive writing on astronomical topics. Effectively, we can begin to observe people arguing with one another. While we must remember that the people who wrote or edited our sources cannot be assumed to be neutral recorders of what happened and what was said, the picture that can be constructed is much richer and more complex than what went before.

We look first at the situation in the early years of the restored Han dynasty. Despite the strong associations of Liu Xin's Triple Concordance system with Wang Mang's seizure of power, it continued in use for more than half a century. Then, around 85 CE, change finally came, and Liu Xin's system was replaced. We have records of the practical and theoretical grounds on which the old system was rejected, and of the creation and implementation of a new system, constructed by Bian Xin and Li Fang. Not long after this, we are able to follow the story of how in c. 92 CE Jia Kui successfully advocated a fundamental innovation in both theory and practice: he insisted on the ecliptic—the path of the sun against the stellar background—as central to astronomical observation and calculation.

Thus it was essential to add an ecliptic ring to the armillary sphere used by the Grand Clerk's officials to provide data for the construction and testing of astronomical systems.[1] The richness of records from this period makes it easy to tell a detailed story of technical innovation in its fullest context, leading up to the work of Zhang Heng (78–139 CE), for whom astronomical calculation was just one of several fields in which he gained a reputation for exceptional originality. This chapter will enable the reader to appreciate the full vigour and creativity of the early Chinese tradition of learning and practice about the motions of the heavens.

6.1 Adopting a new astronomical system: from the Triple Concordance to the Han Quarter Remainder

We shall now begin to use a resource of unparalleled value for the historian of astronomy in early imperial China. This is a section of the *Hou Han shu* that we know to be in major part the work of two great experts in astronomical systems, Cai Yong 蔡邕 (132–192 CE) and Liu Hong 劉洪 (*c.* 130–*c.* 210 CE). Sima Biao 司馬彪 (c. 240–c. 306 CE), who was responsible for the monographs that now appear in the *Hou Han shu*,[2] tells us

> 光和元年中, 議郎蔡邕, 郎中劉洪補續律曆志, [...] 今考論其業, 義指博通, 術數略舉, 是以集錄為上下篇, 放續前志, 以備一家.
>
> In the first year of the Guanghe period [AD 178], the Gentleman for Consultation Cai Yong, and the Palace Gentleman Liu Hong expanded and extended the [*Han shu*] Monograph on Harmonics and [Astronomical] Systems [...] Now I have examined and evaluated their work: its ideas are broad and general, and [the details of] numerical procedures are succinct and complete. Therefore I have collected their material into a first and second chapter, and have let this serve as a continuation of the former Monograph [in the *Han shu*], in order fully to represent their point of view. (*Hou Han shu, zhi* 3, 3082; Cullen 2017, 234)

The 'first and second chapter' referred to here are now the second and third parts of the *Lü li zhi* 律曆志 'Monograph on mathematical harmonics and astronomical systems' in the current *Hou Han shu*. Part 1 of the *Lü li zhi* deals

[1] In chapter 5, section 5.7, we discussed the evidence for simple armillary instruments in the late Western Han.

[2] On the general issue of the monographs, see Mansvelt Beck (1990).

with harmonics, and will not be discussed in this book. Part 2 contains a series of official documents – memorials, edicts and so on, linked by detailed narrative sections that are also clearly based on official sources. Part 3 is a description of the Han Quarter Remainder astronomical system, prefaced by an essay on the observational basis of such systems. While engaged in their work of compilation, Cai and Liu were allowed to consult the archives of the Dong guan 東觀 'Eastern Pavilion', in which confidential state papers were kept.[3] The result is a rich repertoire of documents that gives us a much deeper insight into the events of the Eastern Han that concern us than we could ever hope to have for the Western Han. In this chapter and the next we shall not have space to do more than present the essential features of some of the events that they describe in such detail.[4]

But let us begin at the beginning. Right at the start of the first of the two chapters by Cai Yong and Liu Hong, it becomes clear that early on in the Eastern Han there was a consciousness that something was noticeably wrong with the astronomical system in use—the Triple Concordance, created by Liu Xin on the basis of the Grand Inception system of 104 BCE.

> 自太初元年始用三統曆, 施行百有餘年, 曆稍後天, 朔先於曆, 朔或在晦, 月或朔見. 考其行, 日有退無進, 月有進無退.
> 建武八年中, 太僕朱浮, 太中大夫許淑等數上書, 言曆朔不正, 宜當改更. 時分度覺差尚微, 上以天下初定, 未遑考正.

> The Triple Concordance system began to be used from the first year of the Grand Inception period [104 BCE]. It had been in use for more than a century, and the system lagged slightly behind Heaven, [so that] the conjunction fell in advance of the [predictions of] the system, [with the] conjunction sometimes falling on the [predicted day of] the end of the lunation, and the moon sometimes being visible on the [predicted day of] conjunction. Examining its movements, the sun was retarded [from its predicted position] and never advanced [beyond it], while the moon was advanced [from its predicted position] and was never retarded.

> In the eighth year of the Jianwu period [32 CE] the Chamberlain for the Imperial Stud Zhu Fu, and the Superior Grand Master of the Palace Xu Shu, with others, memorialized that the conjunctions [given by] the system were incorrect, and that it would be appropriate to revise it. [But] at that time the observable discrepancies

[3] *Hou Han shu, zhi* 3, 3083, commentary, quoting a memorial of Cai Yong. This was a privilege that had been sought 40 years earlier by Zhang Heng 張衡, who twice held the office of Grand Clerk and was a major literary figure as well as a constructor of instruments and expert on celestial calculation. He was never allowed the access he sought: *Hou Han shu* 59, 1940. On Zhang Heng's work, see section 6.5.

[4] A full translation of this material is given in Cullen (2017), chapter 5. The reader who wants to follow the story past the end of Han may do so in chapter 4 of Morgan (2013).

> of the fractions and the degrees were still quite subtle; since the empire had only just been settled, the emperor felt he did not yet have the leisure to examine and correctly determine [such matters]. (*Hou Han shu, zhi* 2, 3025; Cullen 2017, 375)

Thirty years later, the topic was renewed. The striking feature of the account we have, apart from its high degree of detail, is the obvious assumption that the only way to get things right is to observe carefully over an extended period, and compare observation with prediction:

> 至永平五年, 官曆署七月十六日月食. 待詔楊岑見時月食多先曆, 即縮用筭上為日, 因上言「月當十五日食, 官曆不中」. 詔書令岑普候, 與官曆課. 起七月, 盡十一月, 弦望凡五, 官曆皆失, 岑皆中. 庚寅, 詔書令岑署弦望月食官, 復令待詔張盛, 景防, 鮑鄴等以四分法與岑課. 歲餘, 盛等所中多岑六事. 十二年十一月丙子, 詔書令盛, 防代岑署弦望月食加時. 四分之術, 始頗施行. 是時盛, 防等未能分明曆元, 綜校分度, 故但用其弦望而已.
>
> When it came to the fifth year of the Yongping period [62 CE], the official system [i.e. the Triple Concordance] set out that the moon would be eclipsed on the 16th day of the seventh month [8 September 62 CE]. [But] the Expectant Official Yang Cen saw that at that time lunar eclipses where generally in advance of the [predictions of] the system, so he [simply] counted one back to find the day, and hence sent up a memorial saying "The moon should be eclipsed on the 15th day [7 September]; the official system is off target."[5] An edict ordered Cen to observe carefully, and to check with the official system. Starting from the seventh month, and going up to the 11th month, there were in all five [cycles of] crescents and full moons, all of which were missed by the official system, and all of which were hit by Cen.[6] On a *gengyin*.27 day[7] an edict ordered Chen to set out crescents, full moons and lunar eclipses [omit 官], and further ordered the Expectant Officials Zhang Cheng, Jing Fang, Bao Ye and others to check the quarter remainder methods with [those of] Cen. After more than a year, the instances where Cheng and the others were on target exceeded Cen's by six. In

[5] My spreadsheet version of the Triple Concordance predicts a lunar eclipse at the full moon of this month which is day *dingwei*.44. The month begins on day *renchen*.29, so that the prediction is indeed for the 16th day, commencing at Chang'an at JD 1,743,953.19750, 8 September. However, according to Starry Night Pro, the eclipse actually began with penumbral contact at about 18:00 local Chang'an time on 7 September, just before moonrise, reaching maximum umbra at about 20:45, JD 1,743,953.06278. So the claim made here was correct.

[6] The predictions and actual phases are as follows: eighth month: Conjunction: *renxu*.59, 23 September, 06:31 (actual conjunction at 21 September, 21:17); ninth month: Conjunction: *xinmao*.28, 22 October, 19:16 (actual conjunction at 21 October, 09:34); tenth month: Conjunction *xinyou*.58, 21 November, 08:00 (actual conjunction at 20 November, 02.33); 11th month: Conjunction *gengyin*.27, 20 December, 20:44 (actual conjunction at 19 December, 22:58).

[7] No month is mentioned, but presumably this was the first such day after the end of the 11th month, which was the second day of the first month of the next year.

the 12th year, the 11th month, day *bingzi*.13 [27th day of month; 29 December 69 CE] an edict ordered Cheng, [Jing] Fang and the others to replace Cen in setting out the instants of occurrence of crescents, full moons and lunar eclipses. [Thus] quarter remainder methods first began to be used to some extent. At that time Cheng, Fang and the others were not yet able to distinguish clearly what the system origin should be, or to comprehensively compare the fractions and degrees, so only their crescent and full moon [predictions] were used. (*Hou Han shu, zhi* 2, 3025; Cullen 2017, 375–6)

Here we see for the first time that, at the same time that the Triple Concordance remained in official use, an alternative system was being developed, here called the *si fen* 四分 'Quarter remainder'. As we have seen in section 1.3, to say that a system is of the quarter remainder type only means that the length of the solar cycle is 365 ¼ days, but does not define it uniquely. To do that, we need a system origin—which is precisely what Zhang Cheng and his colleagues did not yet have. Perhaps they had a system origin for solar and lunar phenomena, which would have enabled them to make predictions of lunar phases, but did not yet have a 'High Origin' which could act as the starting point for planetary phenomena as well?

Further discussions took place in 66 CE, but led to no definitive result. We do not have any account of what passed in the next two decades, but eventually the discrepancy between prediction and observation became intolerable:

至元和二年, 太初失天益遠, 日, 月宿度相覺浸多, 而候者皆知冬至之日日在斗二十一度, 未至牽牛五度, 而以為牽牛中星, 後天四分日之三, 晦朔弦望差天一日, 宿差五度. 章帝知其謬錯, 以問史官, 雖知不合, 而不能易, 故召治曆編訢, 李梵等綜校其狀.

When it came to the second year of the Yuanhe period [85 CE], the Grand Inception was missing the Heavens by an increasing distance, and the [predicted] lodge degrees of the sun and moon were perceived as increasingly excessive, while the observers all realized that on the day of the winter solstice the sun was [still only at] at 21 *du* of [the lodge] Dipper, and failed to reach Ox by five *du*, having taken [the start of the lodge as marked by] the middle star of the Ox [asterism].[8] [Predicted lunar phases] lagged ¾ day behind heaven. So the [predicted] last days, conjunctions, crescents and full moons were a day different from heaven, and the lodges were five *du* different. Emperor Zhang became aware of these errors, and enquired of the Clerk's officials concerning them, but

[8] The bright middle star of the six composing this asterism, β Capricorni, actually marks the start of the lodge for purposes of measurement. See *Kai yuan zhan jing*, 61, 2b. In its annual eastwards motion relative to the stars, the sun has to pass through Dipper before reaching the start of Ox.

> even though they knew things did not fit, they could not change anything. So he ordered the experts in [astronomical] systems Bian Xin and Li Fang to make a comprehensive comparison of all the circumstances. (*Hou Han shu, zhi* 2, 3026; Cullen 2017, 377)

Some of the discrepancies referred to here would have been easily detectable. Take for instance the conjunction beginning the first civil month of the year that began in spring of 85 CE. The Triple Concordance predicted that the conjunction would be on 14 February, at 26/81 of a day after midnight, i.e. at about 08:00. In fact, the moon was in conjunction with the sun (i.e. at the same longitude) at about 07:00 on 13 February. The difference in this case was 25 hours, so that the first crescent moon would have become visible a day before the expected date.

As for the winter solstice of that year, the Triple Concordance predicted this for 25 December, with the sun being at the first *du* of Ox. The text states, however, that on that date the sun was in fact about five *du* to the west of the star marking the start of Ox, that is, it was at 21 *du* of the preceding lodge, Dipper, which was 26 *du* wide. Modern calculations agree with the five *du* figure to the nearest *du* for the given date. This discrepancy was the result of the slow westwards shift of the solsticial and equinoctial points relative to the stars over a cycle of about 26,000 years, a phenomenon that had not yet been systematically recognized in the Eastern Han.[9] In fact the position of the sun amongst the lodges at any moment could not have been observed directly. A five *du* discrepancy in the sun's position would, however, have produced a five-day shift in the sequence of dawn and dusk centred stars, which could well have been noted. An alternative way of finding the sun's position might have been to observe the transit of a given star at night and then find the time elapsed to the next noon. Detecting a five *du* shift would have required a clepsydra running to within an accuracy of only a few minutes over 12 hours.

Bian Xin and Li Fang either worked quickly or had already drafted most of what was needed, since on 18 March of the same year, the emperor was able to issue an edict ordering that a new system based on quarter-day constants should be put into use.[10] The content of the edict may be compared with that issued by Wudi in 104 BCE at the time of the Grand Inception reform (see chapter 3,

[9] Calculations also show that the true winter solstice actually fell close to 18:00 Luoyang time on 22 December, when the sun would have been about eight *du* to the west of the star referred to. The five *du* shift noted in the text would correspond to about 350 years, while the eight *du* shift would have taken more like 550 years. Such calculations are only meaningful if we can assume that the sun's real winter solstice position was accurately observed and recorded at both ends of the timescale – which in the current instance was clearly not the case.

[10] See *Hou Han shu, zhi* 2, 3026–7; Cullen 2017, 377–9.

section 3.3.1). Wudi's edict made reference to the emperor's responsibility to follow the example of ancient sage rulers in overseeing an accurate calendar, and referred to the cosmic significance of the use of 81 as the day factor divisor. Zhangdi certainly invoked the pattern set by the ancient sages, and also mentioned the problems that could occur if the calendar was incorrect—pointing out that if, for instance, the day of Establishment of Spring was wrongly predicted, then amnesties might be granted at a time when spring had not actually begun, and one would therefore be out of step with the cosmos. Thus, he noted, there had been a series of disasters, including cattle plagues and poor harvests.

But a wholly new element appears in this edict, and that is the reference to a number of the so-called 'apocrypha', an English term used to refer to texts that in Chinese went under names such as *chen* 讖 'prognostications' *wei* 緯 'weft[threads]' (as opposed to the *jing* 經, literally 'warp[-threads]' which were the received classics), or *tu* 圖 'diagrams/charts' or various combinations of such terms.[11] Typically, such texts were taken as embodying an esoteric tradition that Confucius and his successors had not seen fit to include in the received classics, but had handed down by other means. Modern scholarship holds that such texts were not in fact ancient, but originated around the beginning of the Common Era, probably in the period of political and ideological conflict from the end of Western Han, through the rule of Wang Mang, to the early Eastern Han.[12] Their titles are usually based on those of the received classics, and their contents may contain references to the calendar, heavenly bodies, or matters relevant to astronomical systems, as well as words of prophecy that were clearly meant as references to current events.[13] Thus at one point in his edict, Zhangdi says:

> 春秋保乾圖曰: 『三百年斗曆改憲. 』史官用太初鄧平術, 有餘分一, 在三百年之域, 行度轉差, 浸以謬錯.

[11] The use of 'apocrypha' (from a Greek word meaning 'hidden') to refer to these texts seems to have been prompted by its use by some Christian writers, particularly Protestants, to refer to certain books of the Bible that are known only in Greek translation, and which are thus thought by some to be of lower scriptural status than those also found in the Hebrew canon.

[12] See for instance Jack L. Dull (1966). 'A historical introduction to the apocryphal, *ch'an-wei*, texts of the Han Dynasty.' University of Washington, PhD, and Twitchett, Loewe and Fairbank (1986), 759. These texts are only known today through fragments quoted by various authors, the texts themselves having been destroyed under the Sui 隋 dynasty in the early seventh century CE. One of the most comprehensive collections of fragments is Yasui Kōzan 安居香山 and Nakamura Shōhachi 中村璋八 (1959). Isho shūsei 緯書集成 (Comprehensive collection of [fragments of] weft books; 8 vols.). Tokyo, Kangi Bunka Kenyūkai 漢魏文化研究会.

[13] On the connection of the apocryphal texts with astronomical systems, see for instance Yabuuti Kiyosi (1974) 'The calendar reforms in the Han dynasties and ideas in their backgrounds.' *Archives internationales d'histoire des sciences* 24: 51–65.

> The *Chun qiu bao qian tu* ('The Spring and Autumn annals: chart showing the preservation of Uranic power') says 'in 300 years, the Dipper sequence changes its order'. [If in] the Grand Inception system of Deng Ping used by the Clerk's officials there was to be a single fraction left over, then in the space of 300 years the degrees of motion would move into error, and [the whole system] would be imbued with falsity ... (*Hou Han shu, zhi* 2, 3027; Cullen 2017, 379)

Hence it was clear to the emperor that the time for change had come.

The emperor might quote such texts, but not all scholars were convinced of their value. Huan Tan had argued against their use, and in his old age narrowly escaped execution when he angered Guangwudi by denying the relevance of the apocrypha to the question of where the *ling tai* 靈臺 'numinous terrace' (i.e. the observatory) should be placed in relation to the new capital at Luoyang.[14] We shall see in this chapter that those Eastern Han scholars who were experts in astronomical systems were usually prudent enough not to declare the apocrypha to be irrelevant to the judgments they had to make; indeed, they might cite them when occasion arose. On the other hand they were well able to give decisive weight to factors that did not depend on these texts. This becomes evident when we look at the description of the Han Quarter Remainder system set out in the second of the chapters compiled by Cai Yong and Liu Hong. To this we now turn.

6.2 The Han Quarter Remainder system and its observational basis

The basic constants underlying the new system are ones we have already seen when we reviewed the astronomical practice of the period before the Grand Inception reform of 104 BCE: see chapter 2.5.1. The computational structures of the system are similar to those we have already seen in Liu Xin's Triple Concordance system—the removal of complete cycles to reduce the size of numbers encountered during computation, and the provision of auxiliary quantities derived from the basic data in order to speed up the process. There is a change in the order in which the different sections of the system are arranged: whereas in the Triple Concordance all constants are specified before computation procedures,

[14] *Hou Han shu* 28a, 961. The remains of the mound where the *ling tai* stood are still to be seen, south of the walls of the ancient city, at latitude 34° 42′ N, longitude 112° 37.8′ E. See Figure 1.1. This shows a rectangular mound about 41 m north-south and 31 m east-west, with a height of about 8.5m, with two levels of terracing. There appear to have been built structures on all four sides of both levels and an open level area on top. For a detailed description, see Anon. (1978).

in the Han system constants relating to planets are not given until after computation procedures for sun and moon have been specified. The point of departure for planetary computations has also been revised, being taken as the moment of conjunction with the sun rather than the supposed date of first dawn visibility.

6.2.1 Choosing a new system origin

From the Han point of view, perhaps the major technical change introduced is a new system origin, or rather a pair of origins, one for basic calendrical phenomena, and one for planets and lunar eclipses.

> 當漢高皇帝受命四十有五歲, 陽在上章, 陰在執徐, 冬十有一月甲子夜半朔旦冬至, 日月閏積之數皆自此始, 立元正朔, 謂之漢曆. 又上兩元, 而月食五星之元, 並發端焉.
>
> In the 45th year after the High Sovereign Emperor of Han received the mandate, when the Yang was at *shangzhang* [the seventh of its ten positions, corresponding to geng 庚], and the Yin was at *zhixu* [its fifth position out of 12, corresponding to chen 辰, so the sexagenary name of the year was *gengchen*.17], in winter, the 11th month, day *jiazi*.1, at midnight, the moment of conjunction [occurred at the same time as] the winter solstice. The reckonings of the sun, moon and accumulation of intercalations starts from this point. It establishes the origin and standardizes the conjunction. This is called the Han System. Going back a further two Origin [periods of 4560 years],[15] then that is the [system] origin [used for predicting] lunar eclipses and the [movements of the] five planets, which all start from this point. (*Hou Han shu, zhi* 3, 3057; Cullen 2017, 150)

The instant of system origin first defined here is 25 December 162 BCE, at midnight Luoyang local time, Julian Day Number 1,662,610.18750. The first civil year of the system (according to the practice followed in 85 CE, which was that introduced in 104 BCE) thus began in early 161 BCE, when 45 complete years of Han rule, beginning in 206 BCE, had indeed elapsed up to the start of that year. Two Origin periods take us back to a High Origin *shang yuan* 上元 at Luoyang midnight beginning 25 December 9282 BCE, Julian Day Number −1,668,469.81250.

As we have noted before, setting a High Origin date in this way did not amount to a claim that system origin conditions had actually been observed and recorded at that date—the purpose was simply to find a starting date for cycles

[15] As explained in chapter 1, note 49, this period ensures complete repetition of the system origin conditions of a quarter remainder system: coincidence of winter solstice and conjunction of the first Celestial month at midnight beginning a *jiazi*.1 day, in a year whose sexagenary year name is the same as the year of origin.

that produced predictions of contemporary phenomena that were close enough to observation to be satisfactory. While the 162 BCE origin was adequate for predicting solstices, equinoxes and mean luni-solar conjunctions, it was necessary to push further back in order to find an origin that could be used to predict lunar eclipses and planetary movements.

But why exactly was the new system origin placed when it was? The main reason for the choice appears to have been the need to deal with the fact that the Triple Concordance had been predicting lunar phases about ¾ day too late during the early Eastern Han. This problem was, however, simple to solve. Box 6.1 shows how this was done—by finding a system origin for the Han Quarter Remainder that shifted lunar phase predictions ¾ day earlier, while also giving the new system origin the usual property of falling at midnight beginning a *jiazi*.1 day. This was possible because there was in fact a year, 162 BCE, when the Triple Concordance predicted that winter solstice and conjunction would fall on a *jiazi*.1 day (25 December) about ¾ day after midnight – so all that was needed to correct the timing was to shift the prediction for solstice and conjunction ¾ day earlier to midnight starting that day, thus producing the change required and defining a system origin with the usual properties. This, then, is when the new origin was placed (see Box 6.1).

Box 6.1: The choice of the Han Quarter Remainder system origin, and its consequences

The Han Quarter Remainder system origin

The Han Quarter Remainder system used a system origin that predicted that winter solstice and luni-solar conjunction coincided at midnight commencing the first Celestial month (11th Xia month), at the start of a *jiazi*.1 day which in the Julian calendar would have been 25 December 162 BCE. Let us compare this to the predictions of the Triple Concordance, whose system origin for solar and lunar cycles was the coincidence of solstice and conjunction at midnight commencing the *jiazi*.1 day 25 December 105 BCE.

Now

$$162 - 105 = 57 = 3 \times 19.$$

Thus, since the winter solstice of late 162 BCE was a whole number of 19-year Rule Cycles away from the Triple Concordance system origin in late 105 BCE, the Triple Concordance predicted that solstice and conjunction should coincide in 162 BCE, as they always do at the start of a Rule. Like the Han Quarter

continued

Box 6.1: *Continued*

Remainder, the Triple Concordance predicted that solstice and conjunction would both fall on the *jiazi* day 25 December 162 BCE, with the solstice $\frac{3,420}{4,617}$ of a day after midnight, and the conjunction beginning the first celestial month at $\frac{60}{81}$ day after midnight on that day.

Now, since $\frac{60}{81}$ day = (57 × 60)/(57 × 81) day = $\frac{3,420}{4,617}$ day = 17 hours 47 minutes, we see that winter solstice and conjunction indeed coincided according to the Triple Concordance at about 18:00, 6 p.m. The Han Quarter Remainder, in effect, simply shifted the prediction 17 hours 47 minutes earlier to the preceding midnight – almost exactly the ¾ day shift needed to repair the situation seen in the early Eastern Han.

Actual conditions at the system origin, and their effect on predictions of phases c. 85 CE

According to modern calculations, the actual conjunction of late December 162 BCE fell on 26 December, 01:11 local Luoyang time, JD 1,662,611.23694. This was, however, the true conjunction, not the mean conjunction, which is the position predicted by equally spaced conjunctions best approximating on average to recent true conjunctions. Making an estimate of the position of the mean conjunction based on the conjunctions of the 6 preceding and 6 following lunations, we arrive at 25 December, 15:00, JD 1,662,610.833. So in fact the Triple Concordance made a prediction of the mean conjunction that was only about 3 hours too late, while the Han Quarter Remainder system made a prediction of the mean conjunction that was 15 hours too early! How could that result in better predictions by the Han Quarter Remainder system around 85 CE? The answer lies in the value used for the mean length of a lunation (the 'mean synodic month' in modern terms). The value used by systems of the Quarter Remainder type is, as we saw in chapter 1:

$$29\ \tfrac{499}{940} = 29.53085 \text{ days (7 sig. figs.)}$$

The modern mean synodic month value is 29.53059 days, which is 0.00026 days less. So by 85 CE, which is around 3,000 lunations later, the total error will amount to 3,000 × 0.00026 day = 0.78 day, and as a result we find that the conjunction in December 84 CE is predicted by the new system as being on 16 December at $\frac{450}{940}$ through the day, i.e. about 11:30. The mean conjunction was at about 08:00 on that day, with the true conjunction at about 10:50. The combination of the shift of the Han Quarter Remainder system origin to an earlier moment in 162 BCE and the error in the lunation length have combined to produce a quite good prediction in the first century CE.

It may well have seemed that simple to Bian Xin and Li Fang, who would have been responsible for choosing the new system origin. However, as Box 6.1 also shows, the astronomical conditions at the time of their chosen system origin actually fitted the Triple Concordance considerably better than the new system, since the Triple Concordance prediction of the conjunction was about 3 hours too early, whereas the prediction of the new system was 15 hours too early. Good conjunction predictions for the first century CE resulted from the fact that the new system used a lunation length that was slightly too long, so that after about two centuries the error in system origin was cancelled out. We cannot tell whether Bian Xin and Li Fang were aware of this complication. But the record shows that one of their contemporaries, Jia Kui, was very well aware of it. To his work we shall soon turn our attention (in section 6.3).

6.2.2 The origin of the Han Quarter Remainder system's constants

In chapter 5, we saw how Liu Xin exerted his skill in the use of classical texts and his abilities as a mathematician to show how all the basic constants of his system could be derived from what in his day were seen as the fundamental numerical structures of the cosmos. In no case does he appeal explicitly to observational evidence, even if in the case of the planetary constants it is clear that he must have made use of such evidence.

The Han Quarter Remainder system justifies itself on a completely different basis. Before listing constants and specifying calculation procedures, the monograph gives us an essay informing us, amongst other things, exactly how each constant functions, and how it relates to other parts of the structure of the system. For instance, we are told:

> 歲首至也，月首朔也. 至朔同日謂之章，同在日首謂之蔀，蔀終六旬謂之紀，歲朔又復謂之元.
>
> The start [literally 'head'] of the *sui* [solar cycle] is the [winter] solstice, and the start of the *yue* [lunation] is the conjunction. When the solstice and the conjunction fall on the same day, that is called a Rule [Head]. When this takes place at the start of a day [i.e. midnight], this is called an Obscuration [Head]. When an Obscuration has terminated the six decades [of the sexagenary cycle], that is called an Era [Head]. When the sexagenary numbers of the solar cycle and the conjunction both return, that is called an Origin [Head]. (*Hou Han shu, zhi* 3, 3056; Cullen 2017, 144)

But how do we know the lengths of these cycles? The opening passage of the essay sets the tone:

> 昔者聖人之作曆也，觀琁璣之運，三光之行，道之發斂，景之長短，斗綱(之)[所]建，青龍所躔，參伍以變，錯綜其數，而制術焉.
>
> When in former times the sages created astronomical systems, they watched the turning of the *xuan ji* [in this context the term refers to the Pole Star of that day], the motions of the Three Luminaries, the expansion and contraction of the Roads [i.e. the equator and ecliptic], the growing and shrinking of the shadow, the establishment of retaining cord the Dipper, and the orbit of the Caerulean Dragon [i.e. the Jupiter cycle]. The Three [Luminaries, i.e. the sun, moon and stars] and the Five [Planets] changed [their positions], and they wove together the numbers, so as to make them into methods [for calculation]. (*Hou Han shu, zhi* 3, 3055; Cullen 2017, 142)

The key to establishing methods of predictive calculation is, we are told, observation. This is stressed more than once in the rest of the text:

> 孔壺為漏，浮箭為刻，下漏數刻，以考中星，昏明生焉.
>
> The perforated vessel performs its dripping, and the floating arrow makes its divisions *ke*. One lets the drips go down and counts the divisions, in order to examine the centred stars, and [the times for] dusk and dawn are thereby produced. (*Hou Han shu, zhi* 3, 3056; Cullen 2017, 145)

> 曆數之生也, 乃立儀, 表, 以校日景. 景長則日遠, 天度之端也. 日發其端, 周而為歲, 然其景不復, 四周千四百六十一日, 而景復初, 是則日行之終. 以周除日, 得三百六十五四分度之一, 為歲之日數. 日日行一度, 亦為天度.
>
> In the process of producing the numbers of the astronomical system, one sets up [armillary] instruments and gnomons, in order to compare the solar shadows. When the shadow is longest the sun is most distant; this is the starting point for the degrees of heaven. The sun sets out from that starting point, and makes a solar cycle when it has circuited. However, the [noon] shadow does not return: the shadow returns to its starting point after four circuits, 1461 days, and so this represents the termination of the sun's motion. Dividing days by circuits, one obtains 365 ¼, which is the number of days in a solar cycle.[16] The sun moves one *du* in a day, so [this figure] is also the *du* in [a circuit of] the heavens. (*Hou Han shu, zhi* 3, 3056; Cullen 2017, 151)

[16] Pliny says something similar about the length of the year and the return of the shadow: '…the sun's course is divided into 360 parts, but in order that an observation of the shadows that it casts may come round to the starting-point, five and a quarter days *per annum* are added … in order to make our chronology tally with the course of the sun.' *Natural history*, II.vi.35, pages 190–191.

This emphasis on observation as a basis for deriving constants shows that a considerable change of epistemological emphasis has taken place since the cosmological ambitions of Liu Xin.

6.3 The work of Jia Kui

Jia Kui 賈逵 (30–101 CE) was a major figure in the intellectual life of his day, and also had the good fortune of enjoying more or less unbroken imperial favour. He is recorded as having written a number of works on the classics in the 'Old text' tradition of learning, but none of them has survived, apart from a few fragments. He is said to have taken part in the great debates held in the White Tiger Hall *Bai hu guan* 白虎觀 in 79 CE, though no record of his specific contributions has been preserved.[17] He was given the rank of Court Gentleman by Mingdi 明帝 some time after 58 CE, and worked with the historian Ban Gu in editing materials in the imperial library. In 91 CE he was given the rank of *Zuo zhong liang zhiang* 左中郎將 'Leader of the Gentlemen of the Household of the Left', which gave him responsibility for a corps of talented cadets designated as *dai zhao* 待詔 'Awaiting an edict [of summons to office]'.[18]

The first mentions of Jia Kui relevant to our story are found in the account by Cai Yong and Liu Hong of the establishment of the new astronomical system in 85 CE. Bian Xin and Li Fang had originally specified that the first month after the start of an Origin cycle should be a long (30-day) month, whereas it would have normally been a short (29-day) month. Jia Kui was given charge of a working group of ten scholars to look into the matter, and his conclusion that the first month should, as usual, be short was accepted.[19] In 92 CE, Jia Kui submitted a report on the question of changes in the astronomical system, of which Cai and Liu say:

> 逵論集狀, 後之議者, 用得折衷, 故詳錄焉.
>
> [Jia] Kui's discussion collects all the documentation, and those who discuss this matter in later times may use this to weigh opinions. Therefore we have cited it in detail. (*Hou Han shu, zhi* 2, 3027; Cullen 2017, 381)

Jia Kui begins by noting that the decision to change the winter solstice position of the sun in the new system from start of the lodge Ox to 21 ¼ *du* of the lodge

[17] *Hou Han shu* 79b, 2582; on these debates, see Twitchett, Loewe and Fairbank (1986), 763–764.

[18] See his biography in *Hou Han shu* 36, 1234–1241. According to the commentary at *Hou Han shu* 11, 340, *dai zhao* refers to 諸以材技徵召, 未有正官 'all those summoned to serve by reason of their talent, but who do not yet have a regular official post'.

[19] *Hou Han shu, zhi* 2, 3027.

Dipper was the result of careful observation over the long term. He was clearly able to review the observation notes at first hand, since he speaks of himself as *an xing shi shi guan zhu* 按行事史官注 'referring to the notes of the [Grand] Clerk's officials in charge of this matter'. He tells us that a special programme of observation of the solsticial position was mounted by the Grand Clerk from 85 to 89 CE, which confirmed the newly adopted datum. This position, he notes, is in accord with the *Xing jing* 星經 'Star Canon' ascribed to Shi Shen 石申 of the Warring States period,[20] and also with the *Shang shu kao ling yao* 尚書考靈曜 'The Book of Documents: examining the numinous luminaries', one of the 'apocryphal texts' in favour at court during this period, but now only existing in fragments. The change in winter solstice position is, in modern terms, the result of precession, though this phenomenon was not systematically understood at the time. The fact that the works referred to used 21 ¼ *du* of Dipper rather than the old position at the start of Ox is of course further confirmation that they were not in fact ancient texts, as they were widely thought to be in Jia Kui's day.

Jia Kui then moves on to a matter directly relevant to the choice of a new system origin for the Han Quarter Remainder system, as discussed in Box 6.1. When we looked at this question earlier in the present chapter, it was not clear whether Bian Xin and Li Fang realized that the conditions specified at the new system origin, the winter solstice of 162 BCE, were not representative of actual conditions at the time. Jia Kui, however, is quite explicit on this point: at the time of that system origin, he states, the Han Quarter Remainder system was actually further from the truth that the Triple Concordance, the system that it eventually replaced. Modern calculations confirm that this view was correct, as we saw in Box 6.1, but how did Jia Kui know when conjunctions of the early Western Han actually fell? While accurate calculation on the basis of modern celestial mechanics was certainly not then possible, Jia Kui had another resource, since the Grand Clerk's offices had detailed lists of solar eclipses covering the entire time from the beginning of the dynasty in 206 BCE. And in Jia Kui's time it was acknowledged that a solar eclipse could only happen at the time of solar-lunar conjunction—indeed, that it was the best example of a conjunction that one might see, since the moon passed directly in line with the sun, rather than simply aligning with it in longitude from some distance away. Based on the recorded data, he observes:

> 以太初曆考漢元盡太初元年日食二十三事，其十七得朔，四得晦，二得二日；新曆七得朔，十四得晦，二得二日.

[20] See *Shi ji* 27, 1343, and chapter 5, section 5.7.

> 以太初曆考太初元年盡更始二年二十四事,　十得晦;　以新曆十六得朔,
> 七得二日, 一得晦.
> 以太初曆考建武元年盡永元元年二十三事, 五得朔, 十八得晦; 以新曆十
> 七得朔, 三得晦, 三得二日.
>
> Looking into 23 instances of solar eclipses from the first year of Han [206 BCE] to the first year of the Grand Inception period [104 BCE] using the Grand Inception system, 17 come out as on first day of the month, four come out as on the last day of the month, and two come out as on the second day.[21] With the new system,[22] seven come out as on the first day of the month, 14 come out as on the [second] (last) day, and two come out as on the [last] (second) day.[23]
>
> Looking back to 24 instances of solar conjunctions from the Spring and Autumn Annals [8th to 5th centuries BCE] using the new system, in 23 cases it misses altogether.
>
> Looking into 24 instances from the first year of the Grand Inception period [104 BCE] to the second year of the Gengshi period [24 CE] using the Grand Inception system, ten come out as on the last day. With the new system 16 come out as on the first day, seven come out as on the second day, and one comes out as on the last day.
>
> Looking into 23 instances from the first year of the Jianwu period [26 CE] to the first year of the Yongyuan period [89 CE] using the Grand Inception system, five come out as on the first day, and 18 come out as on the last day. With the new system, 17 come out as on the first day, three come out as on the last day, and three come out as on the second day. (*Hou Han shu, zhi* 2, 3028; Cullen 2017, 383)

It is worth pausing a moment to reflect on the impressive fact that Jia Kui had access to the dates of seventy solar eclipses over the three centuries preceding his own day, from the beginning of the Western Han up to the time he wrote his memorial. Ptolemy, by contrast, cites no records of solar eclipses at all.[24] What is more, from Jia Kui's earlier reference to the 'notes of the [Grand] Clerk's officials' it seems that he would have had direct access to the observation notes of

[21] The references here are to the days of the month in the calendar calculated according to the Grand Inception system, whose core data were later re-used in the Triple Concordance system. As explained in section 2.1.1, the first day of the month is *shuo* 朔 'beginning', when it was expected that conjunction of the sun and moon would occur, and the last day is *hui* 晦 'darkening'.

[22] That is, calculating the calendar according to the Han Quarter Remainder system, which may start and end months on different days from the Grand Inception.

[23] There is obviously an error here: I emend as in the square brackets. We know that the Han Quarter Remainder system placed conjunctions about ¾ day too early in 162 BCE, so any eclipses that occurred would have tended to fall on the second day of the calendar month rather than on the first day; comparison between actual eclipses and the conjunctions predicted by the new system bears this out. Presumably the figures for the last day and second day have been reversed by a copyist.

[24] Toomer (1998), 654 & 310–320. He does, however, cite 22 lunar eclipses, three of which he says that he observed himself.

eclipses in the records of the Grand Clerk. Many of these records appear have been copied into the historical sources that have come down to us, such as the *Han shu* and the *Hou Han shu*. It is clear that some of the eclipse records in those sources have reached us in garbled form, perhaps after multiple processes of copying numerical data, but the consensus of recent research is that the details given are in nearly every case based on real events that were actually observed.[25] If we check the statements made by Jia Kui against modern computations of solar eclipses that might have been observed in Han times, and compare the predictions of the Triple Concordance and Han Quarter Remainder systems, the results are close to the account he gives of the performances of the two systems.

From the data presented by Jia Kui, the picture is clear: for the century from the foundation of Han up to the reform of 104 BCE, the Grand Inception system (equivalent in calendrical terms to the Triple Concordance system) gives calculations of conjunction dates close to those that actually occurred. In the next century, it begins to slip behind, while the Han Quarter Remainder system starts to yield more accurate conjunction dates. By the time we are within 60 years of Jia Kui's memorial, the new system is predicting conjunctions very well, and the Triple Conjunction is obviously lagging behind the phenomena actually seen in the sky. Neither system is wrong in itself: it is simply that any given astronomical system has a limited period of validity, and will eventually require revision.

This, Jia Kui points out, is precisely what one would expect from the way that astronomical systems are constructed on the basis of cycles of relatively short duration:

> 天道參差不齊，必有餘，餘又有長短，不可以等齊．治曆者方以七十六歲斷之，則餘分消長，稍得一日．

> The alignments and discrepancies of the way of heaven are not commensurable. There must be remainders, and those remainders [themselves] will themselves differ in ways that cannot be made commensurable through equalizing.[26] Those who manage astronomical systems chop off a period of 76 years [for commensurability of day, lunation and solar cycle], so the remaining fractions gradually increase until they build up to a day. (*Hou Han shu, zhi* 2, 3028; Cullen 2017, 384)

[25] See for instance David W. Pankenier (2012) 'On the reliability of Han dynasty solar eclipse records.' *Journal of Astronomical History and Heritage* 15 (3): 200–212.

[26] The references to *qi* 齊 here are I suggest more technical than might seem, as is the mention of *qi deng* 齊等. I believe that what is referred to here is the aim of the Chinese astronomical system-builder to represent the behaviour of the heavens by a series of (possibly large) whole numbers. Success in doing this requires that all the celestial cycles should be commensurable, i.e. that they can all be expressed as multiples of some given quantity, which may be set as equal to unity to simplify matters. The term *qi* 齊 is used in a similar way by Liu Hui 劉徽around AD 263 in describing what

That there should be changes in the system used is completely in accordance with the teachings of the sages of antiquity:

> 故易金火相革之卦象曰: 『君子以治曆明時. 』又曰: 『湯, 武革命, 順乎天應乎人. 』言聖人必曆象日月星辰, 明數不可貫數千萬歲, 其閒必改更, 先距求度數, 取合日月星辰所在而已. 故求度數, 取合日月星辰, 有異世之術.
>
> So in the *Book of Change*, the image for the hexagram of metal and fire controlling each other says 'The Gentleman clarifies the seasons by regulating the [astronomical] system.' It also says 'When Tang and Wu changed the mandate,[27] they accorded with heaven and responded to Man.' That means the Sage must 'sequence and delineate the sun, the moon and the stellar markers'.[28] However, in clarifying the reckonings he cannot run the same reckonings through thousands of myriads of years, but there must be revisions during that interval. At the first setting of a standard, when one seeks for the reckoning of the degrees, one just gets them by fitting in with the locations of the sun, moon and stellar markers. So in seeking for the reckoning of degrees, and getting them by fitting in with the sun, moon and stellar markers, each separate age has its own practice. (*Hou Han shu, zhi* 2, 3028; Cullen 2017, 384–5)

Moreover, the suggestion that the proper time to have adopted the Triple Concordance system was at the beginning of the Han, three centuries before the adoption of the new Han Quarter Remainder system, is consistent with the message of the 'apocryphal texts', which were seen as authoritative by Eastern Han rulers (see chapter 5):

> 太初曆不能下通於今, 新曆不能上得漢元. 一家曆法必在三百年之閒. 故讖文曰『三百年斗曆改憲』. 漢興, 當用太初而不改, 下至太初元年百二歲乃改. 故其前有先晦一日合朔, 下至成, 哀, 以二日為朔, 故合朔多在晦, 此其明效也. 」

is done to fractions with different denominators which are to be added: 母互乘子知以齊其子也 'the reciprocal multiplication by denominators is what renders the numerators commensurable' (Guo 1990, 184–5, where 知 is glossed as 者). The reference to *deng* 等 'equalized' may refer to the process of finding the common factors of numbers by alternate subtraction until equal differences are found, as in the algorithm set out by Euclid in Book 7, propositions 1 and 2 (see T. L. Heath (1925, second edition of original 1908 publication, revised with additions; repr. Dover, 1956, New York) *The thirteen books of Euclid's elements, translated from the text of Heiberg*. Cambridge, volume 2, 296–300). At any rate, it does not seem likely that Jia Kui is just talking in a general way about the difficulty of representing *particular* cosmic constants exactly, nor is he making any claim that such constants change over time. The real difficulty as he sees it is apparently that of building a system using finite whole numbers.

[27] Tang 湯 and Wu 武 were respectively the founders of the Shang and Zhou dynasties.

[28] This was the expression used by Emperor Yao to the Xi and He families in high antiquity: see chapter 1, section 1.1, and *Shang shu* 2, 9a-10b in *Shi san jing zhu shu*, vol. 1, 21-1 & 21-2.

> The Grand Inception system cannot be extended down to today, and the new system cannot reach back up to the first year of Han. The rules of the system of a given school are [rightly] located within a particular 300-year period. So a text in the Apocrypha says 'In 300 years the Dipper sequence changes its order.'[29] At the rise of Han, it would have been right to use the Grand Inception system but they made no change. The change was only made 102 years later when it came to the first year of the Grand Inception period. So at the start there were conjunctions one day before the last day of the month, and when it came down to the time of Chengdi [r. 32–5 BCE] and Aidi [r. 6–1 BCE] the second day [of the lunation] was taken as the conjunction. So this is a clear explanation of the frequent occurrence of the conjunction on the last day of the month. (*Hou Han shu, zhi* 2, 3028; Cullen 2017, 385)

As we have seen, Jia Kui places great emphasis on the use of solar eclipse records as a way of testing the validity of past and present astronomical systems at different periods. Let us therefore turn to the question of how such events were understood in Han times.

6.3.1 The significance of a physical explanation of eclipses in relation to astronomical systems

From the 92 CE memorial by Jia Kui just discussed, it is clear that when he needs a precise and demanding test of an astronomical system, he looks for the consistently successful prediction of the dates of the first days of lunar months, *shuo*, when luni-solar conjunction ought to occur, as including the days of any recorded solar eclipses. Thus, the fact that the Han Quarter Remainder system fails to do this for the dates of solar eclipses from the time of the Spring and Autumn annals (early eighth to late fifth centuries BCE) is, for him, firm evidence that it is invalid for such an early period.

But as we have seen in earlier chapters, in the Western Han solar eclipses were recorded as falling most frequently on days designated as *hui*, the last day of the lunar month, rather than on the first day, *shuo*, when the moon and sun ought to have been closest together. On the modern view, according to which a solar eclipse occurs when the moon crosses a terrestrial observer's direct line of sight to the sun, and hence is in conjunction with the sun, this would be a highly anomalous state of affairs—and Jia Kui evidently also saw it as problematic. Yet nobody in the Western Han is recorded as having shown signs of seeing the situation as worthy of comment. At the time of the Grand Inception reform of

[29] This text is the *Chun qiu bao qian tu*; see section 6.1.

104 BCE the occurrence of eclipses before the predicted day of conjunction is not mentioned as a motive for the reform. Before the reform, only five recorded eclipses out of 29 (17%) fell on *shuo*, while after the reform it was still only a minority of eclipses, nine out of 25 (36%) that fell on *shuo*. Nobody, however, is recorded as suggesting that this was not a satisfactory state of affairs, nor that the system in use needed to be modified as a result.

A hundred and fifty years before Jia Kui, Jing Fang 京房 (c. 40 BCE) does not seem to have seen any problem in solar eclipses occurring on *hui* as well as *shuo*, as we see from his comment given after a record of a solar eclipse:

> 惠帝七年［...］五月丁卯，先晦一日，日有食之，幾盡［...］京房易傳曰：「凡日食不以晦朔者，名曰薄．人君誅將不以理．或賊臣將暴起．日月雖不同宿，陰氣盛，薄日光也．」
>
> In the seventh year of Huidi, the fifth month, day *dingmao*.4 [17 July 188 BCE], one day before the last day of the month, the sun was eclipsed, almost totally[30] ... Jing Fang's commentary on the Book of Change says 'All eclipses that do not fall on either the first or last days are called 'dimming'. If the ruler is about to carry out an unreasonable punishment, or a treacherous minister is about to foment disorder, then even if the sun and moon are not in the same lodge, the qi of Yin [embodied in the moon] is at the full, and dims the sun's brilliance' (*Han shu* 27B2, 1500.)

So for Jing Fang, it seems that an eclipse can occur when sun and moon simply share the same lodge – which can easily happen a day before conjunction, since the moon moves at about 13 *du* per day, and several lodges are at least as wide as that. However, in Jing Fang's view eclipses do not always occur under those circumstances, and can also occur outside them, as indicated in the passage just quoted. There it is clear that the moon can cause the sun (which as the image of the ruler embodies Yang) to lose its brilliance merely by exerting its Yin influence from some distance away, if that influence is strengthened by the ruler's failure to act as he should, or by the encroachment of a too-powerful subject. From other comments by Jing Fang found in the same section of the *Han shu*, it is clear that he considers moral failings in high places as a major factor in eclipse causation: thus in relation to an eclipse in 14 BCE he is quoted as saying that drunkenness can bring about such events (*Han shu* 27B2, 1505).

The earliest firmly dated statement of the view that a solar eclipse is caused by the moon physically blocking the sun's light is found in the *Lun heng* 論衡 of Wang Chong 王充 (27–c.100 CE), where he writes:

[30] The eclipse was at maximum, and very near totality, at about 15:12 Chang'an local time, JD 1,652,953.83334.

或說: '日食者, 月掩之也. 日在上, 月在下, 障於日之形也. 日月合相襲, 月在上, 日在下者, 不能掩日. 日在上, 月在日下, 障於日, 月光掩日光, 故謂之食也 ... '

Some say 'When the sun is eclipsed, that is the moon covering it. The sun is above, and the moon is below, so it is a barrier to the shape of the sun. When the moon and the sun come together, but the moon is above and the sun below, then [the moon] cannot cover the sun. But when the sun is above and the moon is below the sun, it is a barrier to the sun, so the moon's light covers the sun's light, hence it is called "eating up"[31] ... ' (*Lun heng* 32, 14b)

The view that Wang Chong describes here would imply that a solar eclipse should always fall at the time of a conjunction of sun and moon, even though not every conjunction need be associated with an eclipse. Although Wang Chong goes on to attack this view of solar eclipses, his opposition makes it clear that it was one current in his own day. Since Wang Chong is writing close to the time when Jia Kui first implies that an eclipse ought to fall on a day of conjunction, it seems reasonable to conclude that the idea that the moon might block the view of the sun and hence cause an eclipse could well have originated some time in the first century CE.[32]

As for lunar eclipses, we have seen that as early as 62 CE Yang Cen was using errors in lunar eclipse timing to argue that the system in use was failing to predict lunar phases correctly (see 6.1). However, we have to wait rather longer to find an explicit statement of a physical theory of lunar eclipses. The idea that all the other heavenly bodies derive their light from the sun is found in a fragment ascribed to Jing Fang by Liu Xiang:

月與星至陰也. 有形無光. 日照之乃有光.

The moon and the stars are the ultimate in Yin. They have shape, but have no light. When the sun illuminates them, then they have light. (*Tai ping yu lan* 4, 10a)

But what physical reason could there be for the moon ceasing to shine when it is exactly opposite the sun in the heavens, as during a lunar eclipse? The first clear answer to that question is found in the writing of Zhang Heng (on whom see later in this chapter), some time in the early second century CE:

[31] The literal meaning of the word normally used for an eclipse, *shi* 食, sometimes written 蝕, is 'eat', as in *ri you shi chi* 日有食之 'There was something that ate up the sun.'

[32] See Needham and Wang Ling (1959), 410–417; see also the discussions in Christopher Cullen (1977). 'Cosmographical Discussions in China from Early Times up to the T'ang Dynasty.' School of Oriental and African Studies, PhD, III, 159, 176–180.

故月光生於日之所照，魄生於日之所蔽．當日則光盈，就日則光盡也．眾星被燿，因水轉光．當日之衝，光常不合者，蔽於（他）〔地〕也，是謂闇虛．在星星微，月過則食．

So the light of the moon comes from where the sun shines on it, and its dark portion comes from where the sun is blocked. When it is opposite the sun, its light is at the full, but when it comes up to the sun its light is exhausted. The multitude of stars are also lit up, and reflect back the light through their watery [nature]. Diametrically opposite to the sun, where its light never joins up, since it is blocked by the earth, that is called the 'dark space'. When it coincides with a star, the star fades, and when the moon passes through it, it is eclipsed. (*Hou Han shu, zhi* 10, 3215 comm.)

At first glance this sounds like the modern view, according to which a lunar eclipse occurs when the moon passes through the cone of the earth's shadow opposite the sun. However, in the cosmos described by Zhang Heng the earth occupies most of the lower part of the heavenly sphere (see Fig. 6.7). Thus, when the sun is below the earth at night, it is hard to see quite why the sun's light should only be blocked off for only a small region a degree or two wide precisely opposite the sun. On the contrary, one might have expected that a large area of the celestial sphere opposite the sun would have been darkened, so that the moon would be eclipsed every lunation whenever the moon was more or less opposite the sun, rather than the moon being eclipsed only rarely. Liu Zhi 劉智 (c. 220–290 CE) made this point strongly, though somewhat later Jiang Ji 姜岌 (c. 385 CE) noted the same objection but suggested that the explanation was that the sun's rays pass round the inside of the celestial sphere *you huo zhu xun tu er sheng* 猶火之循突而升 'like flames licking up inside a chimney' so that there is only one small area precisely opposite the sun that they fail to reach.[33] Jiang Ji was noted as having located the position of the sun by observing lunar eclipses:

岌以月蝕檢日宿度所在：為曆術者宗焉

[Jiang] Ji made use of lunar eclipses to check the *du* of the lodge where the sun is located; those who practice the methods of *li* follow his example. (*Jin shu* 18, 570)

The validity of this procedure is dependent on a view of lunar eclipses that locates the small 'dark space' *an xu* 闇虛 diametrically opposite the sun, as implied by the view set out by Zhang Heng and defended by Jiang Ji.

[33] See *Kai yuan zhan jing*, 1, 26a–b and 1, 21b, discussed in Cullen (1977) IV, 290 and 310–311.

6.3.2 Jia Kui and the importance of the ecliptic in observation and calculation

In chapter 5, we pointed to evidence of certain observations being made in the first century BCE, observations that could only have been made by people who no longer thought of the heavens as a relatively flat expanse far above the earth, but instead took it to be a sphere surrounding the earth, and who used observational instruments that were consistent with this view. Such instruments may have consisted of a simple graduated disc or armillary ring provided with sights: see section 5.7. In section 5.8, we saw Yang Xiong and Huan Tan arguing about which view was the correct one – and the conclusion was in favour of the *hun tian* 渾天 'spherical heaven' view, and against the *gai tian* 蓋天 'umbrella/covering heaven' view.

In the next section of his report, Jia Kui discusses the effects of movements of heavenly bodies on the celestial sphere, and the means by which those movements might be measured, in a way that shows that he is completely at ease with the *hun tian* view. Others, however, apparently had less sophisticated views than did he and his colleagues:

> 臣前上傅安等用黃道度日月弦望多近．史官一以赤道度之，不與日月同，於今曆弦望至差一日以上，輒奏以為變，至以為日却縮退行．於黃道，自得行度，不為變．願請太史官日月宿簿及星度課，與待詔星象考校．奏可．

> Your servant has previously submitted a memorial pointing out that when Fu An and his colleagues[34] used the Yellow Road [i.e. the ecliptic] to measure the [positions of] sun and moon at half and full moons, they were mostly correct. But the [Grand] Clerk's officials, who only used the Red Road [i.e. the celestial equator], were not in agreement with the sun and moon. They were often more than a day wrong in comparison with the present system, so they went ahead and memorialized this as being a portent, even to the extent of making out that the sun had suffered a setback and moved retrograde. On the Yellow Road the degrees of motion turn out naturally, and no such 'portents' occur. I requested that the Grand Clerk's records of the lodges of the sun and moon, together with the stellar data should be checked, and compared with the stellar phenomena [observed] by the Expectant Officials. My memorial was approved. (*Hou Han shu*, *zhi* 2, 3028–9; Cullen 2017, 385–386)

[34] Unfortunately we have no direct indication of when Fu An was active. Since no indication of date is given by Jia Kui, one might reasonably infer that the activity referred to was quite recent.

To understand what Jia Kui is saying, we need to consider two important circles on the celestial sphere. The 'Yellow Road' *huang dao* 黃道 is the annual circuit traced out by the sun as it moves against the background of the stars, called the 'ecliptic' in English because an eclipse (whether lunar or solar) can only occur when the moon crosses this line. In modern astronomical terms, the ecliptic is the intersection of the plane of the earth's orbit around the sun outwards with the celestial sphere, the latter being imagined as effectively of infinite radius. If the stars were not blotted out by the light of the sun, then as the earth orbits round the sun, an observer on the earth would see the sun against a different background of stars on each day of the year, and that pattern would repeat every year. This produces the same apparent effect for an observer on the moving earth as if the sun was orbiting a stationary earth on a path that was fixed relative to the stars, and that apparent path is the ecliptic. The 'Red Road' *chi dao* 赤道 is the celestial equator, the circle midway between the north and south celestial poles. It is the intersection of the plane of the earth's equator with the celestial sphere. Both of these lines are 'great circles' of the celestial sphere, meaning that their diameter is that of the sphere itself. Now if these two circles coincided, the sun would move round the celestial equator, and would always be 90° from both the north celestial pole, and from the south celestial pole. But in fact the ecliptic is tilted about 23 ½° to the plane of the equator, so that for half the year the sun is to the north of the equator and half the year to the south.

We may show the situation as in Figure 6.1, which shows the path of the sun on the celestial sphere (seen from the outside, so that motion west to east is shown from left to right) for the half year when it is nearer to the north celestial pole P than to the south celestial pole P′. The position S_2 then corresponds to the summer solstice, and S_4 to the autumn equinox, when the sun is crossing the equator moving southwards. The tilt of the ecliptic relative to the equator has been exaggerated for clarity, and we can only see the half of each circle on the side of the sphere facing towards us.

Let us assume that the sun moves at constant speed round the ecliptic in the direction $S_1S_2S_3S_4$. If we measure the movement with reference to the equator by drawing meridian lines from the poles through each solar position, and seeing where they cross the equator, we can see that the distance E_1E_2 is greater than the corresponding distance S_1S_2 moved along the ecliptic, since the two arcs PS_1E_1P' and PS_2E_2P' are further apart near the equator than they are near the poles. On the other hand, because the ecliptic crosses the equator at an angle at S_4, the distance E_3S_4 is less than the corresponding distance S_3S_4 on the ecliptic. As a result the sun moving at constant speed on the ecliptic will seem to move faster with reference to the equator near a solstice, and slower near an equinox.

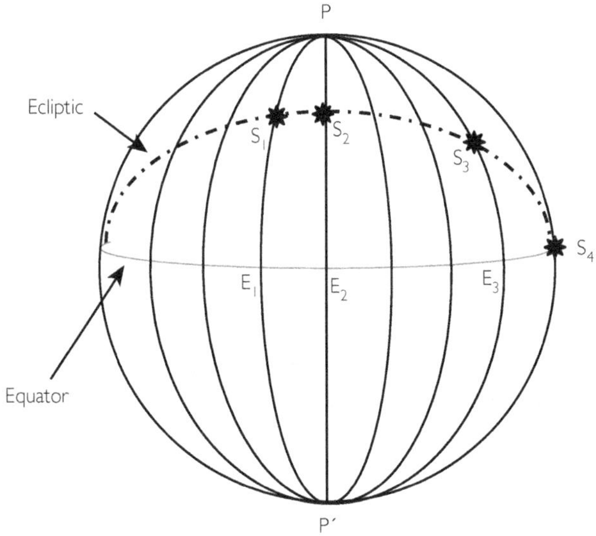

Figure 6.1 Movement of sun on ecliptic, and resultant movement relative to the equator.

According to Jia Kui, the 'Clerk's officials' were calculating the positions of the sun and moon as if they moved at constant speed with respect to the equator, with the result that the actual sun and moon would sometimes be ahead of the predicted equatorial positions and sometimes behind them – so that, as he says, they seemed to have moved retrograde, which would certainly be a portent deserving to be reported to one's superiors, had it occurred.

Jia Kui points out that there is a simple numerical pattern underlying these results:

> 今史官一以赤道為度，不與日月行同，其斗，牽牛，東井，輿鬼，赤道得十五，而黃道得十三度半；行東壁，奎，婁，軫，角，亢，赤道七度，黃道八度；
>
> Now the Clerks all use the Red Road as a scale for measuring the degrees, which is not in accord with the motions of the sun and moon. In Dipper and Ox [near the winter solstice] and Well and Ghost [near the summer solstice] one gets 15 *du* on the Red Road while getting 13 ½ *du* on the Yellow Road. When moving through Wall, Straddler and Harvester [near the spring equinox] and through Axletree, Horn and Gullet [near the autumn equinox] it is seven *du* on the Red Road for eight *du* on the Yellow Road. (*Hou Han shu, zhi* 2, 3029; Cullen 2017, 386)

Or, turning the problem the other way round, consider the moon, which in systems of the quarter remainder type is taken to move 13 $^{7}/_{19}$ *du* a day. Do its actual measured shifts on the heavens from day to day fit in with the idea that it should move that much each day with reference to the equator? The actual observations of the Clerk's officials themselves, whose registers Jia Kui was able to inspect, disprove this:

以今太史官候注考元和二年九月已來月行牽牛，東井四十九事，無行十一度者；行婁，角三十七事，無行十五六度者，如安言.

If one uses the observational records of the present Grand Clerk from the ninth month of the second year of the Yuanhe period [7 October to 5 November 85 CE] onwards to examine 49 instances in which the moon has moved through Ox and Well [49 complete lunations would take us to the end of the intercalary seventh month of Yongyuan 1, 25 August to 22 September 89 CE], there is none in which [the actually observed motion along the path of the moon from day to day] has been 11 *du* [as one would have expected if it had moved 13 *du* on the Red Road]. And for 37 instances in which it has moved through Harvester and Horn, there is none in which [the actually observed motion across the heavens from day to day] has been 15 or 16 *du* [as one would have expected if it had moved 13 *du* on the Red Road].[35] This is just as [Fu] An said.[36] (*Hou Han shu, zhi* 2, 3029; Cullen 2017, 387)

Finally, he points out that observations made a century and a half earlier indicated the same thing:

案甘露二年大司農中丞耿壽昌奏，以圖儀度日月行，考驗天運狀，日月行至牽牛，東井，日過一度，月行十五度，至婁，角，日行一度，月行十三度，赤道使然，此前世所共知也.

I note that in the second year of the Ganlu period [52 BCE] the Assistant of the Superintendent of Agriculture, Geng Shouchang, memorialized that he had measured the motions of the sun and moon with a plotting [literally 'diagram'] instrument. Checking on the appearance of the celestial motions, when in their motions the sun and moon reached Ox and Well, the sun exceeded one *du* [daily] and the moon moved 15 *du*. But when they reached Harvester and Horn, the sun moved one *du* a day, and the moon moved 13 *du*. It was the Red Road [serving as a reference system] that made it thus—this is something everyone in the previous age [i.e. the late Western Han] knew about. (*Hou Han shu, zhi* 2, 3029; Cullen 2017, 388)

Clearly this 'diagram instrument' *tu yi* 圖儀 was some kind of device enabling measurements of equatorial motion to be made. We have previously seen

[35] In fact the actual daily motion of the moon across the celestial sphere does not fall below 11 degrees or rise much above 15.5 degrees, whatever its position on the celestial sphere: see Figure 6.5 and the associated discussion. This corresponds fairly closely to what the officials had observed.

[36] See section 6.3.2.

evidence of a graduated circle or armillary ring placed in plane of the meridian in order to measure north polar distances, but now we have evidence of another circular scale, which must have been in the plane of the celestial equator. It begins to seem possible that a simple armillary sphere with at least two rings might have been put together around 50 BCE. The *Han shu* bibliographical monograph records two texts ascribed to 'Geng Chang 耿昌', who seems likely to be the same person as Geng Shouchang:

月行帛圖二百三十二卷.
月行度二卷.

Charts of lunar motion on silk: 232 rolls
The degrees of lunar motion: 2 rolls (*Han shu* 30, 1766)

The first item may well have contained records of the kind of systematic observations to which Ji Kui is referring.

Jia Kui's summary drives his point home:

如言黃道有驗, 合天, 日無前却, 弦望不差一日, 比用赤道密近, 宜施用.

As has been said, the Yellow Road has its verification, and fits together with Heaven, so that the sun has no advances or laggings, and crescent and full moons are not out by a single day. This is much more accurate than using the Red Road [as a reference], and should be put in practice. (*Hou Han shu, zhi* 2, 3029; Cullen 2017, 388)

Action on his recommendations was not immediate. But eventually—two years after his death in 101 CE—something was done:

案逵論，永元四年也．至十五年七月甲辰，詔書造太史黃道銅儀

Note that Jia Kui's Discussion was [submitted] in the fourth year of the Yongyuan period [92 CE]. When the 15th year [103 CE] was reached, in the seventh month, day *jiachen*.41, there was an edict to construct the Grand Clerk's Yellow Road Bronze Instrument ... (*Hou Han shu, zhi* 2, 3029; Cullen 2017, 388–389)

Evidently this instrument had an ecliptic ring as well as an equatorial ring of the kind that was apparently already in use. A basic armillary sphere was now complete. However, the officials in charge of observation were not eager to change their ways:

史官以部日月行，參弦望，雖密近而不為注日．儀，黃道與度轉運，難以候，是以少循其事．

[When] the Clerk's officials used [this device] to divide up the motions of the sun and moon, and to calibrate against crescents and full moon, even though it was highly accurate they did not use it for their daily record-making. [On this] instrument, the Yellow Road and its graduations rotated, making observation difficult, and so they paid little attention to it. (*Hou Han shu, zhi* 2, 3030; Cullen 2017, 389–390)

6.3.3 The normalization of armillary instruments, and their projection back into the past

Even if the Grand Clerk's officials did not like the idea of having to use an instrument with an unfamiliar new ecliptic ring on it, it is clear that armillary instruments of some kind were no longer a novelty around 100 CE. This is clear from an interesting change in the interpretation of one important ancient classical text that becomes evident at this period. The text in question occurs in the *Shang shu* 尚書 'Book of Documents', in a section that describes the activities of the legendary emperor Shun 舜 of high antiquity on taking over the throne from Yao 堯 (see section 1.1). There we are told:

在璿璣玉衡 以齊七政

He attended to the *xuan ji yu heng* in order to regulate the seven governments/governors. (*Shang shu* 3, 4b, in (Ruan Yuan 阮元 (1764–1849) 1973 reprint of original of 1815: vol. 1, *Shang shu* p.35)

The complex history of the interpretation of this passage has been studied elsewhere.[37] It appears that up to the end of Western Han, it was normal to interpret *xuan ji yu heng*, variously written, as referring to stars near the north celestial pole. By around 100 CE, however, interpretations had changed significantly. This is most directly illustrated by looking at Sima Qian's monograph on the heavens, and comparing it with the later views collected in the commentary. Thus Sima Qian says:

北斗七星，所謂「旋，璣，玉衡以齊七政」

The seven stars of the Northern Dipper are what are called 'the *xuan ji yu heng* that regulate the seven governments/governors'. (*Shi ji* 27, 1291)

The commentary, however, cites two Eastern Han scholars who see things quite differently:

馬融 云「璿，美玉也．機，渾天儀，可轉旋，故曰機．衡，其中橫筩．以璿為機，以 玉為衡，蓋貴天象也」．鄭玄注大傳云「渾儀中筩為旋機，外規為玉衡」也．

[37] See Christopher Cullen and Anne S. L. Farrer (1983) 'On the Term "Hsüan Chi" and the Flanged Trilobate Jade Discs.' *Bulletin of the School of Oriental and African Studies, University of London* 46 (1): 52–76.

> Ma Rong (79–166 CE) says: '*Xuan* [means] a fine jade. The *Ji* is a *hun tian* 'spherical heaven' instrument, which can be rotated, so it is called a *ji* 'device'. The *heng* is its central cross-wise [sighting] tube. The *ji* is made of *xuan*, and the *heng* is made of [common] jade, presumably in order to honour the image of heaven.' The commentary of Zheng Xuan (127–200 CE) on the *Da zhuan* 'Great tradition' [i.e. of interpretation of the *Shang shu*] says: 'The central tube of the sphere instrument is the *xuan ji*, and the outer circle is the *yu heng*.' (*Shi ji* 27, 1291, commentary)

By the time these two scholars were writing, in the second century CE, the use of 'spherical heaven instruments', or armillary spheres, by observers of the heavens was evidently so well established that it was easy to assume that they must always have been used. From this time onwards, the *Shang shu* passage was almost universally interpreted as describing the use of an armillary sphere by Shun, as imagined in the late Qing illustration of the scene reproduced in Figure 6.2.[38]

6.3.4 Jia Kui and the measures of the celestial sphere

It is in Jia Kui's work that we find the first firmly datable set of figures specifying the main measures of the celestial sphere. These are given in what we would today think of as angular terms, using *du* 度 as the unit.[39] Jia Kui would have used the *du* as in systems of the quarter remainder type, in which a complete circuit of the heaven is performed by the sun moving at one *du* per day, in a cycle from winter solstice to winter solstice that is 365 ¼ days long, thus implying that a complete circuit of the heavens is 365 ¼ *du*. Using the sexagesimal angular measures created by the Babylonians, we may say that 365 ¼ *du* are 360°, or that 1 *du* is 0.99°.

As part of the introduction to his arguments in favour of using the ecliptic as the reference circle for solar motion, Jia Kui refers to some of the principal dimensions of the sphere:

> … 前對言冬至日去極一百一十五度，夏至日去極六十七度，春秋分日去極九十一度.

[38] From *Qin ding shu jing tu shuo* 欽定書經圖說 *(The Book of Documents Illustrated and Explained: imperially commissioned)*. (1905). Sun Jianai 孫家鼐 et al. (edited) and Zhan Xiullin 詹秀林 et al. (illustrations), Beijing, Da xue tang bian shu ju 大學堂編書局, 2, 8a.

[39] Some caution is needed here. The general concept of angle as a measure of the amount by which any line is rotated relative to another is not found in China at this period, and is in fact absent from Chinese geometry until the arrival of western mathematics from the 17th century onwards. With reference to the celestial sphere, the *du* is used as a measure of displacement between two points on the sphere, and if the radius of that sphere is thought to be known in linear measure, then the length of a *du* may be given in *li* 里, the common measure of larger terrestrial distances, roughly equivalent to 0.4 km in Han times. On this see for instance Cullen (1996), 185–188.

Figure 6.2 Emperor Shun and his supposed armillary sphere (note 38).

> ... I have previously answered that at the winter solstice the sun is 115 *du* from the pole, that at the summer solstice it is 67 *du* from the pole, and that at the spring and autumn equinoxes it is 91 *du* from the pole. (*Hou Han shu, zhi* 2, 3029; Cullen 2017, 386)

At the equinoxes, the sun should be on the celestial equator, equidistant from the north and south celestial poles. Its distance from either pole should be a right angle, 90° in modern terms, and $(365\ \tfrac{1}{4}\ du)/4 = 91\ \tfrac{5}{16}\ du$ in the measures used by Jia Kui. His 91 *du* is therefore correct as a round number. How far is the sun from the equator at the solstices? We have:

Winter solstice: 115 *du* – 91 *du* = 24 *du*
Summer solstice: 91 *du* – 67 *du* = 24 *du*.

This quantity, which in modern terms represents the inclination of the ecliptic plane to the equatorial plane, is nowadays called the obliquity of the ecliptic; 24 *du* is 23.7°. Ptolemy of Alexandria (c. 100–c. 170 CE), who was an approximate contemporary of Jia Kui, gives data which imply an obliquity of 23.9°.[40] Modern calculations suggest that the obliquity in the first century CE would have been about 23.75°.[41] The figure of 24 *du* implied by Jia Kui is commonly cited by other authors.[42] We may thus sum up his view of the celestial sphere in outline in Figure 6.3.

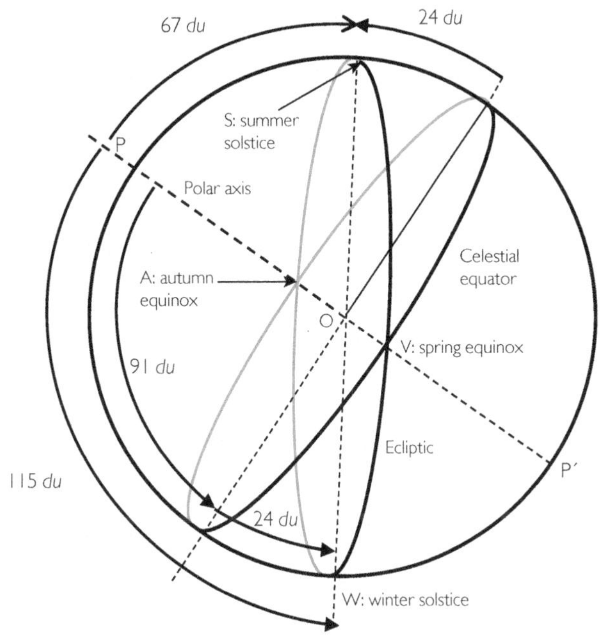

Figure 6.3 Jia Kui's view of the celestial sphere; not to scale.

[40] *Almagest* III.1:12; Toomer (1998), 61–63. In fact Ptolemy gives the meridian arc between the two solstices, which is double the obliquity. It seems likely that Ptolemy's value is not, as he claims, derived from meridian observations, but is simply taken over from Eratosthenes (c. 276–c. 195/194 BCE) and Hipparchus (c. 190–c. 120 BCE), who both claimed that the ratio of the meridian arc between the two solstices to the whole meridian circle is 11:83; see Jones (2002). See also *Geminos's introduction to the Phenomena: a translation and study of a Hellenistic survey of astronomy,* 156, where Geminus states that the equator is four 60ths of the circumference (i.e. 24°) from each of the tropics.

[41] A. Wittmann (1979) 'The obliquity of the ecliptic.' *Astronomy and Astrophysics,* 73 (1–2): 129–131, Table 2.

[42] Somewhat puzzlingly, a few lines later than the passage cited above, Jia Kui states that the equator goes 25 *du* to the north and south of the ecliptic a discrepancy for which I can think of no explanation apart from a possible copyist's confusing of the characters for four 四 and five 五.

Note that Jia Kui himself does not say how much the polar axis is inclined to the observer's horizon; in his discussion he is concerned only with measurements relative to points on the celestial sphere itself.

Another piece of information given to us for the first time relates to the ecliptic circle added to the new instrument constructed in 103 CE in accordance with Jia Kui's recommendation. We are given a new set of widths for the 28 lodges, this time with respect to the ecliptic rather than the equator. The list is shown in Table 6.1, where the widths for the lodges on the equator have been inserted for comparison.[43]

How were the lodges on the ecliptic defined, and how were their widths found? As we shall shortly see (section 6.5.3), it is very likely that measurement on a model sphere was used for this purpose; the results obtained were correct to within about ½ *du*.[44] The ecliptic extensions of the lodges were (in modern terms) defined by finding the positions on the ecliptic which had the same right ascensions as the beginnings of the lodges on the equator. This was done using a simple graphical procedure in which the beginnings and ends of the lodges on the ecliptic were defined by, in effect, drawing meridians from the north celestial pole to the south celestial pole through the positions on the equator where lodges began and ended, and finding where those meridians intersected the ecliptic, similarly to the situation shown in Figure 6.1.

Given this indication of precision in measurements made on the sphere, it would be interesting to have an idea of just how large were the graduations on the device that whoever was responsible for the ecliptic lodge widths used to make his measurements. Unfortunately, we have no record of the size of the new instrument made in response to Jia Kui's initiative. The nearest we can get is a note in the *Sui shu* standard history, six centuries later, which says that the armillary device made by Zhang Heng a few decades after Jia Kui showed each *du* on its scale as measuring 4 *fen* 分 (*Sui shu*, 19, 516;). This is 0.4 *cun* 寸 'inch' or 0.04 *chi* 尺 'foot'. Taking a Han *chi* as 23.1 cm,[45] a *du* graduation would be about 0.9 cm. The overall diameter of the device was therefore about 1.1 m. A precision of about ½ *du* does not seem implausible under those circumstances, so long as the instrument was accurately made. There is, however, a problem with this reference, since the date given for the

[43] For the widths on the ecliptic given in the account of the new instrument, see *Hou Han shu, zhi* 2, 3029; Cullen 2017, 388–389. Both ecliptic and equatorial widths are also given in tabulations that form part of the description of the Han Quarter Remainder system in *Hou Han shu, zhi* 3, 3074–3076, Cullen 2017, 221–223.

[44] See also Cullen (2000), 363.

[45] See for instance Twitchett, Loewe and Fairbank (1986), xxxviii.

Table 6.1 ***Ecliptic width of lodges on Jia Kui's instrument, compared with equatorial widths***

	Lodge	Width on ecliptic /*du*	Width on equator /*du*
1	*Jue* 角 'Horn'	13	12
2	*Kang* 亢 'Gullet'	10	9
3	*Di* 氐 'Base'	16	15
4	*Fang* 房 'Chamber'	5	5
5	*Xin* 心 'Heart'	5	5
6	*Wei* 尾 'Tail'	18	18
7	*Ji* 箕 'Winnower'	10	11
8	*Dou* 斗 'Dipper'	24 ¼	26 ¼
9	*Niu* 牛 'Ox'	7	8
10	*Nu* 女 'Woman'	11	12
11	*Xu* 虚 'Barrens'	10	10
12	*Wei* 危 'Rooftop'	16	17
13	*Shi* 室 'House'	18	16
14	*Bi* 壁 'Wall'	10	9
15	*Kuei* 奎 'Straddler'	17	16
16	*Lou* 婁 'Harvester'	12	12
17	*Wei* 胃 'Stomach'	15	14
18	*Mao* 昴 'Mane'	12	11
19	*Bi* 畢 'Net'	16	16
20	*Zui* 觜 'Beak'	3	2
21	*Shen* 參 'Triaster'	8	9
22	*Jing* 井 'Well'	30	33
23	*Gui* 鬼 'Ghost'	4	4
24	*Liu* 柳 'Willow'	14	15
25	*Xing* 星 'Star'	7	7
26	*Zhang* 張 'Spread'	17	18
27	*Yi* 翼 'Wing'	19	18
28	*Zhen* 軫 'Axletree'	18	17

manufacture of the instrument in question is 164 CE, decades after Zhang Heng's death in 139 CE. The same *Sui shu* passage also tells us that Wang Fan 王蕃 (228–266 CE) made an instrument with 3 *fen* to the *du*. We may perhaps limit ourselves to saying that the precision achieved is not inconsistent with indications of the order of magnitude of the known dimensions of ancient instruments.

6.3.5 Jia Kui and the speed of the moon: the Nine Roads

In his discussion of the advantages of using the ecliptic as a reference system for the motion of the sun and moon, Jia Kui made it plain that he had access to quite detailed and systematic official records of the distance moved by the moon against the background of the stars in the course of a day: see section 6.3.2. He went on to exploit these data to make the first quantitative analysis of the ways in which the apparent speed of the moon varied over time. In this section we shall discuss the basis and significance of his work.

In his analysis of lunar motion, Jia Kui makes reference to the concept of the *jiu dao* 九道 'Nine Roads', references to which begin to be found around the beginning of the Eastern Han, usually (but not always) with reference to the motion of the moon. No single definite explanation of the Nine Roads makes sense of all the occurrences of this term in the early literature, and it is possible that it never had a single consistent meaning.[46] Ancient commentators who discuss the Nine Roads seem themselves to be relying on conjecture to make sense of the way the term was originally used. On the other hand, there are signs that some writers had adopted the Nine Roads concept with a fairly precise technical sense in mind, despite the fact that not all instances of its usage are consistent with the sense they apparently gave it. We shall review the evidence below.

6.3.5.1 THE MOON IN MOTION: ITS PATH AND SPEED

But before we look at the texts, let us look at the moon, and the 'road' (or rather 'roads') that it follows round the sky. It is common in our sources to say that the sun, moon and five planets all follow the Yellow Road – the ecliptic, the great circle round the celestial sphere that is in fact the projection onto that sphere of the plane of the earth's orbit round the sun. Leaving aside the case of the sun, which by definition always lies on the ecliptic,[47] this statement is only

[46] Thus, for instance, Xu Zhentao 徐振韬 (2009) *Zhong guo gu dai tian wen xue ci dian* 中国古代天文学词典 *(A dictionary of pre-modern Chinese astronomy)*. Beijing, China science and technology press, 115–116, cites various early references in the three entries using this term, but ends by saying that its historical significance remains to be determined. Qu Anjing 曲安京 (2008), 332–336, reviews early references, and notes (p. 333) the exasperation of the historian of astronomy and mathematics Qian Baocong 錢寶琮 (1892–1974) at the obscurity of one early attempt at explanation. As Qu points out, the use of the term from the Tang onwards is much clearer—but it appears to carry a rather different meaning.

[47] This statement requires some qualification. If the earth were a point mass at the position now occupied by the centre of the earth, an observer at that point would see the sun exactly on the ecliptic. But since the earth is an extended body, the effect of parallax means that an observer in its northern regions will see the sun a very small distance (about two thousandths of a degree) to the south of the ecliptic, and vice versa.

an approximation for the planets. The planes of their orbits round the sun have small but non-negligible inclinations to the earth's orbital plane, so they can all be seen to depart significantly from the ecliptic. Nor does the plane of the moon's orbit round the earth coincide neatly with the ecliptic. The path of the moon round the celestial sphere is in fact a great circle inclined to the ecliptic at about 5°, so that the moon is sometimes to the north of the ecliptic and sometimes to the south of it. The moon's path crosses the ecliptic at the two points called the ascending node, marked A (when the moon 'ascends' from south to north), and the descending node, marked D (when the moon 'descends' from south to north): see Figure 6.4.[48]

Because the moon moves about thirteen times faster than the sun as seen from the earth, the moon will overtake the sun about every 29 ½ days: this moment of closest approach by the two bodies is a conjunction.[49] For instance, if the moon were at M′ when the sun was at S, that would be a conjunction—actual coincidence of sun and moon is not required for conjunction to occur.

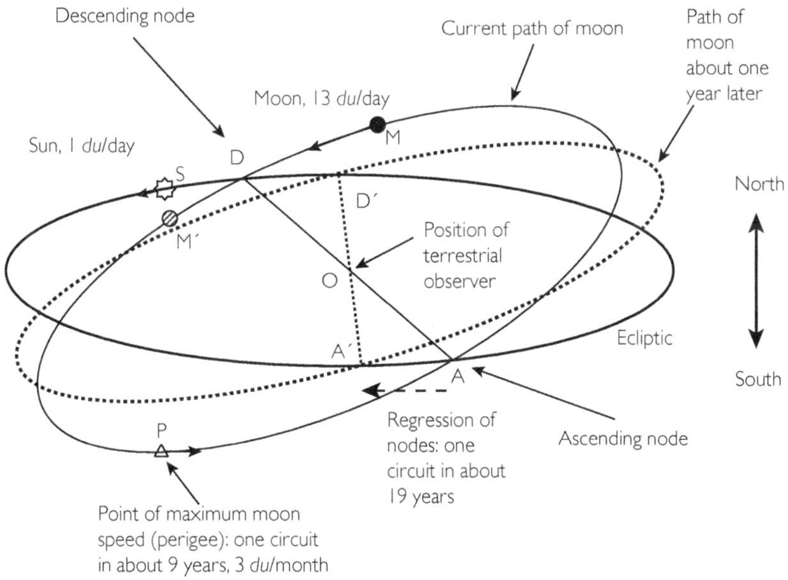

Figure 6.4 The changing path and speed of the moon. The inclination of the moon's path to the ecliptic has been exaggerated for clarity.

[48] Conventional symbols used for these points in the western tradition are ☊ (ascending node) and ☋ (descending node).

[49] In modern terms, the sun and moon are said to be in conjunction when they have the same longitude.

Because the moon is closer to the earth than the sun, only the side of the moon facing away from the observer on earth would be illuminated. If, which happens rarely, the moon caught up with the sun close to either of the nodes D and A, it would block the sun from the view of the terrestrial observer at O, who would experience a solar eclipse. If the moon passed through the node A while the sun was at D (or vice versa) then the shadow of the earth would fall on the moon, and a lunar eclipse would be the result.[50]

However, the moon's path is not fixed relative to the ecliptic: as a result of the complex gravitational interactions within the sun-moon-earth system, the two nodes 'regress' (i.e. they move from east to west, in the opposite direction to the sun and moon). Thus in the diagram, the ascending node will move from A to A′ in about a year, and after about 19 years (more precisely 18.6 years) it will be back where it started.

The speeds of the moon and sun along their paths are marked on the diagram using the typical values of one *du*/day for the sun and 13 *du*/day for the moon. In the early imperial period, the constancy of the sun's speed was not questioned, but as we shall shortly see it was eventually noticed that the speed of the moon varied. In modern terms, this variation is due to the fact that the moon's orbit round the earth is elliptical, and the moon moves fastest when it is at the point in its orbit closest to the earth (perigee).[51] Modern calculation shows that this would produce a variation in daily lunar speed from a minimum of about 12 *du* per day to a maximum of 15 *du* per day in the course of each orbit. As shown on the diagram, the point of maximum speed also moves along the moon's path, completing a circuit in about nine years (more precisely 3233 days or 8.85 years). As a result of this speed variation, the moon will not in general be found precisely at the position calculated on the basis of its mean speed. This discrepancy between the true position and the mean speed predicted position is known in the western tradition as the 'equation of centre', or the 'elliptic anomaly', and can be as large as 6°, a little less than half a day's mean motion of the moon.

Finally, let us bring together the values of some important short-term periods in the moon's motion:

The mean synodic month: 29.53059 days

[50] As noted elsewhere, since a lunar eclipse involves a shadow falling on the moon, it will be seen simultaneously by all observers on the earth who can see the moon. Since a solar eclipse requires the moon to be located very close to the observer's sight-line to the sun, it will only be seen by some observers on the earth.

[51] A similar effect will be seen if one whirls round a small mass attached to a string, and then shortens the string—the weight speeds up, so that the angular momentum of the mass remains constant.

This interval, which we have met before, is the mean interval between two conjunctions of sun and moon—that is, the time taken for the moon to begin from one conjunction with the sun at M′, go all the way round its path, and catch up once more with the sun, which will have moved about 29 ½ *du* from its original position at S.

The mean sidereal month: 27.32166 days

This is the mean interval between passages of the moon past the same point relative to the stars.[52] If it were not for the regression of the nodes, this period would be equal to the time for the moon to get from a point such as the descending node D, round its path once, and back to D again. But because of the steady shift of the nodes, that period has a different value:

The mean draconitic month (or 'nodical month'): 27.21222 days

In fact the moon will have to go a little past the new slightly westwards shifted position of the node to reach an alignment similar to its previous position, so the sidereal month is slightly longer than the draconitic month. But in either case, the time taken will be less than a synodic month. Finally, we have:

The mean anomalistic month: 27.55455 days

This is the mean interval of passages of the moon past P, the point of perigee where the moon is moving fastest. The fact that the perigee moves round the moon's path in the same direction as the moon makes this period longer than either the draconitic or sidereal months.

6.3.5.2 THE NINE ROADS BEFORE JIA KUI

Anybody who watches the night sky systematically with the naked eye may verify that the moon does not always follow the same path relative to the stars. If, for instance, we watch the passage of the moon past the bright star Regulus (α Leonis), which happens to lie almost exactly on the ecliptic, the moon may pass right over it, so as to eclipse the star, or may on occasion go as far as 5°—ten moon-widths—to the north of it, or the same distance to the south. Before the Eastern Han, we do not, however, find evidence that anyone felt that the moon's path through the sky varied in a systematic way. That it did vary, on the other hand, was certainly known. Thus for instance around 100 BCE Sima Qian wrote as follows in his monograph on the heavens:

[52] More precisely, this is the moon's orbital period in a non-rotating frame of reference. Such a frame is only approximately represented by the visible 'fixed' stars, which all have their own ('proper') motions, small but detectable by modern means.

月行中道，安寧和平．陰間，多水，陰事．[…] 陽間，驕恣．陽星，多暴獄．太陽，大旱喪也．

When the moon follows its middle path, all is quiet and peaceful. If [it moves into] the Yin space [to the south], there is much water, and intrigue. [...] If [it moves into the Yang space, [to the north] there is arrogance and wilfulness. [If it approaches] the Yang star, there is much violence and imprisonment. [If it approaches] the Great Yang, there will be great drought and loss. (*Shi ji* 27, 1331; translation modified from Pankenier, David W. 2013: 492)

Sometime during the first century CE, the view arose that the moon's changing path was in some way ninefold. A statement to this effect occurs in its fullest form in the *Han shu*, in the monograph on *tian wen* contributed by Ma Xu 馬續 (70–141 CE).

日有中道，月有九行．中道者，黃道，一曰光道．光道北至東井，去北極近；南至牽牛，去北極遠；東至角，西至婁，去極中．
[…] 月 有九行者：黑道二，出黃道北；赤道二，出黃道南；白道二，出黃道西；青道二，出黃道東．立春，春分，月東從青道；立秋，秋分，西從白道；立冬，冬至，北從 黑道；立夏，夏至，南從赤道．然用之，一決房中道．青赤出陽道，白黑出陰道．若月失節度而妄行，出陽道則旱風，出陰道則陰雨．

The sun has its Middle Road, but the moon has a ninefold motion. The Middle Road is the Yellow Road [i.e. the ecliptic]; one name for it is the Bright Way. The Bright Way goes northward to the lodge Well, which is nearest to the north [celestial] pole, and south to the lodge Ox, which is furthest from the north [celestial] pole. It goes east to Horn, and west to Harvester, whose distances from the pole are midway [between the other two].[53]

[...] [This is how] the moon has a ninefold motion: two Black Roads, which go to the north of the Yellow Road; two Red Roads, which go south of the Yellow Road, two White Roads, which go west of the Yellow Road; two Blue Roads, which go east of the Yellow Road [making a total of nine, including the Yellow Road]. At Establishment of Spring and Spring Equinox, it follows the Blue Roads in the east. At Establishment of Autumn and Autumn Equinox, it follows the White Roads in the east. At Establishment of Winter and Winter Solstice, it follows the Black Roads in the north. At Establishment of Summer and Summer Solstice, it follows the Red Roads in the north. (*Han shu* 26, 1294–1295)

[53] In the late first millennium BCE, these four lodges were thought to contain the positions of the sun at the summer solstice (Well), the winter solstice (Ox), the autumn equinox (Horn) and the spring equinox (Harvester). As we have seen earlier in this chapter, by the time of Jia Kui it was beginning to be noted that this was no longer the case. See also Cullen (1996), 97–101.

The 'Yellow Road' here is clearly the ecliptic as before. What the other coloured 'roads' are intended to be is not clear, but the 'Red Roads' here are certainly not the equator. Around 500 CE, the *Song shu* 宋書 standard history quotes a shorter version of this material with an attribution to Liu Xiang 劉向 (77–6 BCE), which would date it near the end of Western Han; other sources attribute it to one of the apocryphal works that began to appear around the time of Wang Mang.[54] But in any case it is clear that whoever wrote this obscure passage thought that the moon followed a particular road at a particular season of the year, which is certainly not the case, given the periodicities set out in the preceding section.

There is a reference to the 'Nine Roads' in another *Han shu* monograph, the one in which Liu Xin's Triple Concordance is specified, where we read:

> 九章歲為百七十一歲，而九道小終.
>
> The Nine Roads [reach their] Lesser Conclusion in nine Rule Years [19], making 171 years. (*Han shu* 21b, 1007; Cullen 2017, 120)

But the text continues by saying that it is the sun that has nine roads, while the moon has 19 roads. There is no evident connection with the complexities of lunar motion discussed above, but as we shall see, Jia Kui apparently thought that he had found one.

6.3.5.3 JIA KUI ON THE MOON

Jia Kui begins his discussion by pointing to the failure of an attempt made some years earlier by Zhang Long 張隆 to predict lunar motion more accurately, based on the Book of Change. He then continues:

> 梵，統以史官候注考校，月行當有遲疾，不必在牽牛，東井，婁，角之閒，又非所謂朓，側匿，乃由月所行道有遠近出入所生，率一月移故所疾處三度，九歲九道一復，凡九章，百七十一歲，復十一月合朔旦冬至，合春秋，三統九道終數，可以知合朔，弦，望，月食加時．據官注天度為分率，以其術法上考建武以來月食凡三十八事，差密近，有益，宜課試上．
>
> Fan and Tong examined and compared these results with the observation notes of the Clerks. The lunar motion does in fact have slowing and acceleration. This is not consistently [linked to its position] amongst [the solsticial lodges] Ox and Well, [or the equinoctial lodges] Harvester and Horn, nor is it what is called *tiao*

[54] *Song shu* 13, 287; also the Tang commentary on *Li ji* 禮記14, 12a (though it is the sun that is there said to have nine motions) in *Shi san jing zhu shu*, vol. 5, 279A. Commentators make spirited but not obviously successful attempts to deal with the problems presented by this material.

> 朓 or *ce ni* 側匿,[55] but is produced by the fact that the path followed by the moon varies in distance and goes to south and north of [the Yellow Road].[56] The rule is that one lunation shifts the old position of [maximum] speed by three *du*, so that in nine years the Nine Roads make one return, and in nine Rule Periods, which is 171 years [9 × 19 = 171] it returns to month 11, conjunction and winter solstice. This accords with the Spring and Autumn [annals], and with the reckoning of the conclusion [of the cycle] of the Nine Roads according to the Three Concordances, and can be used to know the times of occurrence of conjunctions, crescents, full moons and lunar eclipses. Relying on the Clerks' records of the degrees of heaven for the fractions and the rates, using these methods to look back to 38 instances of lunar eclipses since the Jianwu period [25–55 CE], the differences are very close. [This method] is useful, and is fitting to be checked and submitted to the throne for trial. (*Hou Han shu, zhi* 2, 3030; Cullen 2017, 390–391)

Let us consider this statement in detail. In the first place, what might the Clerks have actually seen if they observed the movement of the moon over a prolonged period, as they are said to have done? We know from another reference by Jia Kui that such observations were actually carried out between 85 and 89 CE (see section 6.3.2). Unfortunately the details of those observations have not been preserved, but modern calculations do allow us to reconstruct what could have been observed. I have used the NASA 'Horizons' online ephemeris program to compute lunar positions at daily intervals from 85 CE onwards, and from these I have calculated the movement along the lunar path from one position to the next. One year's worth of the results is shown in Figure 6.5.

As can be seen, the moon's speed varies from a minimum of 12 *du* per day to a maximum between 15 and 16 *du*: the moon certainly does have 'slowing and acceleration'. This is consistent with what reported as the results of the 85–89 CE observation programme by the Clerk's officials, which concluded that the daily lunar motion was between the limits 11 and 16 *du* (see 6.3.2).

Now what might be the basis of Jia Kui's statement that the position of the moon's maximum speed shifts by three *du* from one lunation to the next? As noted above, the period for the moon to return to the same position amongst

[55] These terms refer respectively to when the moon appears in the west on the last day of a lunar month, or in the east on the first day of a month. See *Wen xuan* 文選 *(Selected literature)*. (1965 reprint of 1936 edition). Xiao Tong 蕭統 (501–531 CE), Hong Kong, Commercial Press, 58, 1249; in the commentary, Zheng Xuan 鄭玄 (127–200 CE) attributes the first to the moon speeding up and the second to it slowing down.

[56] Jia Kui cannot simply be referring to the fact that the moon's path being close to the ecliptic means that it is sometimes to the north of the equator (the Red Road) and sometimes to the south, or he would not have informed us that the effect is not dependent on the position of the moon amongst the lodges.

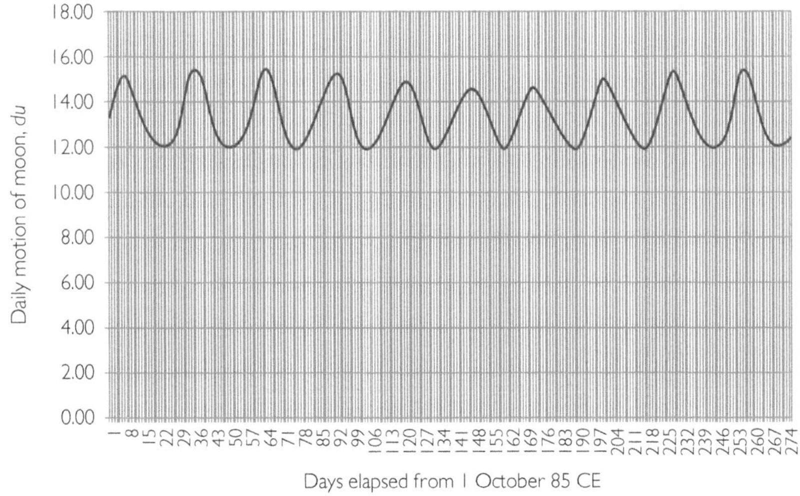

Figure 6.5 Graph showing daily displacement of moon from 1 October 85 CE.

the stars is the sidereal month, for which a modern value is 27.32166 days. However, the time taken for the moon's maximum speed to be reached again is a little longer, the anomalistic month, 27.55455 days. The difference between the two periods is:

$$(27.55455 - 27.32166) \text{ days} = 0.23289 \text{ days.}$$

If we take the moon as moving at about 13 *du* a day, the distance moved by the moon in this interval would be close to:

$$13 \times 0.23289\ du = 3.02757\ du = 3.0\ du, \text{ to 2 significant figures.}$$

This is the distance by which the moon will have shifted relative to the corresponding position a month earlier, and it is precisely the shift in the position of maximum speed identified by Jia Kui when he says 'one lunation shifts the old position of [maximum] speed by three *du*'. It seems therefore that Jia Kui has discovered the variation in the moon's speed due to its elliptic orbit, known in the western tradition as the 'elliptical anomaly', or 'equation of centre'.

However, Jia Kui mistakenly ascribes this variation to the same cause to which he correctly ascribes the annual variation in the sun's speed relative to the equator—the displacement of its path to northwards or southwards, but in the case of the moon with reference to the ecliptic rather than the equator. The fact that the moon moves north and south of the ecliptic is indeed the cause of a smaller part of the complex variation in lunar speed,[57] but the relevant cycle is of the order of half a month long—since for the same reasons that applied in the case with the sun's equatorial motion described earlier, there would be two maxima and minima of speed in each lunar circuit.

What about the reference to 'nine years'? If the position of maximum speed shifts by three *du* a month, then the number of anomalistic months needed for the position to return to the start will be:

$$365.25\ du\ /\ 3.00\ du = 121.75$$

In days, this amounts to:

$$121.75 \times 27.55455 \text{ days} = 3354.77 \text{ days,}$$

which is equivalent to 9.2 solar cycles. This is clearly the origin of Jia Kui's statement that 'in nine years the Nine Roads make one return'. It seems that when Jia Kui refers to the Nine Roads he is thinking of the mean paths taken by the moon in each year of the period. He notes also that this nine-year period recalls the earlier reference to the Nine Roads in the Triple Concordance specification quoted in section 6.3.5, where it was said that the 'Nine Roads reach their lesser conclusion' in 9 × 19 = 171 years – though as we saw, the Nine Roads in that instance seem more likely to have referred to the sun than to the moon.

Jia Kui certainly realizes that his discovery represents a potential advance in the accuracy of predictions of lunar position, since mean positions of the moon can be corrected to allow for the lags and leads produced by the variations in lunar speed. As he says, it 'can be used to know the times of occurrence of conjunctions, crescents, full moons and lunar eclipses'. And he claims that he has tested the results obtained on 38 lunar eclipses recorded since the start of the Eastern Han, and found that it produces a good fit between prediction and observation. The point of using lunar eclipses was presumably the fact that they can only occur when the moon is precisely opposite the sun, whose position is easy to calculate at a given date, so that it was not necessary to have good records of measured lunar positions so long as the date of the eclipse was known.

It seems, therefore, that Jia Kui's 'Nine Roads' procedure may have been a fully developed method for taking account of the moon's speed variation to make

[57] Ptolemy noted the existence of this effect, but decided to neglect it, since he considered the maximum effect on the moon's longitude measured on the ecliptic to be too small, at about 5′ of arc, $\frac{1}{12}°$ (a modern calculation gives 6′). See *Almagest* IV.6, Toomer (1998), 191.

more accurate predictions of the position of the moon. Unfortunately, Cai Yong and Liu Hong were apparently unable to find any records of how the procedure worked. All they can say after their account of Jia Kui's work is:

> 案史官舊有九道術，廢而不修.
>
> Note that previously the Clerks had the Nine Roads method, but it was abandoned before it had been perfected. (*Hou Han shu, zhi* 2, 3032; Cullen 2017, 391)

As we shall see, despite a number of references to the Nine Roads in the years that followed, it was not until the work of Liu Hong himself that a systematic theory of the moon was first laid out in detail: see section 8.2.

6.4 The solar tables of Huo Rong: fitting it all together (but not quite ...)

Not long after the death of Jia Kui, in the fourth year of the Yongyuan 永元 reign period (102 CE), one of the Expectant Officials for whom he had been responsible, Huo Rong 霍融, launched a critique of the practice of official sky-watchers in another area of their responsibilities—timekeeping by means of a water-clock, or clepsydra.

As set out in chapter 5, section 5.2, the simple outflow type clepsydras known from this period indicated time based on the falling level of water in a container that drained slowly through a narrow spout. A graduated indicator rod was attached to a float inside the vessel and protruded through a slot in the lid. As the water level fell, the graduation on the indicator rod opposite a fixed pointer indicated the time elapsed, based on the system of dividing the cycle of day and night into a hundred equal *ke* 刻 'graduations' or 'marks'. It was evidently the practice for those in charge of clepsydras to set up one clepsydra 'run' for the day, and a separate 'run' for the night. Day and night each had their own indicator rods, and their length indicated the expected duration of the run expected: a short rod would be used for a short night, and a long rod (hence carrying more *ke* graduations) would be used for a long night. The question was, given the annual variation in day and night lengths, how was one to know what lengths of rod were appropriate at different times of the year? Huo Rong pointed out:

> 官漏刻率九日增減一刻，不與天相應，或時差至二刻半.

> The rate for official clepsydra graduations is that they are increased or decreased by one *ke* [each] nine days. This does not correspond to the heavens: sometimes the difference in timing is as much as 2 ½ ke. (*Hou Han shu, zhi* 2, 3032; Cullen 2017, 391–2)

This very simple scheme, based on an arithmetical progression with constant differences, has ancient parallels in other cultures. As Neugebauer notes, a scheme based on variation of daylight between 18h and 6h, with a change of 2h each month, is found in an Egyptian papyrus dated from c. 1200 BCE, and other schemes based on a constant change in day-length at fixed time intervals are found in Greek texts from around 300 BCE and 180 BCE.[58] As Huo Rong notes, the crude scheme he criticizes will inevitably differ significantly from observation.[59] By the time of Ptolemy in the second century CE, and possibly earlier, in the time of Hipparchus in the second century BCE, accurate trignometrical methods were available that made it possible to calculate day-lengths to high accuracy, although not everybody made use of them, and arithmetical progressions continued in use for some time.[60]

The Grand Clerk and his officials were ordered to respond. Like good civil servants in every age and clime, they began by pointing out that they had only been acting in conformity with well-established practice:

> 案官所施漏法令甲第六常符漏品，孝宣皇帝三年十二月乙酉下，建武十年二月壬午詔書施行.
>
> We note that method for using clepsydras practised by the officials is modelled on Order A, [section] number 6, category 'Perpetually correct clepsydras', of the third year of the [late] Filial Xuandi[61] [69 BCE], the 12th month, day *yiyou*.22, which was ordered to be followed by an edict of the tenth year of the Jianwu period [34 CE], second month, day *renwu*.19. (*Hou Han shu, zhi* 2, 3032; Cullen 2017, 392)

[58] See Neugebauer (1975), 706–7; see also the review of such methods in *Geminos's introduction to the Phenomena: a translation and study of a Hellenistic survey of astronomy*, 73–82.

[59] As we shall shortly see, values of 65 *ke* and 45 *ke* were assumed for daylight at the summer and winter solstices: these included 2.5 *ke* of twilight at each end of the day. A trial of the '1 *ke* each 9 days' rule based on these extreme values suggests that the resulting day-lengths can differ from values likely to have been observed by over 2 *ke*, as suggested by Huo Rong.

[60] See *Geminos's introduction to the Phenomena: a translation and study of a Hellenistic survey of astronomy*, 74. On Ptolemy's methods for predicting day-length, see *Almagest* II.8 and 9 in Toomer (1998), 99–104; Ptolemy provides tables based on his own trigonometrical calculations that can be used to find (amongst other things) day-lengths, without his readers actually having to do any trigonometry themselves. A modern explanation of the trigonometrical basis of such tabulations is given in Smart and Green (1979 (reprint of 6th edition 1977)), 46–8.

[61] 'Filial' *xiao* 孝 was conventionally prefixed to the posthumous name of an emperor, in this case Xuandi, in formal official references.

Huo Rong's arguments had evidently proved persuasive, since they hastily added:

> 漏刻以日長短為數，率日南北二度四分而增減一刻．一氣俱十五日，日去極各有多少．今官漏率九日移一刻，不隨日進退．
>
> [However,] clepsydra graduations [ought to] take their numbers from the varying length of the day, the rate being that for every 2 ¼ *du* [of movement] southwards or northwards by the sun they increase or decrease by one graduation. One *qi* is 15 days, and [for each *qi*] the sun's polar distance varies. Now the rate for official clepsydras is to shift by one graduation for each nine days. This does not follow the advancing or retreating of the sun. (*Hou Han shu, zhi* 2, 3032; Cullen 2017, 392)

So now instead of equal changes in day-length for equal time intervals, there are to be equal changes in day-length for equal changes in the sun's north polar distance. A little later we shall consider how far this was an improvement in accuracy of prediction. Not long afterwards an edict was issued giving approval to the new way of doing things:

> 其年十一月甲寅，詔曰：「告司徒，司空：漏所以節時分，定昏明．昏明長短，起於日去極遠近，日道周〔圜〕，不可以計率分，當據儀度，下参晷景．今官漏以計率分昏明，九日增減一刻，違失其實，至為疏數以耦法．太史待詔霍融上言，不與天相應．太常史官運儀下水，官漏失天者至三刻．以晷景為刻，少所違失，密近有驗．今下晷景漏刻四十八箭.」[62]
>
> That same year, in the 11th month, day *jiayin*.51, an edict stated: 'Let it be announced to the Minister of Education, and to the Minister of Works: the clepsydra serves to regulate the divisions of time, to fix darkness and daylight. The varying lengths of darkness and daylight arise from the variations of the sun's polar distance. The way of the sun follows round in a circle, and cannot be divided up according to calculated rates: one has to rely on instrumental measurements, and align it with the sundial shadow. Now the official clepsydra divides up darkness and daylight by calculated rates, and adds or subtracts one graduation each nine days, which is contrary to reality, and at the extreme may lead to an inaccurate number derived from an arbitrary method. The Expectant Official under the Grand Clerk, Huo Rong, has sent up a memorial saying 'This is not answerable to the heavens'. The Clerks of the Chamberlain for Ceremonials have operated their instruments and let water flow down, [and found

[62] I follow the suggestion of the Zhonghua edition collation note that the characters '立成斧官府當用者, 計吏到, 班予四十八箭', to which it is hard to attach any clear meaning, are intrusive and should be omitted.

> that] the officials' clepsydra misses the heavens by as much as three graduations. Using the sundial shadow to make the graduations diminishes the error, and its accuracy can be verified. Now We decree 48 indicator-rods [lit. 'arrows'] for the clepsydra graduations [from the] sundial shadow.' (*Hou Han shu, zhi* 2, 3033; Cullen 2017, 392–3)

And there the account in the first of the two chapters by Cai Yong and Liu Hong ends. But more is to be found elsewhere, as the following note by the two editors tells us:

> 文多，故(魁)取二十四氣日所在，并黃道去極，晷景，漏刻，昏明中星刻于下.
>
> The text is lengthy, so we have set out below [i.e. in the next chapter of *Hou Han shu*] the positions of the sun at the 24 *qi*, together with the distance of the ecliptic from the pole, the solar shadow, the graduations of the clepsydra, and the graduations of darkness and daylight and the centred stars. (*Hou Han shu, zhi* 2, 3033; Cullen 2017, 393)

We find this, and more, in the next chapter (*Hou Han shu, zhi* 3, 3077–9; Cullen 2017, 224–31).[63] Here, in Table 6.2, is a rendition of the table there given:

6.4.1 The structure of the table

The entire table is structured around the division of the solar cycle from one winter solstice to the next into 24 equal intervals, the *qi* 氣. In the presentation of the table in pre-modern Chinese editions, the names of the *qi* are given as headings to groups of data for each *qi*; there is no attempt at aligning each type of datum into rows and columns: see Figure 6.6. In the translated version, the table has been re-oriented in accordance with modern western usage. But since the way data are presented is inseparable from the ways in which they were understood and used, it is as well to be aware that the way the data were traditionally presented gave the reader an impression quite different from the highly correlated modern style used in Table 6.2. The ninth column from the right in the right-hand page of the two shown in Figure 6.6 has the words *er shi si qi* 二十四氣 'The 24 *qi*', in large characters, followed at the head of the tenth column by the name of the first *qi*, *dong zhi* 冬至 'Winter Solstice'. The rest of the column is taken up by commentary in smaller characters on the traditional significance of this *qi*. The first column on the left-hand page begins (slightly lower that the preceding heading) with the words *ri suo zai* 日所在 'position of the sun', followed in small characters

[63] Parts of the following discussion were first set out in Cullen (2007a).

Table 6.2 ***The solar table of 102 CE***

The 24 *qi* (number 1 is winter solstice,13 is summer solstice)	Solar position		Shadow of 8 *chi* gnomon/ *chi*	Polar distance of ecliptic/ *du*	Day clepsydra/ *ke*	Night clepsydra/ *ke*	Dusk centred star		Dawn centred star	
	Lodge number	*Du* in lodge (advance/ retard)					Lodge number	*Du* in lodge (advance/ retard)	Lodge number	*Du* in lodge (advance/ retard)
1	1	21 8/32 (−2)	13	115	45	55	8	5 11/12 (0)	23	2 4/12 (−1)
2	3	2 7/32 (+1)	12 30/100	113 1/12	45 8/10	54 2/10	9	6 7/12 (−1)	24	7 2/12 (−2)
3	4	5 14/32 (+2)	11	110 8/12	46 8/10	53 2/10	10	11 7/12 (−1)	26	6/12 (−3)
4	5	10 21/32 (+2)	9 60/100	106 4/12	48 6/10	51 4/10	12	5 2/12 (−3)	27	7 5/12 (−3)
5	6	8 28/32 (+3)	7 95/100	101 1/12	50 8/10	49 2/10	14	6 5/12 (−4)	28	8/12 (−3)
6	7	8 3/32 (+1)	6 50/100	95 1/12	53 3/10	46 7/10	15	17 2/12 (−3)	1	3/12 (−2)
7	8	14 10/32 (0)	5 20/100	89 4/12	55 8/10	44 2/10	16	4 (0)	1	10 11/12 (−2)
8	10	1 17/32 (−1)	4 15/100	83 2/12	58 3/10	41 7/10	18	4 9/12 (−1)	1	21 6/12 (−2)
9	11	2 24/32 (−2)	3 20/100	77 10/12	60 5/10	39 5/10	19	17 (1)	2	6 6/12 (0)
10	12	6 31/32 (−3)	2 52/100	73 2/12	62 4/10	37 6/10	20	17 9/12 (2)	3	10 3/12 (1)
11	14	4 6/32 (−4)	1 98/100	69 8/12	63 9/10	36 1/10	22	0 8/12	5	8/12 (2)
12	15	10 13/32 (−3)	1 68/100	67 2/12	64 9/10	35 1/10	23	5 9/12 (−1)	5	14 1/12 (2)
13	15	25 20/32 (−3)	1 50/100	67 1/12	65	35	24	12 2/12 (−2)	6	12 2/12 (3)
14	17	3 27/32 (0)	1 70/100	67 10/12	64 7/10	35 3/10	27	1 10/12 (−3)	8	2 10/12 (0)
15	18	4 2/32 (+1)	2	70	63 8/10	36 2/10	27	15 5/12 (−3)	9	3 9/12(−1)

continued

Table 6.2 ***Continued***

16	19	$12\frac{9}{32}$ (+1)	$2\frac{55}{100}$	$73\frac{1}{12}$	$62\frac{3}{10}$	$37\frac{7}{10}$	28	$9\frac{10}{12}$ (−3)	10	$9\frac{8}{12}$ (−1)
17	20	$9\frac{16}{32}$ (+2)	$3\frac{33}{100}$	$78\frac{7}{12}$	$60\frac{2}{10}$	$39\frac{8}{10}$	1	$10\frac{3}{12}$ (−2)	12	$3\frac{9}{12}$ (−3)
18	21	$6\frac{23}{32}$ (+1)	$4\frac{35}{100}$	$84\frac{4}{12}$	$57\frac{8}{10}$	$42\frac{2}{10}$	1	$21\frac{1}{12}$ (−2)	14	$5\frac{5}{12}$ (−4)
19	22	$4\frac{20}{32}$ (0)	$5\frac{50}{100}$	$90\frac{7}{12}$	$55\frac{2}{10}$	$44\frac{8}{10}$	2	$5\frac{3}{12}$ (0)	15	$16\frac{4}{12}$ (−3)
20	23	$8\frac{5}{32}$ (−1)	$6\frac{85}{100}$	$96\frac{10}{12}$	$52\frac{6}{10}$	$47\frac{4}{10}$	3	$7\frac{9}{12}$ (1)	16	$3\frac{4}{12}$ (0)
21	24	$14\frac{21}{32}$ (−2)	$8\frac{40}{100}$	$102\frac{4}{12}$	$50\frac{3}{10}$	$49\frac{7}{10}$	4	$6\frac{9}{12}$ (2)	18	$3\frac{10}{12}$ (1)
22	27	$4\frac{19}{32}$ (−3)	10	$107\frac{4}{12}$	$48\frac{2}{10}$	$51\frac{8}{10}$	5	$8\frac{1}{12}$ (2)	19	$15\frac{10}{12}$ (1)
23	28	$1\frac{26}{32}$ (−3)	$11\frac{40}{100}$	$110\frac{11}{12}$	$46\frac{7}{10}$	$53\frac{3}{10}$	6	$3\frac{7}{12}$ (3)	20	$15\frac{10}{12}$ (2)
24	1	$6\frac{1}{32}$ (−2)	$12\frac{56}{100}$	$113\frac{10}{12}$	$45\frac{5}{10}$	$54\frac{5}{10}$	7	$0\frac{7}{12}$ (1)	21	$15\frac{3}{12}$ (1)

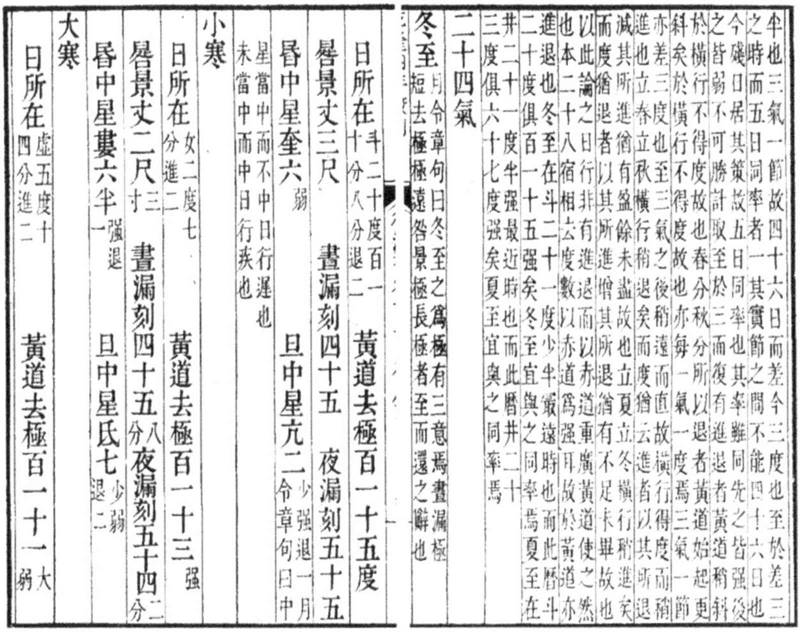

半也三氣一節故四十六日而差今三度也至於差三
之時而五日同率者一其實節之間不能四十六日也
今䟽日居其策故五日同率也其率雖同先之皆強後
之皆弱不可勝計取至於三而復有進退者黃道稍斜
於橫行不得度故也春分秋分所以退者黃道始起更
斜矣於橫行不得度故也亦每一氣一度焉三氣一節
亦差三度也至三氣之後稍遠而直故橫行得度而稍
進也立春立秋橫行稍退矣而度猶云進者以其所退
減其所進猶有盈餘未盡故也立夏立冬橫行稍進矣
而度猶退者以其所進增其所退猶有不足未畢故也
以此論之日行非有進退而以赤道重廣黃道使之然
也本二十八宿相去度數以赤道為強耳故於黃道亦
進退也冬至在斗二十一度少半最遠時也而此曆斗
二十度俱百一十五強矣冬至宜與之同率焉夏至在
井二十一度半強最近時也而此曆井二十
三度俱六十七度強矣夏至宜與之同率焉

二十四氣

冬至 月令章句曰冬至之為極有三意焉晝漏極短去極極遠晷景極長極者至而還之辭也

日所在斗二十度百一十分八分退二 黃道去極百一十五度
晷景丈三尺 晝漏刻四十五 夜漏刻五十五
昏中星奎六弱 旦中星亢二少強退一月令章句曰中
星當中而不中日行遲也
未當中而中日行疾也

小寒

日所在女二度七分進二 黃道去極百一十三強
晷景丈二尺三寸 晝漏刻四十五八分 夜漏刻五十四分二
昏中星婁六半強退一 旦中星氐七少弱退二

大寒

日所在虛五度十四分進二 黃道去極百一十一大弱

Figure 6.6 Page from the 1739 Wu ying dian 武英殿 edition of *Hou Han shu, zhi* 3, 21a–21b, showing part of the table of solar data.

by the name of the lodge in which the sun is found at winter solstice, the position of the sun within that lodge, in *du*, and the note *jin er* 進二 'advance: 2'. These data are given in the second and third columns of Table 6.2. Further down we read *huang dao qu ji bai yi shi wu du* 黃道去極百一十五度 'the ecliptic is 115 *du* from the pole.' The next two columns give the length of noon shadow, the number of *ke* in day and night, and the stars that will be seen on the meridian at dusk and dawn. All these data are given in following columns of Table 6.2. The fifth column of the left-hand page is headed with the name of the second *qi*, *xiao han* 小寒 'Lesser Cold' in large characters, and is followed by corresponding data for that lodge; the name of the third *qi*, *da han* 大寒 'Great Cold' heads the ninth column, and data for this and subsequent *qi* follow on later pages.

6.4.2 The observational basis of the table

Apart from the position of the sun amongst the lodges, to which we shall return, there are two columns in the translated version of the table that appear to be observed data. These are:

(a) The third column from the left, which gives the shadow cast at noon by an 8-*chi* gnomon on the day when a given *qi* begins.
(b) The fourth column from the left, which gives the distance from the north celestial pole (north polar distance, often abbreviated as NPD) of the point on the ecliptic where the sun is located at the moment of inception of a given *qi*.

My analysis suggests that the shadow lengths given are extremely plausible for the likely place of observation, which is the Eastern Han capital of Luoyang 洛陽[64]—if one assumes, which seems highly probable, that the dates for the various *qi* dates are those calculated on the basis of the Han Quarter Remainder system in the year the table was created, rather than the actual dates of those *qi* in modern terms.[65] The lengths listed are a little less than those that result from calculating the distance from the foot of the gnomon to the point where a line from the centre of the sun's disc to the top of the gnomon intersects the ground. The shadows given are mostly within 5⁄100 *cun* (about 1 mm) of the distance to the end of the umbra of the gnomon, which corresponds to the point where a line from the sun's upper limb to the top of the gnomon strikes the ground. That seems a reasonable enough measurement of shadow length for any observer to have made. As for the solar north polar distance values, these suggest the presence of a systematic error resulting from the axis of the observing device, such as an armillary ring, being aligned about 0.7° too high for the observer's latitude.[66] Allowing for that error, the figures given are within about a ¼ *du* of the actual values on the calculated dates.[67]

6.4.3 Calculated quantities

The fact that the north polar distance and shadow length values in the table were observed rather than calculated is consistent with the edict of 102 CE (see 6.4): 'The way of the sun follows round in a circle, and cannot be divided up according to calculated rates: one has to rely on instrumental measurements, and align

[64] More precisely, the *ling tai* 靈臺 observatory, whose remains are at latitude 34° 42′N, longitude 112° 37.8′ E. See above, footnote 14.

[65] That is why the shadow lengths, the polar distances and the day and night clepsydra runs at the spring and autumn equinoxes are not equal.

[66] The view that there was a systematic error of this magnitude is strengthened by the fact that an independent analysis suggested a 0.86° misalignment in a similar case. This is the misalignment of the axis of the observer's instrument calculated by Sun and Kistemaker in analysing the north polar distance values of stars that they concluded were determined c. 78 BCE: see Sun and Kistemaker (1997), 64.

[67] See Cullen (2007a), 84–95.

it with the sundial shadow.' But immediately before that, we read 'The varying lengths of darkness and daylight arise from the variations of the sun's polar distance'. And what is more, the description of the Han Quarter Remainder system tells us exactly how to find the times in question from solar NPD:

> 黃道去極，日景之生，據儀，表也．漏刻之生，以去極遠近差乘節氣之差．如遠近而差一刻，以相增損．
>
> For producing the distance of the Yellow Road from the pole, and the solar shadow, one relies on the [armillary] instrument and the gnomon.
>
> For producing the clepsydra *ke*, multiply the differences in distance from the pole by the differences for the Nodal Qi, with a difference of 1 *ke* being counted as this accords with the varying distance, then let them be added and subtracted. (*Hou Han shu, zhi* 3, 3075–6; Cullen 2017, 224)

It is easy to show that following this prescription generates the precise day and night lengths given in the table: see Box 6.2. Once we have found the lengths of day and night for any given *qi*, it is not difficult to calculate the figures given in the last four columns of the table, which are the points in the relevant lodges

Box 6.2: Finding day and night lengths from solar north polar distance

Looking at the figures for day and night length at the solstices (which are what are meant by 'the Nodal Qi' in the prescription for day and night lengths given in the main text), we have:

Winter solstice:	day 45 *ke*	night 55 *ke*
Summer solstice:	day 65 *ke*	night 35 *ke*

Thus the '[clepsydra] difference for the nodal *qi*' is 20 *ke* for both day and night.

The lack of symmetry in these data points to the fact that, in accordance with early Chinese practice, daytime is not counted as running from sunrise to sunset, but from dawn to dusk, with a conventional 'twilight' of 2.5 *ke* (36 minutes) at each end of the day. Thus the underlying solsticial day:night ratio is 40:60, equivalent to winter solstice sunrise at 7:12 and sunset at 16:48. At Luoyang the actually observed local times of rising and setting at date of the winter solstice of 101 CE as calculated by the Han Quarter Remainder system (on 24 December, which was about a day later than the actual solstice), would be 7:09 and 16:57 respectively. Since, as we have already seen (section 5.5), a nine-hour run by

continued

Box 6.2: *Continued*

a simple clepsydra might have given a timing error of the order of 8 minutes, the conventional ratios could be used without any obvious discrepancy being revealed.

Proceeding as specified, we see from the table:

Winter solstice north polar distance of ecliptic:	115 *du*
Summer solstice north polar distance of ecliptic:	67 $\frac{1}{12}$ *du*

So the difference between these (which is what is meant by "the [difference between] nearest and furthest") is 47 $\frac{11}{12}$ *du*. This corresponds to the difference in day-length of 20 *ke* as found earlier, so to 1 *ke* there corresponds to a solar north-south displacement of 2 $\frac{5}{12}$ *du*, to the nearest $\frac{1}{12}$ *du*, not far from the approximate value of 2 $\frac{1}{4}$ *du* mentioned by the Grand Clerk's officials in their response to Huo Rong.

Suppose we want to predict the day clepsydra for the second *qi*, we note from the table the ecliptic north polar distance for that *qi*, which is 113 $\frac{1}{12}$ *du*, a change of 1 $\frac{11}{12}$ *du* since the winter solstice. We now obey the instruction to 'multiply the differences in [ecliptic] distance from the pole by the [clepsydra] difference for the nodal *qi*', obtaining:

$$(1\tfrac{11}{12}\ du) \times (20\ ke) = 460/12\ du.ke$$

We are now told that 'a difference of 1 *ke* [is] counted as this accords with the [difference between] nearest and furthest', in other words we are to divide this figure by the 47 $\frac{11}{12}$ *du* found above, to obtain:

$$(460/12\ du.ke)\ /\ (47\tfrac{11}{12}\ du) = (460/12\ du.ke)\ /\ (575/12\ du)$$
$$= 460/575\ ke = 8/10\ ke \text{ precisely}$$

This is the difference that must be added to the previous day clepsydra value of 45 *ke* to obtain 45 $\frac{8}{10}$ *ke*, which is in fact the tabulated value for day length at the second *qi*. Proceeding similarly, using the north polar distances of the sun for each *qi* in turn, the clepsydra values for day and night at all other *qi* may be obtained to the nearest $\frac{1}{10}$ *ke*, exactly as tabulated, and the dusk and dawn stars thus follow as explained.

which will be on the meridian at dusk and dawn. We know the location of the sun amongst the lodges, and the sun will be on the meridian at noon. It is therefore possible to find what *du* of which lodge will cross the meridian a known time earlier (at dawn) or later (at dusk). The results given in the table can thus be reproduced precisely, showing that we have indeed understood the thinking on which it is based.[68]

Of course, this procedure can only produce results that approximate to observation. It is, however, a great improvement on the previous simple arithmetic progression method of '1 *ke* for each 9 days'. A check against modern calculations for the expected day-length for the tabulated north polar distance of the sun for each *qi* at the latitude of Luoyang suggests that the error of the new method is less than 1 *ke* of day-length for 20 *qi* out of the 24.

6.4.4 Advances and retardations

Finally, what is to be done about the figures for 'advances and retardations' given after some of the positions of the sun in the table? In the translated version, these are shown as respectively positive and negative. Identical 'advances and retardations' were given in the listing of the equatorial widths of the 28 lodges cited earlier in the *Hou Han shu* system specification.

Again, we may look to the *Hou Han shu* system specification to find out how to apply these figures:

> 其以赤道命度，進加退減之．其步以黃道．
>
> You are to count off the *du* along the Red Road. Add to this for advance and subtract for retardation, and the motion is [reckoned] according to the Yellow Road. (*Hou Han shu, zhi* 3, 3074; Cullen 2017, 219–220)

So what do we have here? It seems at first sight as though we are being given a means of converting motion with respect to the equatorial co-ordinate system to the ecliptic. If so, this would be a considerable advance in the consistent representation of solar motion, and would put an end to the problems signalled by Jia Kui. It would represent the inverse of what in modern terms is called the 'reduction to the equator', the difference between the longitude of the sun and its right ascension.[69]

[68] See Cullen (2007a), 89–90.
[69] See Smart and Green (1979 (reprint of 6th edition 1977)), 148.

In fact, however, it is nothing of the kind. The Han Quarter Remainder system simply calculates the motion of the sun from one *qi* to another with respect to the equator by counting off equal increments of 15 7⁄32 *du* (one 24th of 365 ¼ *du*) from the winter solstice position at 21 ¼ *du* of the lodge Dipper. The quantities in the third column of the translated table are the result. If we add the advances to these positions and subtract the retardations, the result is a new set of 24 positions – which turn out to be exactly those positions relative to the placing of the lodges on the ecliptic that correspond to the sun moving 15 7⁄32 *du* along the ecliptic from one *qi* to the next![70] It seems that the only lasting result of Jia Kui's careful arguments against the idea of constant solar speed along the equator, and in favour of constant solar speed along the ecliptic (thus producing the observed changes in solar speed with respect to the equator) has been to force the Grand Clerk's officials into devising a scheme that did the impossible—postulating constant solar speed along the equator and ecliptic simultaneously!

Fortunately, a solution was eventually found to this impasse, and we shall now consider the work of the person who found it.

6.5 Zhang Heng: a reputation, its origins, and its consequences

Zhang Heng 張衡 (78–139 CE) already had an impressive reputation not long after his death, and it has certainly not diminished since. In a recent English language biographical encyclopaedia, he was termed 'one of the finest intellectuals of Later Han, a brilliant scientist and mathematician'.[71]

His biography in the *Hou Han shu* says of him:

> 衡善機巧，尤致思於天文，陰陽，歷筭.
>
> [Zhang] Heng excelled in ingenuity,[72] and devoted much attention to the patterns of heaven, Yin and Yang, and calculations [in] relation to astronomical systems (*Hou Han shu* 59, 1897).

70 See Cullen (2007a), 87–88.

71 De Crespigny (2007), 1049.

72 The most common sense of the expression *ji qiao* 機巧 is 'cunning, ingenious'. In some instances it clearly refers to ingenuity that is mechanical as well as literary (as a parallel, English 'engineer' is of course cognate with 'ingenuity')—but it may be that this interpretation originated from its occurrence in this particular passage, in connection with a person, Zhang Heng, who made mechanical devices, as well as displaying purely intellectual talent.

An inscription on his tomb stele claimed;

> 數術窮天地，制作侔造化.
>
> With the arts of number, he exhausted heaven and earth; as a maker[73] he was equal to the creative [power of nature]. (*Hou Han shu* 59, 1940)

His talents are said to have extended to making a device that could detect the tremors caused by earthquakes;[74] he was also said to have excelled in poetry and classical learning from an early age. Many fragments of texts by him have survived in quotations and commentaries;[75] we shall consider below the contents of two of these that are particularly relevant to our concerns. He is also cited by Liu Hui, the writer of the great commentary on the *Jiu zhang suan shu* 九章算術 'Mathematical procedures in nine sections' for work on the volume of a sphere.[76]

He twice held the office of Grand Clerk. He was first appointed under Andi (r. 107–125 CE); some time after taking office he left the post for five years, and took it up again in 126 CE on the accession of Shundi. He accompanied his return to office with a lengthy and elegant justification for his sabbatical, *Ying xian* 應閒 'Answering for leisure'.[77] The post of Grand Clerk was by no means a minor one, although it was not one of the highest offices of state; under the Eastern Han, it carried a salary of 600 measures, *shi* 石, of grain.[78] The official job description says of his responsibilities:

[73] The expression *zhi zuo* 制作 can refer to making something physical or to creating a literary work; hence I choose the expression 'maker' to reflect this ambiguity.

[74] See *Hou Han shu* 59, 1909; also Needham and Wang Ling (1959), 626–635. The description of the device that has come down to us is sufficiently detailed to have led to numerous attempts at reconstruction, but not quite clear enough to enable any of the attempts to be taken as definitive. A memorial submitted by Zhang Heng on the occasion of an earthquake in 133 CE does not mention this device, but makes it plain that an earthquake is a portent warning of faults in government just as much as signs in the heavens—hence perhaps the need to be sure that they are detected. See *Hou Han ji* 後漢紀 *(Annals of the Eastern Han dynasty)*. Yuan Hong 袁宏 (328–376 CE), Shanghai, Commercial Press, Si bu cong kan 四部叢刊 series, 18 15a: *yao xing xian yu shang, zhen lie zhu yu xia: tian jie xiang yi* 妖星見於上, 震裂著於下: 天誡詳矣 'Baleful stars appear on high, while earthquakes and fissures are seen below: Heaven's warnings are explicit'.

[75] See those collected and collated in Yan Kejun 嚴可均 (1762–1843) (1965 (repr. of 1958 edn, plus indices)) *Quan shang gu san dai Qin Han San guo Liu chao wen* 全上古三代秦漢三國六朝文 *(Complete texts of High Antiquity, Three Dynasties, Qin, Han, Three Kingdoms and Six Dynasties)*. Beijing, Zhonghua Shuju vol. 1, 759–79.

[76] See Guo Shuchun 郭書春 (2004) *Hui jiao jiu zhang suan shu* 匯校九章算術 *(Comprehensively collated [edition of the text] Mathematical procedures in nine sections]*. Shenyang 瀋陽, Liao ning jiao yu chu ban she 遼寧教育出版社 (Liaoning educational publishers), vol 1, 143; Chemla and Guo (2004), 383.

[77] See *Hou Han shu* 59, 1897–8.

[78] One *shi* as a volume measure of grain was equivalent to about 20 litres: see M.A.N. Loewe (1961) 'The measurement of grain during the Han period.' *T'oung Pao* 49: 64–95. At UK prices in 2013, good quality rice sells for about £1.50/kg, so given that a litre of rice weighs close to a kilogram,

> 掌天時，星曆．凡歲將終，奏新年曆．凡國祭祀，喪，娶之事，掌奏良日及時節禁忌．凡國有瑞應，災異，掌記之．
>
> He takes charge of the times of heaven, and the sequence of the stars. When the year nears its end, he submits the calendar for the new year [to the emperor]. If there are state sacrifices, mourning or marriage ceremonies, he takes charge of submitting a favourable date, and [noting] the prohibitions appropriate to the season. If there are favourable omens for the state, or disasters and prodigies, he takes charge of recording them. (*Hou Han shu* 35, 3572)

The Grand Clerk's staff consisted of some seventy persons, ranging from what might be called 'back office staff' such as the six people whose job was said to be *zhi li* 治曆 'dealing with the [astronomical] system', who were presumably calculators, and others responsible for various forms of divination, including the *Book of Change*, and nine archivists. Staff who served on the observatory, *ling tai* 靈臺 'Numinous Terrace' itself included '14 persons who observe the stars, two who observe the sun, three who observe the winds, twelve who observe the vapours, three who observe the gnomon shadow, and seven who observe the [notes of] the bells and pitchpipes'. There was also a doctor, and a janitor, the former presumably being essential to treat those who fell ill after watching the sky through long, cold winter nights.[79]

Slightly surprisingly, therefore, Zhang Heng's appearance in the materials collected by Cai Yong and Liu Hong is only a brief one, and refers to events in 122 CE, when he held the office of *shang shu lang* 尚書郎 'Gentleman of the Secretariat'. He and one of his colleagues were slightly tersely described as *neng li* 能曆 'capable in [astronomical] systems', and were therefore consulted about a proposal to change the system origin in use. Zhang Heng and his colleague both came in for some criticism in the submission that finally won imperial approval.[80] There is even a slight note of exasperation in that document, which says of the fact that the winter solstice was no longer in the lodge Ox: 'That things obviously are this way is common knowledge amongst the Clerk's officials—not

a *shi* would be worth about £30, and the Director's salary rating would be worth about £6000. Even allowing for the great difference between ancient and modern values of commodities, such calculations make Han officials look badly off, but as pointed out in Hans Bielenstein, *The bureaucracy of Han times* (Cambridge [etc.], 1980) 125–131, by the period we are discussing the nominal salary in grain was little more than a grade ranking. Part of the salary was paid in cash equivalent, and there were additional allowances and regular gifts in kind. As a result, the Director and his family were able to enjoy a standard of living far higher than that of most imperial subjects.

[79] See *Hou Han shu* 35, 3572, commentary. For more on the Grand Clerk and his staff, see Morgan (2013), 76 ff.

[80] *Hou Han shu zhi* 2, 3034–5.

just [Zhang] Heng and [Zhou] Xing.' (*zhao ran ru ci, shi guan suo gong jian, fei du Heng, Xing.* 昭然如此, 史官所共見, 非獨衡, 興: *Hou Han shu zhi 2*, 3035). Could it have been that Zhang Heng was seen by those with less literary reputation and more technical expertise as 'too clever by half', a person who was good at impressing others but did not always take his responsibilities as seriously as he should? In that connection, we may note that Liu Hui's discussion of Zhang Heng's views on the volume of a sphere ends with the words:

> 欲協其陰陽奇偶之說，而不顧踈密矣．雖有文辭，斯亂道破義，病也．
>
> He aimed to unite the notions of *yin* and *yang*, and of odd and even, but did not pay attention to accuracy. Even though he had literary elegance, this merely disordered his Way and harmed his meaning – and that was his defect. (See note 372.)

It may perhaps be that the survival of such works by Zhang Heng as have come down to us has given us a disproportionate impression of his importance in the field of celestial observation and calculation, an impression that experts later in the Eastern Han did not necessarily share, at least so far as their specialist area of interest was concerned.

Or it may be that there were other factors apart from professional skill: we are told that around 110 CE a request for him to be allowed to work with scholars who were researching historical records in the Dong guan 東觀 confidential archives was refused, and that later in his life he was denied access despite repeated requests on his part (note 3). This is said to have been related to his belief that the succession of recognized emperors of the Han after Wang Mang should commence with the Gengshi 更始 (r. 23–25 CE) emperor rather than with Guangwudi 光武帝 (r. 25–57 CE) who replaced him. He is also known to have opposed the imperial practice of treating the apocryphal books *chen wei* 讖緯 as authoritative. All this may have added a political taint to any problems about his technical expertise on the part of those who, rightly or wrongly, did not see him as entirely serious in the areas in which they worked.

6.5.1 The nature and size of heaven and earth, and of the heavenly bodies

There are two texts by Zhang Heng that are of particular interest to us. One of them, *Ling Xian* 靈憲, whose title may mean something like 'The Numinous Explained', gives an account of the origin of the universe from its first emergence from undifferentiated chaos, and goes on to set out the shapes and sizes of heaven and earth, and the nature, properties and movements of the heavenly bodies.[81] The other, which bears the title *Hun tian yi* 渾天儀 'The spherical heaven instrument', is in fact a detailed account of the layout of the celestial sphere in terms of the celestial equator, the ecliptic, and other important features, and makes no direct reference to the construction of an observational device.[82]

The account given by the *Ling xian* is the earliest surviving description of the overall layout of heaven and earth according to the *hun tian* view, and of their dimensions. The account of the *gai tian* universe given in the *Zhou bi* has prepared us for large numbers: the observer assumed in the *Zhou bi*, supposedly somewhere within the territory of Han empire, was located 103,000 *li* from the pole, and heaven was 80,000 *li* above his position. Since one *li* of the Han period was about 0.415 km, the first distance was more than three times the modern value for the diameter of the earth, which would be about 30,000 *li*. But shapes come before sizes, so first we are told why heaven and earth are as they are. First of all, Zhang Heng tells us, there was nothing at all in existence, until after long ages 'Existence was born of non-existence' *zi wu sheng you* 自無生有, and a mass of undifferentiated *qi* appeared 'vastly immense' *mang hong* 庬鴻. Then this *qi* began to subdivide:

> 於是元氣剖判，剛柔始分，清濁異位．天成於外，地定於內．天體於陽，故圓以動；地體於陰，故平以靜．
>
> At this stage, the original *qi* differentiated, hard and soft first divided, pure and turbid took up different positions. Heaven formed on the outside, and Earth became fixed within. Heaven took its body from the Yang, so it was round and in motion; Earth took its body from the Yin, so it was flat and quiescent. (*Hou Han shu, zhi* 10, 3215, commentary)

So we have a round and turning heaven on the outside, and a flat and stationary earth within it. Some words attributed to Zhang Heng in a quotation by Ge Hong 葛洪 (283–343 CE) give us a more concrete picture of the cosmos:

> 天如雞子，地如中黃，孤居於天內，天大而地小．天表裏有水，天地 各乘氣而立，載水而行．
>
> Heaven is like a hen's egg, and the earth is like the yolk inside. It is alone within; heaven is the larger, and earth the smaller. There is water inside and outside heaven; Heaven and earth both stand by riding on *qi*, and move borne up by water. (*Sui shu* 19, 509)

[81] Fragments collated in Yan Kejun 嚴可均 (1762–1843) (1965 (repr. of 1958 edn, plus indices)), vol. 1, 776–7.

[82] Fragments collated in Yan Kejun vol. 1, 777–9

How large are they? A little later, Zhang Heng's text tells us:

> 八極之維，徑二億三萬二千三百里，南北則短減千里，東西則廣增千里．自地至天，半於八極，則地之深亦如之．通而度之，則是渾已．將覆其數，用重[差]鉤股，懸天之景，薄地之義，皆移千里而差一寸得之．過此而往者，未之或知也．未之或知者，宇宙之謂也．宇之表無極，宙之端無窮．
>
> The tie-line of the eight poles has a diameter of 232,300 *li*. It is a thousand *li* shorter from north to south, and a thousand *li* wider from east to west. From earth to heaven is half of the eight poles, and so the depth of earth is equal to that. Measuring it out overall, then this is a sphere. In order to find these numbers, use repeated [differences] and [the relation between] base and altitude. The brightness of heaven suspended [above], and the goodness of the dark earth [below], may both be [understood because] a difference of one *cun* of shadow follows from a movement of 1,000 *li*. Beyond this, no-one has ever known of it. That which no-one has ever known is called vast space. The boundaries of vastness are unfathomable, and the limit of space is unattainable. (*Hou Han shu, zhi* 10, 3216, commentary)

'Half of the eight poles' is 232,300 *li* ÷ 2 = 116,150 *li*; so this cosmos is clearly comparable in size with the *gai tian*.

Putting these dimensions together with the other elements in the description, we obtain the picture (Figure 6.7) of a spherical heaven surrounding a flat earth.

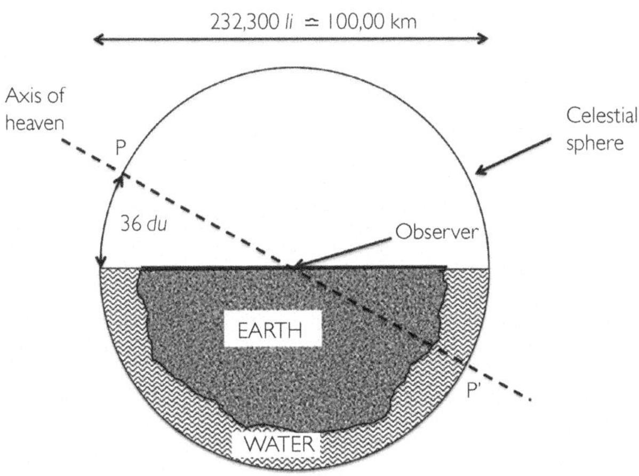

Figure 6.7 Zhang Heng's *hun tian* cosmos; the figure of 36 *du* for the altitude of the north celestial pole is taken from a text quoted in section 6.5.2. The Han empire was about 6,000 *li* (2,500 km) north to south, and is therefore too small to show to scale. All its inhabitants were thus effectively at the centre of the cosmos.

There is no immediately obvious reason why Zhang Heng should have chosen these particular figures for the dimensions of the cosmos.[83] However, a little later in the *Ling xian*, we are told:

> 懸象著明，莫大乎日月．其徑當天周七百三十六分之一，地廣二百四十二分之一．
>
> Of the images hung up [on high], or the brilliances displayed, none is greater than the sun and moon. Their diameter corresponds to 1⁄736 of the circumference of heaven, which is one part in 1⁄242 of the breadth of earth. (*Hou Han shu, zhi* 10, 3216, commentary)

Zu Gengzhi 祖暅之 (c. 450–c.520 CE) discusses the implications of these numbers, and in certain problems that arise from his criticisms of Zhang Heng we can perhaps find the rationale for the figures given above. He states:

> 張衡[曰]日，月(其)徑,[84] 當周天七百三十六分之一，地廣二百四十二分之一．按此而論，天周分數圓周率也．廣分數圓徑率也．以八約之，得周率九十二，徑率二十九：其率傷於周多徑少，衡之疏也．衡以日月之徑居一度之半，又言八極之維徑二億三萬二千三百里．
>
> Zhang Heng states that the diameters of the sun and moon correspond to one part in 736 of the circumference of heaven, and one part in 242 of the breadth of earth. I observe that according to this statement, the number of parts in the circumference of heaven corresponds to the proportion of the circumference of a circle, and the number of parts in the width of earth correspond to the diameter of the circle. Simplifying by a factor of 8, we get 92 for the proportion of the circumference, and 29 for the proportion of the diameter. These proportions are faulty in making the circumference too large and the diameter too small. This is [Zhang] Heng's error. Heng took it that the diameters of sun and moon each took up half a *du*, and further that the tie-line of the eight poles has a diameter of 232,300 *li*. (Qutan Xida 瞿曇悉達 c. 725 CE: 1, 17)

There are two problems here.

(a) Although 736/8 = 92, 242/8 = 30.25, not 29 as Zu Gengzhi states.
(b) Since 736/242 = 3.041 (4 significant figures) the error is in the opposite sense to that which Zu Gengzhi complains of—with a circumference smaller than it should be, for a given diameter.

[83] His references to the use of 'repeated [differences] and [the relation between] base and altitude [… and the fact that …] a difference of one *cun* of shadow follows from a movement of 1,000 *li*' are simply borrowed from the *Zhou bi*, and do not seem to lead to the figure of 1,000 *li* by any plausible route.

[84] The text seems slightly corrupt at this point. The character *qi* 其 'its' certainly makes little sense.

Now Zu Gengzhi is unlikely to have made two such elementary errors; for (b) in particular he was already in possession of a quite good value for the ratio in question, which is of course nowadays called *pi* π.[85] However, if we read 242 二百四十二 as 232 二百三十二, then we have:

(c) $232/8 = 29$, as Zu Gengzhi states.
(d) $736/232 = 3.172$, which is indeed larger than Zu Gengzhi's value for π, as he claims to be the case.

Furthermore, if we take $\frac{1}{232}$ of the diameter of the cosmos rather than $\frac{1}{242}$, we obtain for the diameter of the sun and moon:

232,000 *li* ÷ 232 = 1,000 *li*, just as stated by Zu Gengzhi.

A possible solution seems to be that Zu Gengzhi had access to a better text of the *Ling xian* than what we now have available, which actually did read 232 rather than 242, and originally wrote the first figure rather than the second. However, the *Kai yuan zhan jing* editors, who were perhaps not bothering to follow the calculation, 'restored' Zu Gengzhi's correct figure to the faulty *Hou Han shu* reading, without, however, changing Zu's 'simplified' value of 29 to correspond with it. If this is the case, we have discovered an explanation of Zhang Heng's cosmic dimensions, which would then based in effect on a value of $\pi = \frac{92}{29}$, combined with the postulates that the linear diameters of the sun and moon are both 1,000 *li*, and that their angular diameters are half a *du*—a figure that is close to observational values.

However, if half a *du* is worth 1,000 *li*, that would imply that one *du* is 2,000 *li*, so that the circumference of heaven would contain $\frac{736{,}000}{2{,}000} = 368$ *du*. This is clearly an error. It is perhaps not surprising that Zu Gengzhi sums up this part of his discussion by saying:

> 不明其理, 飾辭華說
>
> ... [Zhang Heng] failed to understand the underlying principles, but used elaborate expressions and over-elegant explanations.

Returning to the *Ling xian*, there are a number of other interesting statements about the nature of the heavenly bodies. In connection with Jia Kui's discussion of the use of eclipses to check on the functioning of *li*, we have already mentioned Zhang Heng's notion that the moon and stars shine by re-emitting the

[85] His father, Zu Chongzhi 祖沖之 (429–500 CE), had set upper and lower limits for the ratio equivalent to 3.1415927 and 3.1415926; see *Sui shu* 16, 387–8.

light of the sun, and also the unilluminated 'dark space' opposite the sun, which causes a lunar eclipse when the sun passes through it: see section 6.3.1.

As for the stars, we are told that there are a large but not uncountable number:

> 中外之官，常明者百有二十四，可名者三百二十，為星二千五百，而海人之占未存焉．微星之數，蓋萬一千五百二十．庶物蠢蠢，咸得繫 命．不然，何以總而理諸！夫三光同形，有似珠玉，神守精存，麗其職而宣其明；及其衰，神歇精斁，於是乎有隕星．然則奔星之所墜，至〔地〕則石〔矣〕．
>
> Of the [asterisms corresponding to] officials within and without the palace, there are a hundred and twenty-four which are always shining, and three hundred and twenty which can be named, making 2,500 stars—and this does not include those by which the peoples of the sea divine. The reckoning of the faintest stars comes to 11,520. The multitude of creatures pullulates, and all of them manage to link their fate [to a star]. If this [i.e. the vastness of the star-count] was not so, how could there be any overall ordering of them? The three luminaries are of similar shape, resembling jade beads. While their spirit is conserved and their essence is protected, they fulfil their function in splendour and spread forth their brightness, but when decay comes, their spirit ceases and their essence is wearied—then there is a falling star. So where a shooting star falls, if one reaches the spot there is a stone. (*Hou Han shu, zhi* 10, 3217, commentary)

A total of 2,500 stars distributed over a total of 444 asterisms, comes to about 6 stars per asterism, which seems reasonable. Sima Qian's *Tian guan shu* 天官書 'Treatise on the heavenly offices' listed 89 asterisms and named 412 individual stars, which gives above five each.[86] It seems a little improbable that Zhang Heng had actually listed 2,500 stars, let alone 11,520 of the 'faintest stars'. At any rate, no later writer claims to have seen a star-list or star-chart by him, let alone one listing over eleven thousand. In fact, in the light of our examination of the work of Liu Xin based on the *Book of Change*, we can see where the figure of 11,520 comes from: it is the number known in the Book of Change as *wan wu zhi shu* 萬物之數 'The number of the myriad creatures' which represents all phenomena corresponding to the hexagrams, and hence every separate thing that can exist: see chapter 4, section 4.4.3. Since, as Zhang Heng points out, everything is linked to a star, there must be 11,520 stars available, even if they are not all visible. This figure is thus a theoretical prediction made on the grounds of cosmology, and is certainly not an observed datum.

[86] *Shi ji* 27; see Pankenier (2013), 444.

The *Sui shu* notes that any star-chart that Zhang Heng might have made was lost at the end of Han, and goes on to mention that in the Three Kingdoms period (c.220–265 CE) Chen Zhuo 陳卓 created comprehensive star-charts incorporating three previous listings, not including that of Zhang Heng, with a total of 283 asterisms, amounting to 1565 individual stars, which seems a somewhat more likely total than those ascribed to Zhang Heng (*Sui shu* 19, 504).

6.5.2 Constructing and using instruments

We know from the account by Cai Yong and Liu Hong that an instrument on the lines laid down by Jia Kui, with an ecliptic as well as an equatorial ring, was constructed in 103 CE, when Zhang Heng would have been in his mid-twenties. Neither Cai and Liu, nor any other Eastern Han source gives details of any celestial observational instrument constructed by Zhang Heng.[87] More than six centuries later, however, the *Sui shu* gives a high level of detail stating clearly that Zhang Heng made a bronze *hun tian* instrument with a circumference of 14.61 *chi*, approx. 3.4 m, making the diameter about 1.1 m (*Sui shu* 19, 516). Unfortunately the date given corresponds to 164 CE, many years after Zhang Heng's death in 139 CE, so little reliance can be placed on this account, which may be the result of confusing Zhang Heng with someone else.

Perhaps the earliest explicit statement that we can find to the effect that Zhang Heng made an armillary sphere out of bronze is by Ge Hong 葛洪 (283–343 CE), writing about two centuries after the time of Zhang Heng and recorded in the *Jin shu* three centuries later. But this statement leads on to what has been interpreted as a much more ambitious claim:

> 張平子既作銅渾天儀於密室中以漏水轉之，令伺之者閉戶而唱之．其伺之者以 告靈臺之觀天者曰：「璇璣所加，某星始見，某星已中，某星今沒」，皆如合符也．
>
> When Zhang Pingzi [i.e. Zhang Heng] had made his bronze *hun tian* instrument, he turned it in accordance with the water of a clepsydra in a closed room. He told the person watching it to shut the door and call out. So the person watching announced to those observing the heavens on the *ling tai*, saying 'According to the

[87] His biography in the *Hou Han shu* says that he *zuo hun tian yi, zhu ling xian, suan wang lun, yan shen xiang ming* 作渾天儀, 著靈憲, 筭罔論, 言甚詳明 'Made the *Hun tian yi* [and] wrote the *Ling xian* and *Suan wang lun* [no work of this name has survived], [of which] the words are extremely detailed and clear.' (*Hou Han shu* 59, 1898). But the words *zuo* and *zhu* used here can both be used to refer to writing texts, and since one of the texts left to us is called *Hun tian yi*, the words used do not necessarily mean that Zhang Heng made a physical *hun tian* instrument.

indication given by the *xuan* ji [i.e., the armillary sphere], such and such a star is making its first appearance, such and such a star is just culminating, and such and such a star is now setting.' Everything matched up perfectly. (*Jin shu*, 11, 281–2)

There is a long tradition of interpreting this and related passages as describing a device that was a precursor of the great construction project of Su Song, c. 1090 CE, who made a large astronomical clock turned by a water-wheel fed from a constant-head apparatus, and incorporating an escapement mechanism operated by the weight of water to divide time into equal intervals—so far as the flow of water remained constant. Su Song himself certainly took this view.[88] However, it seems to me that there is room for some scepticism. Firstly, there are no clear contemporary or near contemporary references to what would have been an elaborate and impressive mechanical device, which it would have been fascinating to observe in operation, nor do the *Sui shu* or *Jin shu* say anything explicit about Zhang Heng making such a device.[89] The mechanism of Zhang Heng's seismoscope, which for most of the time would simply have done nothing at all, is on the other hand described in detail in his biography in *Hou Han shu* 59, 1909. That well-known account seems to be a literary reworking of the older account in *Hou Han ji* 19, 5b–6a.[90] Secondly, there is a much more

[88] See for instance Needham and Wang Ling (1959), 359–366; Joseph Needham, Ling Wang and Derek J. de Solla Price (1986) *Heavenly clockwork: the great astronomical clocks of medieval China*. Cambridge Cambridgeshire; New York, Cambridge University Press, 100–113 makes every possible argument in favour of this interpretation. For Su Song's interpretation of Zhang Heng's work, see Needham, Wang and Price (1986), 21.

[89] After a passage that repeats the text just quoted in abbreviated form, *Jin shu* (11, 285) adds the words 因其關戾, 又轉瑞輪蓂莢 於階下, 隨月虛盈, 依曆開落 'Based on the *guan li*, he also turned an "auspicious wheel" [with a] *ming jia* below the steps [of the throne], which waned and waxed following the moon, and [whose pods] opened and fell according to the calendar.' But this is clearly a description of a quite different device, set up below the steps of the throne, with a rotating display mimicking the *ming jia* plant that grew near the thrones of legendary emperors, developing one pod a day as the moon waxed, and losing one pod a day as it waned: see for instance *Bai hu tong* 白虎通, chapter 5, *feng shan* 封禪. The term *guan li* is entirely obscure; this text is its only known usage in ancient China, and the one instance of its use in a Tang text may well be based on what we have here. It is notable that two later quotations from a work ascribed to Zhang Heng entitled *Lou shui zhuan hun tian yi zhi* 漏水轉渾天儀制 'Rules for turning an armillary sphere in accordance with the water of a clepsydra' make no mention of a mechanism for doing the turning by water power, but simply refer to the operation of an ordinary clepsydra with separate vessels for day and night runs – which would hardly be appropriate for a continuously rotating armillary. See *Wen xuan* 文選 *(Selected literature)*, 56, 1218 and *Chu xue ji* 初學記 *(Records for initial study)*. (1962 repr.). Xu Jian 徐堅 (c.700 CE), Beijing, Zhonghua Press, 35, 595.

[90] Significantly, this earlier account does not include the well-known story of the device indicating an earthquake when no tremor had been perceived at the capital. This is said to have puzzled courtiers, until a report came in of an earthquake in Longxi 隴西, 700 kilometres away. It may be that this story was an embellishment added when the original story was polished for inclusion in the *Hou Han shu* See Robinson, Andrew (2016) for a more detailed outline.

economical interpretation of the text available to us, which is simply to read the phrase *yi lou shui zhuan zhi* 以漏水轉之 as in the translation above, where it refers to a human operator turning the apparatus in accordance with the flow of water in a clepsydra, rather than by the water powering the motion of a mechanism independently of human action. The procedure described would then be a test of the kind of data set out in Huo Rong's tables for the 24 *qi* described earlier, as well as of the *hun tian* view of the heavens: on the basis of the date and the known position of the sun from the tables, one simply turns the armillary sphere, placed in a closed room, to the position that accords with the time shown by the clepsydra, and reads off predictions of the positions of stars that will be rising, setting, or making a meridian transit. The observer outside on the observing platform then checks the actual phenomena and says whether they match the predictions. Accurate predictions by such a method would have been quite impressive enough.[91]

6.5.3 The celestial sphere: solving the problem of ecliptic and equatorial motion

Jia Kui's work raised the problem of how one might change reference systems between the ecliptic and the equator; as we have seen (section 6.4.4), the Han Quarter Remainder system gave a list of 'advances and retardations' that seem to be designed to deal with this issue, but which in fact do not. The solution was, however, finally provided by Zhang Heng, as part of his account of the celestial sphere.

The description of the celestial sphere given by Zhang Heng does not differ in any major way from that given by Jia Kui. As for the observer's view of the sphere, we have the following from material in the eighth century *Kai yuan zhan jing* that may be a part of Zhang Heng's writing, or at worst comes from a commentator. It describes how the sphere is oriented relative to a terrestrial observer, and adds some other detail to the picture:

> 周天三百六十五度四分度之一； 又中分之則一百八十二度八分之五覆地上，一百八十二度八分之五繞地下．故二十八宿，半見半隱，其兩端謂之南北極．北極，乃天 之中也．在正北，出地上三十六度，然則北極上規，

[91] Needham, who was heavily committed to the view that Zhang Heng made a water-powered mechanism, recognized the possibility of such an interpretation of the description of Zhang Heng's device being adopted 'from a position of extreme skepticism': see Joseph Needham, Lu Gwei-djen, John H. Combridge and John Major (1986) *The Hall of Heavenly Records: Korean astronomical instruments and clocks, 1380–1780*, 75. But when the evidence is so scanty, it is not clear to me that it is the sceptic who carries the burden of proof.

徑七十二度，常見不隱．南極，天之中也．在南，入地三十六度，南極下規七十二度，常伏不見．

The circumference of heaven is 365 ¼ *du*. Dividing it in half, 182 ⅝ *du* are above the earth and 182 ⅝ *du* are below the earth. So half of the 28 lodges are visible, and half invisible. The two ends [of the celestial axis] are called the south and north poles. The north pole is the centre of heaven in the north, and it rises 36 *du* above the earth. So the upper circle of the north pole has a diameter of 76 *du*, [within which bodies are] always visible and never hidden. The south pole is the centre of heaven in the south, and it goes 36 *du* below the earth. So the lower circle of the south pole has a diameter of 76 *du*, [within which bodies are] always hidden and are never seen. (*Kai yuan zhan jing*, 1, 3; see Figure 6.8)

We are also told that at the winter solstice the sun spends close to 146 *du* of its daily circuit above the earth and 219 *du* below it, with the figures being reversed at summer solstice.[92] This is consistent with the day and night clepsydra runs of 45 and 55 *ke* given in the solar table of 102 CE (see Table 6.2), since the day clepsydra run begins 2 ½ *ke* before sunrise, when dawn is taken to begin, and does not end until the end of dusk, place 2 ½ *ke* after sunset. We thus have sunrise to sunset as 40 *ke*, and sunset to sunrise as 60 *ke*.

$$^{146}/_{219} = ^{2}/_{3} = ^{40}/_{60}$$

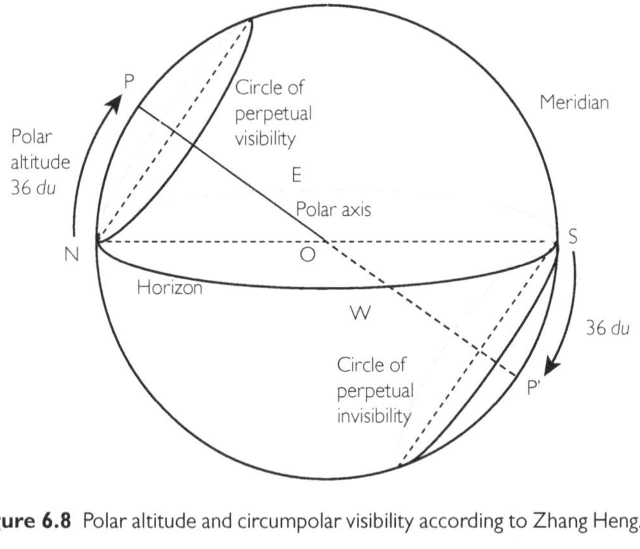

Figure 6.8 Polar altitude and circumpolar visibility according to Zhang Heng.

[92] The figures given in the original text include the two fractions *qiang* 強 and *shao qiang* 少強. On the interpretation of such fractions normally adopted, these would be 1/12 and 4/12 respectively, but that makes no sense in this context. The figures are presumably corrupt.

So the figures given are consistent. The picture given here, as shown in Figure 6.7, is simply the view of an observer in a given position on the earth; the altitude of the pole is, in modern terms, the observer's latitude. The latitude of the observatory on the southern outskirts of Luoyang was 34.7° N which should have yielded a polar altitude of 34.7 × (365.25/360) *du* = 35.21 *du*. The altitude of the pole was therefore overestimated by about 0.8 *du*, very close to the 0.86° overestimate suggested by the analysis of Sun and Kistemaker (see section 5.7). We may also note that the seasonal values of the north polar distance of the sun given in the solar tables of 102 CE suggest that the polar axis of the measuring instrument used was about 0.7 *du* too high.[93] At the winter solstice of 101 CE, the predicted times of sunrise and sunset would have been quite close to observation: see Box 6.2.

Finally, we come to Zhang Heng's ingenious solution to the problem first raised by Jia Kui. It is found in the version of his *Hun tian yi* in the *Hou Han shu* commentary. The relevant part of the text, which is preceded by a short description of the positions of the equator and ecliptic on the celestial sphere, begins somewhat strangely:

> 上頭横行第一行者，黃道進退之數也．本當以銅儀日月度之，則可知也．以儀一歲乃竟，而中閒又有陰雨，難卒成也．
>
> As for the first row running horizontally at the top, that is the amount of advance or retardation on the Yellow Road. Basically one ought to measure it from the sun and moon using the bronze instrument, and then it can be known. Using the instrument, it is concluded in one year. But in that interval, there will be obscuration and rain, which make it hard to get [a] complete [set of data.] (*Hou Han shu, zhi* 3, 3079)

From the references to 'the first row' and *jin tui* 進退 'advance or retardation' it seems that Zhang Heng is describing a table, now lost, which contained data designed to enable one to transform between ecliptic and equatorial motion. Fortunately, this is not all he says on the matter. He begins as above by pointing out that these data could be derived from observing the sun and moon throughout the year using an armillary sphere, and presumably noting their equatorial and ecliptic positions at daily intervals. He then goes on:

> 是以作小渾．盡[94]赤道黃道．乃調賦三百六十五度四分度之一．從冬至所在始起．令之相當值也．取北極及衡各鍼穿[95]之為軸．取薄竹篾穿其

[93] See Cullen (2007a), 94.

[94] I suggest this should be read as *hua* 畫.

[95] Following the reading in *Kaiyuan zhanjing*.

> 兩端．令兩穿中間與渾半等以貫之．令察之與渾相切摩．乃從減半起以為八十二度八分之五[96]．盡衡減之半焉．又中分其篾拗去其半．令其半之際正直與兩端減半相直．令篾半之際從冬至起一度一移之．視篾之半際多少黃赤道幾也．其所多少則進退之數也．從北極數之則去極之度也．
>
> Thus one makes a small sphere,[97] with the Red Road and the Yellow Road, and one marks out on them 365 ¼ *du*, beginning from the winter solstice, so that they are mutually aligned [at that point]. Take the north pole and its opposite [pole], and bore through them to make pivots. Take a thin bamboo strip and pierce holes in both its ends. Let the space between the two holes be equal to half of the sphere in order that it may be journalled [onto the pivots]. Make it so that if you check it, [you can see that] it rubs against the sphere. Then starting from the halfway mark make [divisions of] 91 5⁄16 *du*, so that these occupy the entire half [of the strip]. Now divide the strip down its mid [-line] and remove half of it, making it so that the [inner] edge of the half [-strip] aligns exactly with the middle of the two ends. Now let the half edge of the strip be moved along one degree at a time starting from the winter solstice, and look at the difference in the measurements on the ecliptic and equator [indicated by] the mid-line of the strip. This difference is the 'advance or retardation' quantity. If one measures from the north pole, then that is the polar distance.

Zhang Heng is evidently about to take an extremely practical route to obtain the data he needs—measurements on an actual model sphere with a graduated ecliptic and equator marked on it. It is at first sight frustrating that the resulting tabulation is not given; instead he comments on the pattern revealed by the data in a way that enables us to reconstruct the complete set of data for a quarter of the solar cycle, beginning at a solstice and ending at the next equinox. Clearly these data can be applied to subsequent quarters with an appropriate change of sign. I have explained the reconstruction in detail in (Cullen, Christopher 2000) from which Figure 6.9 is taken.[98] The vertical bars given in the graph show values of the 'advance or retardation' given for each day; Zhang Heng changes this value every three or four days. The solid line shows the results of modern calculations, which are in every case within ¼ *du* of the value given, except near the maximum.

[96] The editors of Zhong Hua edition of the *Hou Han shu* propose to emend this to 百八十二度八分之五, 182 5⁄8 *du*. But since the graduations are to start from halfway along a strip that already goes only halfway round the sphere, this seems unlikely. Since (365 ¼) / 4 = 91 5⁄16, the preceding figures should clearly be 九十一度十六分之五 if the text is to be comprehensible in the only obvious sense. I translate accordingly.

[97] It is worth noting that this phrase is (so far as I know) the first clear usage of *hun* 渾 in a sense that means 'sphere'.

[98] See also the slightly different interpretation in Y. Edmund Lien (2012) 'Zhang Heng's *Huntian yi zhu* Revisited.' *T'oung Pao* 98 (1–3): 31–64.

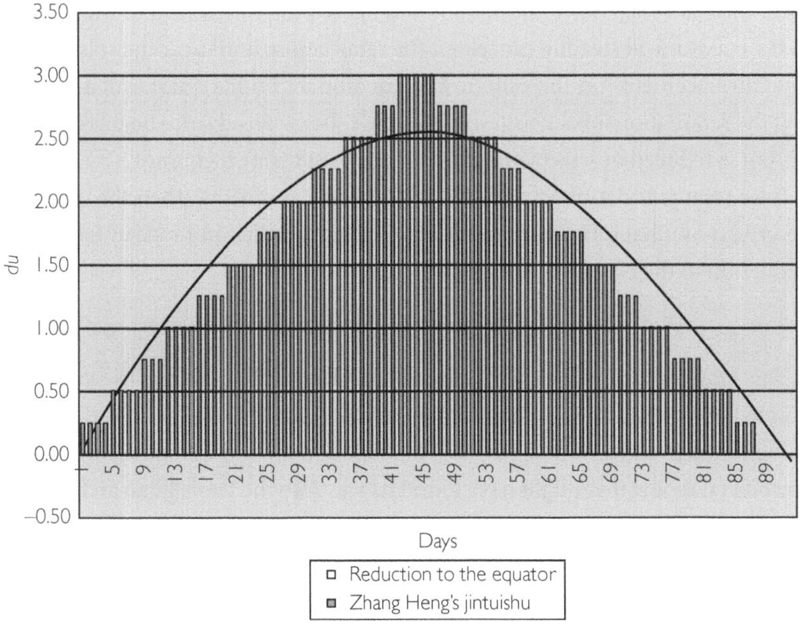

Figure 6.9 Zhang Heng's 'advance or retardation' data.

Zhang Heng explains why the 'advance or retardation' varies as it does in some detail. If we begin with the sun on the ecliptic at the spring or autumn equinox, then the corresponding position on the equator will be on the same line of right ascension through the two poles, so we begin with an 'advance or retardation' of zero. However, as shown in Figure 6.1, every *du* of motion on the ecliptic near an equinox, when the ecliptic is at maximum inclination to the equator, corresponds to a considerably smaller displacement on the equator. As Zhang Heng tells us:

> 春分，秋分所以退者，黃道始起更斜矣，於橫行不得度故也.
>
> The reason why there is retardation [after] the spring and autumn equinoxes is that when the ecliptic starts out [from those positions] it is more inclined, so that the motion cross-wise [on the equator] does not get to the *du* [on the ecliptic]. (*Hou Han shu, zhi* 3, 3076, commentary)

So the retardation will build up, since the measured movement along the ecliptic (longitude in modern terms) will be greater than the corresponding movement along the equator (corresponding to modern right ascension). However, Zhang continues, as the sun moves towards the following solstice, the combination of increasing distance from the equator and decreasing inclination of the ecliptic to the equator will steadily cancel out the retardation, until near the solstice the total displacements on the ecliptic and equator are both a quarter of a circuit, and the retardation is back to zero. After the solstice, an advance begins to build up, but is reduced to zero by the time the sun is at the next equinox.

If we treat retardation as negative and advance as positive, then the quantity described by Zhang Heng is equivalent to what is called in modern terms the 'reduction to the equator', defined as:

Sun's right ascension – Sun's longitude[99]

This notable technical success by Zhang Heng in resolving the problem pointed out by Jia Kui is not mentioned in any records of discussions that have come down to us from the Han; it is possible that he did not incorporate it in any official report that might have found its way into the Dong guan archives by the time that Cai Yong and Liu Hong worked on them. However, in Liu Hong's account of his new *Qian xiang* 'Supernal manifestation' astronomical system created near the end of the dynasty he refers briefly (as we shall see in section 8.1.2) to the use of 'advance or retardation' data in a way that bears a close resemblance to what we have seen here. Perhaps he had read a version of the *Hun tian yi*? We cannot tell.

But now it is time to pass on to survey the record of the controversies that the archive reveals for the rest of the Eastern Han.

[99] Smart and Green (1979 (reprint of 6th edition 1977)), 148. Since the quantity is defined by Zhang Heng for a single quadrant, it must obviously be given the appropriate sign for the quadrant of the ecliptic to which it is applied.

CHAPTER 7

The age of debates

In this chapter, we look at a further aspect of the story revealed by the documents collected by Cai Yong and Liu Hong: the records of debates on technical matters during the first and second centuries CE, debates which were often held in the presence of large audiences of officials. We look at the issues that were raised in such debates, and at the varieties of evidence and arguments that the participants attempted to use to support their positions. Nothing of this kind is known from anywhere else in the ancient world, whether from the Hellenic cultures within which Ptolemy of Alexandria worked, or from the Mesopotamian cultures which produced a mass of cuneiform documents recording the observations and calculations of the 'scribes of Enūma, Anu and Enlil'. First, however, we look at two instances where outsiders came to court to display their expertise, with somewhat different results. We shall see more of this kind of intervention in the next chapter, where Liu Hong becomes directly involved in evaluating technical proposals made by those without official standing.

7.1 Whose voices were heard? Two outsiders

In the great set-piece debates that will be discussed in the central part of this chapter, the only people who were expected to take part were those holding official rank of some kind—whether as regular office-holders or as *dai zhao* 'expectant' officials (chapter 6: 6.3 and note 18). As we have seen (and shall continue to see), there was no expectation that only those with job descriptions giving them responsibilities connected with the heavens might have something worth saying on that subject. Official rank was, however, essential for those who wanted to be heard in formal debate.

But before going any further, it is worth recalling the evidence we have seen that indicates the presence and influence of experts on the observation, interpretation and prediction of the motions of the heavenly bodies who were not—or at least not originally—members of the bureaucracy based in the Han capital. One of the first to appear in our story was Gongsun Qing, who set in motion the events that led to the Grand Inception *Tai chu* 太初 reform of 104 BCE. The reform itself called on the services of Luoxia Hong, Tang Du, and more than twenty other experts from among the people, *min jian* 民間, evidently somewhat to the discomfiture of the Grand Clerk, Sima Qian, who had to stand by while the advice of outsiders was followed in preference to his own. All this was set out in chapter 3.

7.1.1 The case of Lang Yi

In the period with which we are now dealing, the pattern continues. A striking example of the respect that might be accorded to the expertise of non office-holders is shown by the case of Lang Yi 郎顗 from Beihai 北海 (on the coast to the north of the Shandong peninsula) who was summoned to court by Shundi during the first month of 133 CE (23 February to 24 March) to give his advice on a series of disturbing portents. Lang Yi's father, Lang Zong 郎宗, had enjoyed a considerable reputation as an interpreter of the Book of Change, but also as an expert in divination by wind direction, calculations relating to the stars, hemerology and the interpretation of vapours; he seems to have enjoyed a successful and profitable practice as a diviner in addition to his career as a local official. His son followed in his footsteps, and ran what amounted to a private college of classical learning and celestial prognostication:

> 兼明經典, 隱居海畔, 延致學徒常數百人. 晝研精義, 夜占象度, 勤心銳思, 朝夕無倦.
>
> He understood the classics thoroughly, and lived in retirement by the sea, but usually had several hundred students staying with him. By day they researched the essential meaning [of the classics] and by night they prognosticated from the degrees of the [celestial] phenomena. They were completely devoted to sharpening their understanding, taking no rest by morning or evening. (*Hou Han shu* 30b, 1053)

It is clear from the account of Lang Yi's activities that he and his students could not only interpret the *xiang du* 象度 'degrees of the [celestial] phenomena', but could also calculate in advance what they should be. His biography fills twenty pages of a modern edition of the *Hou Han shu*, and contains detailed citations of the memorials he submitted during his time in the capital. Some parts of this material show that he was an expert observer of the heavens. For instance, we read:

> 去年八月二十四日戊辰, 熒惑歷輿鬼東入軒轅, 出后星北, 東去四度, 北旋復還.
>
> Last year [i.e. in 132 CE], on the 24th day of the eighth lunar month, cyclical day *wuchen*.5 [21 September 132 CE], Mars crossed the lodge Ghost and went eastwards into the asterism Xuanyuan, going out to the north of the Empress Star [Regulus, α Leonis], and going four *du* east, then turning back again in the north. (*Hou Han shu* 30b, 1061)

Modern calculations show that the dates given here for the retrogradation of Mars are correct, and so are the positions referred to. Mars did indeed pass over the middle of the asterism 'Ghost' on 21 September 132 CE, and subsequently passed within 4 degrees of α Leonis before beginning a phase of retrograde (i.e. westwards) motion. Later (p.1073) Lang Yi records accurate observations of a close approach of Jupiter and Venus in November of the same year. It seems clear, therefore, that Lang Yi had brought with him to court the results of a regular programme of observations carried out privately, presumably as part of his students' training.

But Lang Yi was not just a sky-watcher: he also claimed to be able to calculate the positions of a planet using the Triple Concordance system. Just before the passage previously cited, he says:

> 熒惑以去年春分後十六日在婁五度, 推步三統, 熒惑今當在翼九度, 今反在柳三度, 則不及五 十餘度.
>
> 16 days after the spring equinox of last year, Mars was at the fifth *du* of the lodge Harvester. Calculating according to the Triple Concordance, Mars should now be at the ninth *du* of Wing. But on the contrary, it is at the third *du* of Willow—so it is falling short by more than 50 *du*. (*Hou Han shu, ibid.*)

The essential details of the quite complex calculations behind these statements are set out in Box 7.1. As noted, the Triple Concordance system predicted an Appearance of Mars 16 days after the spring equinox of 132 CE. According to the Triple Concordance, an Appearance (first dawn visibility of a planet in the east, shortly before sunrise) occurs when the planet is 15 *du* to the west of the sun. Clearly the date Lang Yi mentions was not a fortuitous choice. And at that time, Mars can indeed be calculated to be in the 5th *du* of Harvester according to the Triple Concordance, as Lang Yi states.

Box 7.1: Lang Yi's calculations of the position of Mars

If Lang Yi had the expertise he was thought to possess, and if we have understood the specifications of the Triple Concordance system correctly, it ought to be possible to reproduce Lang Yi's calculations by following those specifications precisely. This is in fact the case, as I have verified. However, the methods used by the Triple Concordance follow the usual principles of avoiding fractions, and 'casting out' *chu* 除 completed cycles whenever possible in order to keep the numbers on the counting board as small as possible, at the cost of multiplying the number of steps needed. For simplicity, this box uses the basic constants on which these procedures are based, but exploits the possibilities of electronic calculations to do in one step what the Triple Concordance splits into a number of separate steps. The following three data are all that are needed to do the calculation (artificially high precision is used, to avoid rounding errors)

1. High Origin of Triple Concordance: Local Chang'an midnight beginning 2 December 143,232 BCE (proleptic Julian calendar), Julian Day Number (JD) −50,593,729.8.
2. Length of solar cycle: since one Concordance Cycle of 1,539 years (counted from one winter solstice to the next) contains precisely Circuits of Heaven [562,120] days, one solar cycle is precisely $^{562,120}/_{1,539}$ days (365.2501624431449 days to 16 significant figures).
3. Interval between Appearances (in modern terms, synodic cycle) of Mars: there are Appearance Number [6,469] appearances in Year Number [13,824] solar cycles. So the interval between Appearances is precisely $(^{562,120}/_{1,539}) \times (^{13,824}/_{6,469})$ days = 780.5253123533831 days (to the same precision as above).

According to the Triple Concordance, the spring equinox of 132 CE fell on 26 March, whose midnight fell at JD 1,769,355.2. At midnight 16 days later (11 April), the time elapsed since High Origin was therefore 1,769,355.2 +16 − (−50,593,729.8) days = 52,363,101 days. But this is very close to 67,087 Appearance cycles of Mars, which amount to 52,363,101.6 days (to the nearest 1/10th of a day). So according to the Triple Concordance 11 April was the day of an Appearance of Mars (first visibility in the east before dawn).

Dividing the time since High Origin by the length of the solar cycle leaves a remainder of slightly over 107 days. So 11 April midnight is 107 days since the last winter solstice, and since the sun moves 1 *du* per day, it is 107 *du* from its winter solstice position, which according to the Triple Concordance is the start

continued

Box 7.1: *Continued*

of the lodge Ox. Since a planet is close to 15 *du* to the west of the sun at an Appearance, Mars must be 92 *du* from the start of Ox. Counting off the widths of subsequent lodges as specified by the Triple Concordance takes us to the fifth *du* of Harvester, precisely as stated by Lang Yi.

Lang Yi states that Mars is 'now' predicted to be at the 9th *du* of Wing, which is 237 *du* from winter solstice, 145 *du* to the east of its Appearance position. But when in 133 CE should the planet have reached that position according to the Triple Concordance? We may find out by using the data given in the Triple Concordance for the motions of Mars during the phases that follow its Appearance. The first phase is said to last 276 days, during which the planet moves 159 *du* eastwards. It is then stationary for 10 days, after which it commences a 62-day phase of retrogradation, moving $^{17}/_{62}$ *du* westwards daily. We achieve the desired shift after 51 days of retrogradation, since $159 - 51 \times (^{17}/_{62}) = 145$. This gives a total of $276 + 10 + 51$ days $= 337$ days, which takes us to 14 March 133 CE, well within the first civil month (23 February to 24 March), when Lang Yi appeared at court to present his memorial.

We are not quite sure when 'now' is for Lang Yi, but if we follow the Triple Concordance we can predict that Mars should be at the 9th *du* of Wing on March 14 of 133 CE, which is well within the lunar month when Lang Yi answered the imperial summons. However, as Lang Yi notes, Mars was in fact observed to be a long way from the predicted position on that date. The actual departure from prediction was more like 40 *du* than the 50 *du* stated in the text,[1] but the error is still obvious to any naked-eye observer who knows the constellations and can recognize Mars by its distinctive reddish appearance. Modern calculations show that Mars will in fact be seen at about the fifth or sixth *du* of Willow, rather than the third *du*, which is the position that Lang Yi stated that he had observed. But given that finding the location of the planet in the lodge would probably have involved clepsydra determination of the difference in transit times of Mars and the determinative star of Willow (δ Hydrae), which would have amounted to only about 23 minutes, the discrepancy is not serious, quite apart from the fact that we do not know exactly when the observation was made. In any case, the large difference between prediction and observation remains.

[1] I suspect that a copyist has read 五十 'fifty' for the graphically similar 四十 'forty'.

Two questions arise out of all this:

(a) What did Lang Yi make of the large discrepancy between observation and his predictions?
(b) What is the significance of the fact that he is still using the Triple Concordance, half a century after it was replaced in official practice by the Han Quarter Remainder system?

As for the first question, Lang Yi makes his attitude quite plain: such discrepancies between prediction and observation are to be taken as portents. In his first address to the emperor, Lang Yi had summarized the types of omen that in his view called for changes of policy (*Hou Han shu* 30b, 1056), and these included the fact that Mars had frequently departed from its predicted position—literally 'missed its *du*' *shi du* 失度. It is notable that he does not suggest that there might be a fault with the way the position of Mars had been calculated, but rather seizes on the major discrepancy with prediction as an omen that demands interpretation. This is a pattern that is already evident in the text known as the *Wu xing zhan* 五星占 'Prognostics of the five planets', which dates from the third century BCE, and that David Brown has identified in ancient Mesopotamia: a relatively simple planetary theory that does not succeed in predicting all details of the planet's actual motion can actually be of service to a diviner by providing portents that enable him to display his expertise.[2]

After the emperor requested more detail, Lang Yi produced a further memorial listing seven major items of concern, ranging from infelicitous winds to anomalous weather and strange vapours in the night sky, and parhelia (atmospheric phenomena producing areas of brightness on both sides of the sun, often associated with haloes). This was followed by another short memorial, and then a more substantial one listing four further items. The observations of the motions of Mars from September 132 CE onwards that we discussed earlier are given in the fourth item of the seven-item memorial. But to understand why Lang Yi thinks they need to be brought to the emperor's attention, we need to continue reading a little further:

> 軒轅者, 後宮也. 熒惑者, 至陽之精也 ...
>
> Xuan Yuan [corresponds to] the rear palace [i.e. the women's quarters]. Mars is the essence of the culmination of Yang. (*Hou Han shu*, 30b, 1061–2)

[2] Cullen (2011b), 248–9, and David Brown (2000) *Mesopotamian planetary astronomy-astrology*. Groningen, Styx, 146–56. See also section 4.2.

Lang Yi explains that Mars is the celestial counterpart of the Yang male principle; its unusual behaviour obviously relates to the emperor's excessive indulgence in the pleasures of his harem. The motion of Mars was in close correspondence with the predictions of the Triple Concordance up to the point where it approached the Empress Star: after that point, however, the Triple Concordance predicted that the planet should continue eastwards until pausing and turning back in mid-January, rather than doing so at the start of December. The fact that it unexpectedly hangs around the star corresponding to the women's quarters suggests that the emperor's behaviour is not as all that it should be.

On the second question—there seem to be two possible reasons why Lang Yi may still be using the Triple Concordance. One may be that he simply does not have access to the details of the official system, the Han Quarter Remainder. While calendars calculated using this system were promulgated annually, there was no reason to make public the technical details of how they were constructed. The systems for planetary motions would have been amongst the most arcane parts of those details, given that the calendars issued to the people did not give information on the planets. Secondly, as we have already seen and shall see again, there appear to have been some who argued that the Triple Concordance should not have been abandoned at all. The seventh item in the first of Lang Yi's longer memorials is in effect a call for a change of *li yuan* 曆元 'system origin', on the grounds that the Han has been in power for three centuries, and that this would be beneficial for the dynasty. Could Lang Yi have been suggesting that a return to the Triple Concordance was advisable? We may note, however, that Lang Yi evidently did not feel able to take his use of the Triple Concordance past a certain point: when he gives a date for the November 132 CE observations of Jupiter and Venus already mentioned, he states that the 20th day of the tenth month fell on cyclical day *guihai*.60. That cyclical day was the one predicted by the Han Quarter Remainder system, whereas the Triple Concordance would have placed the 20th day on cyclical day *jiazi*.1. Lang Yi's use of the official date was a wise move, since open departure from the published imperial calendar might have been interpreted as rebellious, with potentially fatal consequences.

Finally, one further point must be mentioned, although it is not raised anywhere in the account of Lang Yi's appearance at court. If one uses the Han Quarter Remainder system to predict the motion of Mars at the times referred to by Lang Yi, the results seem considerably less anomalous and out of step with observation. The most striking example is the anomaly seen with the Triple Concordance in March 133: whereas the Triple Concordance placed Mars in Wing, the planet was, as Lang Yi correctly noted, actually several tens of *du* to the west, near the first few *du* of Willow: this was a gross departure from prediction. However, according to the Han system, it should have been found in the 14th *du* of Willow, a position much closer to observation.

It seems quite surprising that no representative of the Grand Clerk's office pointed this out, especially given that the Grand Clerk at this time was Zhang Heng himself.[3] Perhaps nobody with the relevant expertise was invited to comment on Lang Yi's advice to the emperor? Or perhaps it is even possible that the Han Quarter Remainder system was not yet equipped with its own planetary prediction system at the time Lang Yi worked.

7.1.2 The case of Xiang Kai

Xiang Kai 襄楷, whose biography follows that of Lang Yi, seems to have been in a similar situation. He came from the region of Linyi 臨沂 in Shandong, and it was said of him that:

> 好學博古, 善天文陰陽之術.
>
> He was an excellent scholar with a wide knowledge of antiquity; he was good at the arts of [interpreting] celestial phenomena and *yin-yang*. (*Hou Han shu* 30b, 1075)

He came to the capital and presented two memorials to the throne in 166 CE, though holding no official position. Unlike Lang Yi, who was summoned to come to court in an official vehicle, he seems to have done so on his own initiative, without having received an official request to offer advice.[4] Like Lang Yi, he cites a number of records of observations involving the planets, which correspond to what would have been actually observed—and since he is not said to have headed a group of students as did Lang Yi, we can only assume that these were observations he had gathered himself. There is also a reference to his having made a calculation to predict a planetary position:

> 臣又推步, 熒惑今當出而潛, 必有陰謀.
>
> Your servant has further calculated that Mars should have now have appeared, but remains hidden; there must be secret plotting. (*Hou Han shu* 30b, 1081)

[3] He is given that title in the record of a memorial he submitted in 133 CE: chapter 6, note 74.

[4] See *Hou Han shu* 30b, 1075–85. Xiang Kai's memorials are discussed in full historical and political context in Rafe De Crespigny (1976) *Portents of protest in the later Han dynasty: the memorials of Hsiang K'ai to Emperor Huan in 166 A.D.* Canberra, Australian National University Press in association with the Faculty of Asian Studies: see pp. 21–33 for a translation of his biography.

Unfortunately, since we do not know what was the date referred to as 'now' by Xiang Kai, it is not easy to judge what system he was using. The Han Quarter Remainder predicted an Appearance of the planet Mars for 4 May 166 CE, and Mars would indeed have been observed about 15 *du* west of the sun at that date. The Triple Concordance, on the other hand, made the prediction for 21 June, over a month later, when Mars was already 27 *du* from the sun, and thus would have been considerably more difficult to miss in the eastern sky before dawn—hence it seems unlikely that Xiang Kai would have complained that it could not be seen when expected. It does therefore seem possible that, unlike Lang Yi, Xiang Kai might have been using the Han system for planetary calculation—which would at least reassure us, in the face of Lang Yi's silence, that the Han system did include a planetary system by the middle of the second century CE.

So now we have two examples, three decades apart, of people who were not office holders of any kind, let alone office-holders with responsibilities for celestial matters, who evidently made detailed quantitative observations of celestial phenomena and had the ability to make the relatively complex calculations required to predict planetary positions. Both of them lived some distance from the capital. There seems no reason to assume that they were the only people under Eastern Han who fulfilled these conditions. Indeed, later in this chapter we shall see that there were others. We may note that Cai Yong and Liu Hong made no mention of Lang Yi's memorials in their collection of documents on astronomical controversies. Could it have been that the discussions that concerned them took place in a different milieu from that in which Lang Yi spoke? Or were Lang Yi and Xiang Kai ignored because what they had to say did not bear on the technical matters of celestial calculation with which Cai Yong and Liu Hong were mainly concerned? We cannot tell.

Despite their similarities, the two men were treated quite differently at court. Lang Yi was offered significant, if minor, official rank (which he did not accept), while Xiang Kai was given a sentence of two years' convict labour, presumably for his criticism of the influence of eunuchs at court. No one seems to have suggested, however, that their studies were in any way illegitimate in themselves. An interest in the heavens, whether for purposes of prediction of phenomena or for the interpretation of what was observed, does not seem to have been seen as out of place, even for commoners with no official responsibility for such matters.

But let us now turn to the activities of those with a higher rank in Han society.

7.2 The debates

We have already noted the existence of the records of discussions on the subject of celestial calculation, collected by Cai Yong and Liu Hong from the archives of the Dong guan office in the late second century CE.[5] This material is unusual, even unique, in the context of ancient science. This uniqueness may be characterized in two ways: by the detailed nature of the documents themselves, and also by the situation of those who generated these documents.

7.2.1 Styles and records of argument in ancient western and eastern Eurasia

Firstly, for the documents: as will shortly be made clear by the examples I shall cite, they enable us to trace the details of the arguments between the disputants quite closely. If we look at the literature of astronomy in the ancient Hellenic and Mesopotamian worlds, this cannot be done. We know that Greek discussion of natural philosophy was frequently agonistic, and we can see signs of that in Ptolemy's writing, where he takes considerable trouble to justify his procedures in the face of a potential attack,[6] but we cannot see the actual *agōn* proceeding before our eyes, with the two sides having their say.[7] In ancient Mesopotamia, we can see signs of struggle, even shrillness, in the efforts that the *ṭupšar Enūma Anu Enlil* (Scribes of [the gods] Enūma Anu Enlil), who recorded their astronomical observations and calculations on clay tablets in cuneiform, made to ensure that the king attended to their interpretations of omens rather than those of their rivals. As one seventh or eighth century BCE clay tablet has it, in a report to the king of Assyria:

> [He who] wrote to the King, my lord 'Venus is visible' is a vile man, an ignoramus, a cheat![8]

If, however, we look for accounts of open discussion or of discursive writing or advocacy about the ways in which the technical details of astronomy were to

[5] See section 6.1; I first discussed this material in Cullen (2007b).

[6] See for example his somewhat nervous defence of the fact that he has not been able to explain the motions of all the planets in the same way, in *Almagest* IX.2, Toomer (1998), 421–3.

[7] Of course there is little room for doubt such public events did take place. In some cases we have what looks like one side of an *agōn*, as in the Hippocratic *Nature of Man*; see Hippocrates and G. E. R. Lloyd (1983) *Hippocratic writings*. Harmondsworth, Penguin, 260–71. But what we do not seem to have is the record of the actual confrontation between opponents.

[8] Brown (2000), 240.

be validated, we do not find much. The elaborate mathematical schemes of late Mesopotamian astronomy can hardly have evolved without technical discussion, but there is little direct evidence of this in the written record. One example of what we do have runs:

> Concerning Mercury, about which the king my lord wrote to me: yesterday Issar-sumu-eres had an argument with Nabu-ahhe-eriba in the palace. Later, at night, they went and all made observations; they saw (it) and were satisfied.[9]

Beyond a few other examples of this kind, we can only proceed by conjecture. However, the material provided by Cai Yong and Liu Hong provides us precisely with what the western records lack, in that it gives us explicit records of who debated with whom about astronomical calculation, and what each side said. In one case at least we even know where the debate took place, who was in the audience, and where they sat.

Further, the situation of the Chinese debaters makes them particularly interesting. For a start, they are officials. They do their astronomy as respected if not exalted members of a bureaucracy, in a strong institutional setting. Who did Ptolemy work for, or with, if anyone? We do not know. What relation if any did he have to existing institutions such as the Library or Museum at Alexandria? Again we mostly draw a blank. One of the things that has been held to make ancient Hellenic science so different from modern science as to make it an entirely different phenomenon is its lack of an institutional setting: in ancient China the institutional setting of astronomy is there by the very nature of the material, and its workings are accessible to us in considerable detail.

We know rather more (but still not a great deal) about the institutional setting within which the Scribes of Enūma Anu Enlil lived, learned, and practised their craft. Francesca Rochberg has traced the way the site of the Scribes' activity appears to have shifted from palace to temple around the middle of the first millennium BCE, although exactly when and why this shift took place is not clear.[10] We know, too that the Scribes appear to have functioned to some extent as a group that, at least in principle, kept its professional skills to itself.[11] We have some evidence that they took care to validate the skills of new entrants to the group. One clay tablet from the Esagila Temple of Babylon, written about 110 BCE, records the settlement of a claim for property and income made by the

[9] Hermann Hunger (2002) *Astrological Reports to Assyrian Kings, State Archives of Assyria, Vol VIII* Helsinki, Helsinki University Press, 50.

[10] Rochberg (2004), 209.

[11] Rochberg (2004), 216–19.

Scribe Bēl-Uṣuršu (himself the son of a Scribe). Recording the decision in his favour, the tablet notes that he:

> ... having appeared in court before us persuaded(?) us that he is able to make all the astronomical observations. We have seen that he is capable of carrying out the activity of keeping watch to its fullest extent [...] He will carry out the celestial observation. He will provide *tersētu* [computed] tablets and almanacs [...] with the other scribes of *Enūma Anu Enlil.* (Rochberg, Francesca 2004: 234–5)

Our picture of the lives of the sky-watchers of Han is much clearer. They had regular salaries, offices, the chance of promotion, guaranteed time off, and even uniforms.[12] In theory all this was at the emperor's whim, but in the Eastern Han many emperors were too weak for this to matter much, and established custom gave considerable security. In the only instance in the material before us where the emperor does intervene in disciplinary proceedings taken against officials on the losing side of an important dispute, he acts to suspend judicial action, on the grounds (it is said by a commentator) that everyone should have their say: see section 7.2.3. And further, even though all their debates were in theory conducted under the emperor's eye, he was by no means always present, and the actual decisions often seem to have been made by their fellow-officials. In the Eastern Han this was a group whose sense of collegiality was notoriously seen by the emperor as a threat sufficient to justify him in developing an alternative corps of political advisers and drafters of policy, the palace eunuchs.[13] And although (as already mentioned) ancient China has been described as having a form of intellectual life that was consciously non-agonistic and concerned to produce consensus,[14] there does seem have been not only a fair amount of tolerance for what has been called 'the disturbing spectacle of experts disagreeing in public',[15] but even a sense that such spectacles were necessary if technical problems were to be effectively settled.

[12] See the account of the recruitment, duties and employment conditions of Han officials in Loewe (2006), chapter 5, particularly 76–80.

[13] Twitchett, Loewe and Fairbank (1986), 287–90.

[14] See for instance G. E. R. Lloyd and Nathan Sivin (2002) *The way and the word: science and medicine in early China and Greece*. New Haven; London, Yale University Press, 249.

[15] This phrase was used by Simon Schaffer in a BBC Radio 4 broadcast in 2001 with reference to controversies about lightning rods as safety features of powder stores in which Benjamin Franklin was involved around 1,750. The fuller context as noted by me and since checked with him (private communication, March 2007) was 'You can get a team of experts to agree if you restrict the range of the enquiry sufficiently. How far such an enquiry actually is restricted by those in authority will depend on the degree to which they can tolerate the disturbing spectacle of experts disagreeing in public'. Of course in the case of the Chinese officials we shall discuss, the 'public' did not extend to any curious commoner who might wish to join in—but it does appear to have included a quite wide selection of their numerous colleagues in the central administration.

Different types of material demand different research tools, and one should try to be both open and reflective about the research tools one chooses. Thus in studying the astronomical and astrological activities of the Mesopotamian scribes, David Brown attempted to apply the Kuhnian language of paradigm shifts, and claimed that he could locate a 'Revolution of Wisdom' in the shift from the Enuma-Anu-Enlil (EAE) paradigm of omen interpretation on the basis of idealized outline numerical schemes to the Prediction of Celestial Phenomena (PCP) paradigm using relatively accurate mathematical procedures, a shift he locates somewhere in the seventh and eighth centuries BC. But there are other models for enquiry that may be more applicable in the present context. In the corpus prepared by Cai Yong and Liu Hong, we see a community of actors concerned with the rhetoric of persuasion, situated in a strong institutional setting that defines their roles and is in turn defined and shaped by their actions. The range of actors here is very varied: there are human beings, certainly, but also observational instruments both old and new, systems of official rank, databanks of observations both official and unofficial, sacred texts with the words of the ancient sages, rules of official procedure and even seating plans for the buildings in which debates take place. There are also systems of mathematical software, the *li* 曆 'astronomical systems', with some actors attempting to promote them to effectively 'black box' status as unchallengeable facts, while others work to subject them to more provisional and questioning modalities, meanwhile pushing their own candidates for 'black box' status—instruments as well as software—in turn. How far this tentative borrowing from Latour is justified (and for present purposes I mean specifically the Latour of *Science in Action*) the reader will shortly be in a position to judge.[16]

7.2.2 The institution of the *yi* and the background of controversy

To understand the documents that we shall consider in the first part of this chapter, we need to understand an important Eastern Han institution, the *yi* 議 'deliberation'. There is clear evidence that the high officials of the Eastern Han met together from time to time in large numbers, in a building designated for that purpose, that there they debated forcibly with one another on important issues of policy, and that their collegiate deliberations mattered a good deal in determining the direction taken by the state. They were not elected representatives

[16] See Bruno Latour (1987) *Science in action: how to follow scientists and engineers through society*. Cambridge, Mass., Harvard University Press.

in anything like the House of Commons that became fundamental to modern statecraft in Britain—but British parliaments have never normally consisted of a House of Commons alone, nor did the purely English parliaments that preceded them. When they worked well (which is not always), pre-modern parliaments did so in part because they made strong provision for representing the permanent interests of other estates of the realm—the landholding nobility and the Church—in the House of Lords. It seems to me that the high officials of the Han were just such a self-conscious estate of the Han realm, but in a world with no Church, and no feudal landholding. Like the prelates and nobles of mediaeval England, Han officials lived in a dangerous world, as was shown by such events as the great proscriptions of the late second century.[17] They certainly had no parliamentary immunity. But they could never safely be ignored, not only because of the official power they wielded but also because they were often the representatives of families whose informal influence extended well beyond the limits of the offices they held. And one of the ways they made their influence felt was through the collective deliberations called *yi* 議.

The importance of the *yi* in the polity of Eastern Han has not been widely stressed in modern western studies. But in China it was recognized as long ago as 1226, when Xu Tianlin 徐天麟 presented to the throne the results of his studies of Eastern Han institutions in the form of his *Dong Han hui yao* 東漢會要 'Collected essentials of Eastern Han [institutions]'.[18] His work is divided into 40 chapters, of which seven are devoted to the description of official posts and an outline of how their duties were carried out. The fourth of these chapters is devoted entirely to the topic of *yi*. Xu's sources are mostly the *Hou Han shu*, which he cites explicitly, with some editorial abbreviation. While it is always helpful to check the original source, Xu's quotations are well selected and presented in context, and a reading of them is an excellent introduction to the topic.

From a review of the available instances of its usage, the word *yi* as found in the *Hou Han shu* can be seen to have several possible shades of reference. It can act as a noun referring to a written report made on an issue that has become a matter for discussion, or it can be a verb signifying the making of such a report. Sometimes a *yi* can be an oral presentation, or it can simply be the discussion in which oral presentations are made. Finally, it can refer to the decision resulting from such a discussion.

[17] Twitchett, Loewe and Fairbank (1986), 327–30.

[18] I use the reprinted edition: *Dong Han hui yao* 東漢會要 *(Collected essentials of Eastern Han [institutions])*. (1978). Xu Tianlin 徐天麟 (*jin shi* 1205 CE), Shanghai, Shanghai gu ji chu ban she 上海古籍出版社.

In the same way that a mediaeval English Parliament could not sit unless it was properly summoned, a *yi* needed to be convened by the authority of those charged with conducting high deliberations of state. In formal terms, this meant the trio of officials often designated as the San Gong 三公 'Three Excellencies':

> 太尉 [...] 凡國有大造大疑, 則與司徒, 司空通而論之. 國有過事, 則與二公通諫爭之.
>
> In all cases where there are great national undertakings or great problems, the Defender in Chief investigates and discusses these matters with the Minister of Education and the Minister of Works. Should there be any fault in state affairs, then he joins with the other two Excellencies in investigation, and in admonition and contention about it. (*Hou Han shu, zhi* 24, 3557)

Consistently with the very general remit given above, a *yi* could be about any issue in relation to which government action was thought appropriate. The deliberations for which the Three Excellencies were responsible frequently took the form of great assemblies of officials meeting in a building specifically designed for that purpose in the offices of one of them:

> 司徒府中有百官朝會殿
>
> In the office of the Minister of Education there is a hall for court meetings of all officials. (*Hou Han shu, zhi* 24, 3560)

It was expected that a *yi* would produce a decision during its session, and it is clear from the cases cited by Xu that such decisions did not simply occur as a result of a pronouncement by the emperor or his agents after receiving *yi* from the assembled officials (whether written or spoken). Although a *yi* was evidently expected to produce a decision, it is not explicitly stated how that decision was actually manifested. Numbers were apparently significant: we are told in some cases that a named official made a *yi*, and that he was joined in this by a list of others specifically named, or in some cases by a given number of persons. In the latter case the number might be quite large:

> 愷等八十四人議, 宜從太初.
>
> Kai and others, 84 persons, made *yi* that it was fitting to follow the Grand Inception [astronomical system]. (*Hou Han shu, zhi* 2, 3034)

Here it seems unlikely that such a large group of people simply put forward similar *yi*, whether orally or in writing; perhaps they did something like adding their names to a list? Further, once a *yi* had been put forward, it might be said that numbers of persons 'followed' *cong* 從, the opinion that had been advanced, though how they indicated this is not made explicit. It may be that acclamation (or the lack of it) played a role: in describing the presentation of the annual reports on provincial administration to the assembled officials, it is said:

> 善者同聲稱之,不善者各爾銜枚
>
> In the case of good [reports] they praise them with joined voices, but if [the reports] are bad they all are [as silent as if they were] gagged. (*Hou Han shu, zhi* 24, 3560, commentary)

Perhaps the clearest statement relating to the mode and status of a *yi* deliberation comes from a comment by the emperor after an extremely contentious decision in 85 CE led some officials to send in their seals in resignation as a protest against abuse from other participants:

> 久議沈滯,各有所志. 蓋事以議從,策由衆定,誾誾衎衎,得禮之容
>
> There may be deadlock in a prolonged *yi*, with each [participant] holding to his opinion. But the business is [done] according to the *yi* which is followed, and the policy is settled in accordance with the many. One must remain calm and preserve a polite demeanour. (*Hou Han shu* 45, 1519)

It appears that the majority gets its way, and 'unparliamentary' behaviour is discouraged.

It does not seem that it could be taken for granted that a *yi* would be a dull occasion. On the contrary, a *yi* might be an unpredictable affair, in which the courage and oratory of an individual might swing the meeting either way. It is clear from such instances that it was not open to powerful persons simply to ignore the sense of the meeting, nor was it always easy to rig the deliberations. The fact that attempts at coercion or intimidation were sometimes made indicates that consent mattered, and could be withheld. An example is the proposal in 89 CE that a victorious general and member of the powerful Dou family, Dou Xian 竇憲, should be welcomed back to the capital with cries of *wan sui* 萬歲 '[live for] ten thousand years!' (a salutation reserved for the emperor). This is opposed by Han Leng 韓棱, who silences the Dou sycophants single-handed in what is clearly a verbal exchange:

> 及憲至, 尚書以下議欲拜之, 伏稱萬歲. 棱正色曰: 「夫上交不諂, 下交不黷, 禮無人臣稱萬歲之制.」 議者皆慙而止.
>
> When [Dou] Xian arrived, those from the rank of Masters of Writing downwards made *yi* that they wished to do obeisance to him, and salute him with 'Ten Thousand Years!' while prostrate. [Han] Leng said with a stern countenance:

> 'In their relations with others, those on high should not flatter, and those below should not act wantonly. The *Rites* have no ordinance for a subject to be saluted with "Ten Thousand Years!"'. Those who had made *yi* all desisted through shame. (*Hou Han shu* 45, 1535)

In an extreme case, when at the end of the dynasty the military dictator Dong Zhuo 董卓, with troops at his back, sought to depose the young emperor in favour of his own candidate, he still felt it necessary to call a meeting of the mass of officials to discuss this matter:

> 因集議廢立, 百僚大會.
>
> ... so he assembled a *yi* on the deposition [of the emperor], and there was a great meeting of all office-holders. (*Hou Han shu* 72, 2324)

When despite his hectoring he still met with open opposition, he fell into a rage and brought the sitting to an end. The next day he summoned another meeting, this time having coerced the Empress Dowager into giving support, and got his own way: but it is significant that he still felt the need to summon the meeting and obtain its consent, however forced, before acting.

One of the most dramatic oral confrontations during a *yi*, again involving a member of the Dou clan, took place in the extremely tense discussions concerning the funeral rites of Dowager Empress Dou in 172 CE (see *Hou Han shu* 56, 1832–3). The eunuchs, who hated the Dou, wanted to see her buried with the rites of a palace lady only. A great assembly was convened by edict in the Meeting Hall 詔公卿大會朝堂, and the eunuch Zhao Zhong 趙忠 chaired it. The 太尉 Defender in chief Li Xian 李咸 (who might be thought of as the normal person to head such deliberations) attended though ill, telling his wife he would not return alive if the empress was not buried with proper ceremony. Several hundred persons were present, but none dared to be the first to speak (so clearly there *should* have been speeches from the floor at such an event), and all gazed at the eunuchs. 既議, 坐者數百人, 各瞻望中官, 良久莫肯先言. Zhao Zhong announced that the consultation must be settled in a timely manner 議當時定, but was surprised by the way those present just stared at one another. Despite attempted intimidation by the eunuch chairman, including an attempt to cut short his *yi*, the Superintendent of Trials Chen Qiu 陳球 stood out for the full ceremony; all those present 'followed his *yi*' 公卿以下, 皆從球議. The sick Li Xian then summoned the energy and courage to speak out loudly in Chen's support causing those present to be conscience-stricken. He then confuted an attempt by the eunuchs to compare the case to earlier instances, approaching the throne and saying 'The Empress Dowager treated Your Majesty as her son—how can your majesty not treat her as your mother?' 太后以陛下為子, 陛下豈得不以太后為母. The emperor (Lingdi, then only 16 years old), perhaps encouraged by the evident opposition to the eunuchs, then halted Li's speech, saying to the eunuchs: 'Although the Dou clan did wrong, it was not the Empress Dowager's fault, and she was always good to me'. The eunuchs found nothing to say in return 無復言 and the *yi* was settled 於是議者乃定. It is clear in this case that the 'settlement' of the *yi* depended on the emperor's extraordinary intervention by taking over the chairman's role and expressing what we have already been told was the sense of the meeting. But such imperial intervention does not seem to have been common, and the emperor was often not present at all.

Clearly, therefore, open confrontation between opponents where the arguments were addressed at least as much to one's peers as to one's superiors were not rare in China in the period we are discussing. That is in itself interesting enough, since it is strong counter-evidence to the claim made by some that the model of inter-peer debate seen as typical of some times and places in ancient Greece was rare in China, where advocacy is said to have been typically directed upwards to one's superiors rather than horizontally towards one's equals.[19] More interesting for us is that such debates took place more than once in connection with *li*, astronomical systems.

7.2.3 Cai Yong and the debate of 175 CE

Let us now turn to a case in which a great *yi* with the full body of officials present gave its attention to issues concerned with astronomical systems. We might ask ourselves how representative of the general form of debates such a technically oriented *yi* might be. Interestingly, Xu Tianlin himself claimed in his account of this particular *yi* that it could serve as a type-specimen for all *yi* of the period.[20] Given Xu's deep studies of Eastern Han institutions, his view must carry a great deal of weight.

The theme of the debate was the question of the 'system origin' *li yuan* 曆元, that is the instant of time at which the system should begin to run, and from which its interlocking cycles should be counted off. It will avoid confusion later if we note that although experts in *li* sometimes write as if each astronomical system had just one single year of origin, things can be considerably more

[19] See in particular Lloyd and Sivin (2002).

[20] *Dong Han hui yao* 東漢會要 (*Collected essentials of Eastern Han [institutions]*), 327

complex. Let us therefore pause to remind ourselves what a system origin was for (see also 1.3 and 6.2.1).

Typically a system origin was defined as being an instant when winter solstice and conjunction of sun and moon coincided at midnight beginning the first day of the first celestial month (the 11th month in the Xia count that was used for civil reckoning) of a given year. That day was also specified as being a particular day in the sexagenary cycle, usually *jiazi*.1. In systems of the quarter remainder type, such conditions would recur at intervals of 1,520 years, an Era *ji* 紀 cycle. If we are only interested in defining dates in the luni-solar calendar, the Era cycle is the longest period we need to consider: the *n*th day in a given Era cycle will be the same day of the same month with the same sexagenary day number as the *n*th day in any other Era cycle. If, however, we apply the sexagenary numbering system to years, then since

$$1{,}520 = 25 \times 60 + 20$$

the sexagenary number of the year will increase by 20 from one 'Era head' *ji shou* 紀首 (the first year of an Era) to the next, and the same number will only repeat after three Eras. This period is an Origin cycle, 4,560 years, and the first year of such a cycle is an 'Origin head' *yuan shou* 元首.[21] However, if an Origin head is defined as falling in a *gengchen*.17 year, it follows from the explanation just given that exactly the same calendrical conditions (apart from the year name) will be found in the Era cycles beginning in *gengzi*.37 and *gengshen*.57 years respectively. Shifting the Origin head to those years will make no calendrical difference at all, and thus for calendrical purposes what may seem at first glance to be three different Origins are calendrically equivalent.

Those who argued about the right choice of system origin certainly did base their claims in part on the predictions for contemporary observations of sun and moon that followed from those choices. But as we shall see, there was more to a system origin than its purely astronomical implications. I have discussed above (3.2 and 3.3) the complex motivations behind the adoption of a system origin on 25 December 105 BCE, a *jiazi*.1 day, for the astronomical reform in which the 太初 *Taichu* 'Grand Inception' system was introduced. Rather than being

[21] If we also want to specify that the planets should be at their initial positions (usually in conjunction with the sun), we may need to use an origin in the very remote past, for which the term *shang yuan* 上元 'High Origin' may be used: see the discussion in section 4.2. But it will still be an origin like any other so far as solar and lunar phenomena are concerned, and for calendrical purposes we may simplify calculations by using the nearest origin to our own time.

based on contemporary astronomical observations, it is clear that the main motivation for this choice was the wish of Emperor Wu to attain immortality by imitating, as he thought, the Yellow Emperor of high antiquity. In the Eastern Han a further important factor was the authority given to a particular system origin by what may be called 'scriptural warrant'. The 'scriptures' in question were not the ancient texts commonly known as 'classics' in English but in Chinese termed *jing* 經 ('warp-threads'), such as the Book of Documents (*Shu jing* 書經 or *Shang shu* 尚書) which contained proclamations by sage rulers of high antiquity, but were the so-called 'apocryphal' texts, sometimes known as *wei* 緯 ('weft-threads'). Although these texts are nowadays thought to have originated around the time of Wang Mang, they were regarded by many in the Eastern Han, including emperors and their advisers, as being truly ancient records of esoteric teachings by sages of antiquity such as Confucius. As we shall see, some of these texts made explicit reference to calendrical matters.[22]

Before we consider the material relating to the great *yi* of 175 CE, let us briefly review the history of the topic of system origins up to that date. During the Eastern Han, it was thought that the Han dynasty had begun by using the Zhuan Xu *li* 顓頊曆 'Zhuan Xu system' a quarter remainder type system said to have been used by the preceding Qin dynasty. As we have seen (chapter 3, section 3.4) this was probably not in fact the case. The nearest Origin Head for that system was in 1506 BCE, a *yimao*.52 year. The Grand Inception system, whose constants were not of the quarter remainder type, adopted 104 BCE, a *dingchou*.14 year, as an Origin Head.[23] We have seen (6.2.1) that under the Eastern Han, in 85 CE, the new Han Quarter Remainder system chose 161 BCE, a *gengchen*.17 year, as an Origin Head; in both these cases the actual instant from which calculations started was, as usual, the winter solstice immediately preceding the year in question. However, in the debate we shall see that the new system was said to have used the calendrically equivalent *gengshen*.57 origin. Indeed a contributor to a discussion that took place in 143 CE states that:

> 四分曆仲紀之元, 起於孝文皇帝後元三年, 歲在庚辰.
>
> The middle [of three] Eras of the [Han] Quarter Remainder system began in the third year of the Houyuan period of Emperor Wen [161 BCE] when the Year was at *gengchen*.17. (*Hou Han shu, zhi* 2, 3036; Cullen 2017, 400)

[22] The importance of these texts is discussed in chapter 6, section 6.2. On their astro-calendrical connections, see Yabuuti Kiyosi (1974).

[23] The sexagenary year numbers used here follow the system used in the Eastern Han. See also chapter 2, section 2.1.3.

So 161 BCE was not an Origin Head, but only the second Era Head in an origin period. That would mean the most recent Origin Head was 1,520 years earlier, in 1681 BCE, a *gengshen*.57 year.

During the Eastern Han there were a number of other discussions, before the debate we shall analyze, of the question of what the system origin should be (Cullen 2017, 393–410). However, in 175 CE the discussion not only took place with special solemnity, but was initiated by a pair of officials who demanded, unusually, that their opponents should be subject to criminal sanctions:

> 靈帝熹平四年, 五官郎中馮光, 沛相上計掾陳晃言: 「曆元不正, 故妖民叛寇益州, 盜賊相續為害. 曆當用甲寅為元而用庚申, 圖緯無以庚申為元者. 近秦所用代周之元. 太史[治曆郎](治治曆)[24] 中郭香, 劉固意造妄說, 乞本庚申元經緯明文, 受虛欺重誅.」

> In the fourth year of the Xiping reign period of Lingdi [175–6 CE] the Palace Gentleman for General Purposes, Feng Guang, and the Accounting Clerk to the Chancellor of Pei[25], Chen Huang stated: 'The system origin is incorrect, so that evil folk are rebelling and thieving in Yizhou, and robbers and bandits make endless trouble. Although the system should use a *jiayin*.51 origin, it uses a *gengshen*.57 origin. Among the Charts and Wefts, there is none that uses a *gengshen*.57 origin. It is near to the origin that Qin used in supplanting Zhou. The Palace Gentlemen for Regulating the Calendar, Guo Xiang and Liu Gu, thought up wild doctrines, and urged that one should treat the *gengshen*.57 origin as basic [and that] the Warp and Wefts had clear texts [to this effect]. They should receive heavy punishment for empty deceptions.' (*Hou Han shu, zhi* 2, 3037; Cullen 2017, 404)

The term 'Charts and Wefts' *tu wei* 圖緯 used here is a common designation of the 'apocryphal texts' mentioned earlier in this section. We know nothing about the calendrical officials Guo Xiang and Liu Gu who were the main object of attack, and only a little about Feng Guang and Chen Huang themselves. Since this is imperial China, we can, however, say something about their salaries. In a later reference to this case they are described as:

> 郎中馮光, 司徒掾陳晃

[24] I read following (for instance, among other pre-modern editions) the *Wu ying dian* 武英殿 edition (1739), *zhi* 2, 14a; it seems to me that the modern Zhonghua text contains a misprint which makes no sense.

[25] The Chancellor of Pei was the special official in charge of the affairs of the district of Pei 沛, in modern Jiangsu province, from whence the dynastic founder Liu Bang 劉邦 originated. Although Pei was far from the capital, the importance of this post made the holder part of the central bureaucracy.

> … the Palace Gentleman, Feng Guang, and the Clerk of the Minister of Education, Chen Huang (*Hou Han shu, zhi* 3, 3042; Cullen 2017, 417)

The salary of a Palace Gentleman for General Purposes was 300 measures (*shi* 石) of grain, and we know that Clerks to one of the other Three Excellencies (the Defender in Chief) had salaries of 400 and 300 measures.[26] While such posts are not in the upper ranks of capital officials (who received salaries of a thousand measures or more; see 6.5), these are not negligible people, and this was a significant attack. It is clear that any other officials who supported Guo Xiang and Liu Gu were involved in a serious accusation—that the choice of system origin they supported was contrary to scripture, and, in virtue of the consequent failure of harmony between human activity and cosmic rhythm, was the cause of social disorder on a large scale. The response from the imperial government was to put into action the full deliberative mechanism of a *yi*:

> 乙卯,詔書下三府,與儒林明道者詳議,務得道真.以群臣會司徒府議.
>
> [Commentary]蔡邕集載:「三月九日,百官會府公殿下,東面,(校)〔太〕尉南面,侍中,郎將,大夫,千石,六百石重行北面,議郎,博士西面.戶曹令史當坐中而讀詔書,公議.蔡邕前坐侍中西北,近公卿,與光,晃相難問是非焉.
>
> On [day] *yimao*.52 an edict referred the matter to the Three Offices [i.e. the offices of the Three Excellencies], for there to be detailed *yi* with the most enlightened of the literati, with the aim of getting at the truth of the Way. They met to make *yi* with the whole flock of officials in the office of the Minister of Education.
>
> Commentary: The *Collected Works of Cai Yong* contains this: 'On the third month, the ninth day, all the officials met in the lower part of the public hall of the office [*sc.* of the Minister of Education] facing east. The Commandant faced south, and the Palace Attendants, Leader of the Court Gentlemen, Grand Masters, and those with emoluments of 1,000 and 600 measures faced north in serried ranks. The Court Gentlemen for Consultation and the Erudits faced west. A clerk of the Director of the Civil Affairs Section was placed in the midst of the seats and read out the edict and the public *yi* [*sc.*the reports submitted in writing]. Cai Yong came forward and sat to the north-west of the Palace Attendants, near to the [Three] Excellencies. Then he joined with [Feng] Guang and [Chen] Huang in raising problems and questions with one another on the rights and wrongs [of the matter].' (*Hou Han shu, zhi* 2, 3037; Cullen 2017, 404)

[26] *Hou Han shu, zhi* 25, 3574 and *zhi* 24, 3558.

In terms of the Julian calendar, the fourth year of the Xiping reign period began on 9 February 175 CE, and ended on 28 January 176 CE. The third month of that year began on 9 April, sexagenary day *wuchen*.5, so that the ninth day was 17 April, sexagenary day *bingzi*.13. Assuming that the *yimao*.52 day when the *yi* was commanded was the one immediately preceding the meeting, the instruction would have been issued on 27 March, so that the parties had about twenty days to prepare their presentations. From the description of the arrangement of the participants in the hall, we may construct a seating plan as in Figure 7.1 (north is at the top).

Although our source does not specify their position in the hall, Feng Guang and Chen Huang must presumably have been somewhere near Cai Yong in front of the Three Excellencies, since they are specifically said to have debated with him on this occasion. As for the number of officials present, we can only say that it can have been no greater than could be accommodated in a large audience hall.

Of Cai Yong's career we can only speak briefly here. He was one of the most brilliant intellectual figures of his day, and the possessor of a famous library.

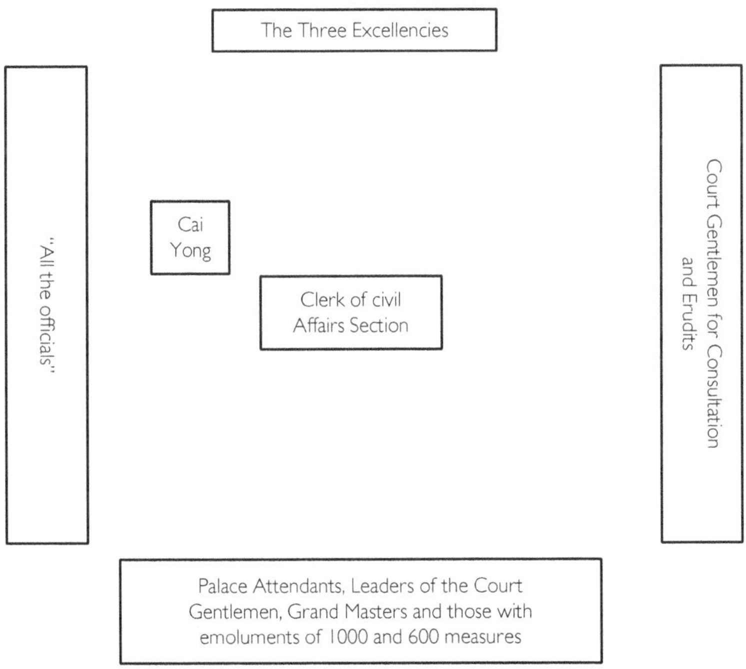

Figure 7.1 Seating plan for the great debate of 175 CE.

Two years after the debate recorded here he suffered a decade of banishment as a result of a conflict with powerful eunuchs at court.[27] At the time of this debate he held the rank of 議郎 *yi lang* 'Court Gentleman for Consultation' ('Consultation' here is a rendering of our familiar *yi*), with a salary of 400 to 600 measures of grain. We have no direct evidence of how he came to be chosen (or to volunteer?) to debate with the two originators of the accusation, but it must have been evident to his colleagues that he was well equipped to think on his feet, as well as being technically expert. Nor do we know whether the words ascribed to him here are a prepared speech that had been committed to writing, or whether it is a later summary of the totality of his points made against Feng and Chen. There certainly was a major element of oral exchange, as is made plain by the statement that the agonists *xiang nan wen shi fei* 相難問是非 'raised problems and questions with one another on the rights and wrongs': in another example where people *xiang nan* during an Eastern Han *yi*, the exchange seems to have ended in a shouting match:

> 安又與憲更相難折. 憲險急負埶, 言辭驕訐, 至詆毀安[...]安終不移.

> An again joined with Xian in further raising problems and confusions with one another. Xian became worked up and lost all restraint, using expressions that were arrogant and accusatory, to the point that he calumniated An [...] but to the end An refused to shift.' (*Hou Han shu* 45, 1521)

After a brief recapitulation of the different system origins used since the start of the Han dynasty, Cai Yong turns directly to the accusation. His first move is an attempt to distance the validity of a system origin from the witness of the apocryphal texts, without attacking the authority of the latter. He also launches the idea that no system origin is right for all time, but that change is inevitable:

> 今光, 晃各以庚申為非, 甲寅為是. 案曆法, 黃帝, 顓頊, 夏, 殷, 周, 魯, 凡六家, 各自有元. 光, 晃所據, 則殷曆元也. 他元雖不明於圖讖, 各自一家之術, 皆當有效於當時. 武帝始用太初丁丑之元, 六家紛錯, 爭訟是非. 太史令張壽王挾甲寅元以非漢曆, 雜候清臺, 課在下第, 卒以疏闊, 連見劾奏, 太初效驗, 無所漏失. 是則雖非圖讖之元, 而有效於前者也. 及用四分以來, 考之行度, 密於太初, 是又新元有效於今者也. [...] 且三光之行, 遲速進退, 不必若一. 術家以筭追而求之, 取合於當時而已. 故有古今之術. 今術之不能上通於古, 亦猶古術之不能下通於今也.

> Now Guang and Huang both say that *gengshen*.57 [origin used by the Han Quarter Remainder] is wrong, and that the *jiayin*.51 [origin] is correct. I would note that each of the astronomical systems of Huangdi, Zhuan Xu, Xia, Yin,

[27] See Mansvelt Beck (1990), 41–6.

Zhou and Lu, six schools in all, had its own origin.[28] The one relied on by Guang and Huang is the origin of the Yin system.[29] Even though the other origins are not made clear in the Charts and Oracles, each one stands in its own right as the method of one school, and all must have had validity at the corresponding time. When Emperor Wu took the initiative to use the *dingchou*.14 origin of the Grand Inception, the six schools were thrown into confusion, and wrangled over the rights and wrongs. The Grand Clerk Zhang Shouwang clung to the *jiayin*.51 origin in order to reject the Han system. After various observations on the Pure Terrace[30], it was checked and put down as a failure, and in the end its inaccuracy lead to it being denounced in a series of memorials.[31] The Grand Inception checked out when tested, and had no deficiencies or errors. This is a case where a system origin was shown to be valid in the past, even though it did not appear in the texts of the Charts and Oracles. When we come to the time from when the Quarter Remainder was used, if one checks the degrees of motion, it is more accurate than the Grand Inception. This additionally is a case of a new system origin being valid today. [...] Moreover, the motions of the Three Luminaries, in their slowings and accelerations and their advances and retardations are not necessarily as one.[32] When experts try to chase after them through calculation, all they can do is to seek a fit at the corresponding time. So there are ancient methods, and also modern ones. The inability of modern methods to be extended back to antiquity is the same as the inability of ancient methods to be extended forward to modern times. (*Hou Han shu, zhi* 2, 3038; Cullen 2017, 405–6).

[28] At least as early as the end of the Western Han, it was asserted that in pre-imperial times there had existed six different astronomical systems, usually identified with those listed here: see for instance *Han shu* 21a, 979 and 30, 1765–6. The Huang Di (Yellow Emperor) and Zhuan Xu systems are named after legendary rulers of high antiquity; Xia, Yin and Zhou refer to the first three dynasties, of which the second two at least are historical; Lu was the state whose annals, supposedly edited by Confucius, are the earliest Chinese historical record from the received tradition. See section 3.4.1.

[29] The earliest indication of the system origin of the Yin system is given by Liu Xin 劉歆 near the beginning of the Common Era. He tells us that the year corresponding to 47 BCE, a *jiaxu*.11 year, was an Era Head for the Yin system (*Han shu* 21a, 1024). If as we are told here the Yin system had a *jiayin*.51 origin, then that Origin Head must have been the previous Era Head, 1,520 years earlier in 1567 BC. As usual, any year a multiple of the Origin period, 4,560 years, away from that date is equivalent.

[30] This was the term used in the first century BCE for the site of the official observatory, referred to under Eastern Han as the 靈臺 *ling tai*, 'Numinous Terrace'.

[31] The story is told in *Han shu* 21a, 978: see chapter 3, section 3.4. Zhang made his protest in 78 BCE; matters were not settled until more than 20 persons had been engaged in an observation programme lasting three years.

[32] This may simply be a reference to the point made earlier by Jia Kui (see chapter 6, section 6.3), to the effect that the value of astronomical constants are not necessarily commensurable, or it may even be a suggestion that there is unpredictable variation from time to time.

Cai Yong now moves to attack his opponents on their central ground, that of the astronomical references said to be contained in apocryphal texts. The historical event to which he refers, the capture of the mythical beast known as a Lin 麟, is mentioned in an entry near the end of the *Chun qiu* 春秋 the 'Spring and Autumn Annals', traditionally said to have been edited by Confucius, under the 14th year of Duke Ai 哀, corresponding to 481 BCE.[33] Confucius is said to have interpreted the arrival of this creature (in effect a Chinese unicorn) as marking the coming end of his life. We need not speculate here on whether Cai Yong believed in the literal truth of this story. What matters for the discussion is that this event is said to have taken place in a year that was a standard reference date in a system of chronology that is recognized even today as having been reliable to the year from as early as 841 BCE.[34] Similarly, modern historians are quite comfortable about using dates based on the Christian (called in this book 'common') era without needing to commit themselves as to how that era may relate to the precise chronology of the life of Jesus of Nazareth.

> 元命苞, 乾鑿度皆以為開闢至獲麟二百七十六萬歲; 及命曆序積獲麟至漢, 起庚午蔀之二十三歲, 竟己酉, 戊子及丁卯蔀六十九歲, 合為二百七十五歲. 漢元年歲在乙未, 上至獲麟則歲在庚申. 推此以上, 上極開闢, 則元在庚申. 讖雖無文, 其數見存. 而光, 晃以為開闢至獲麟二百七十五萬九千八百八十六歲, 獲麟至漢百六十一歲, 轉差少一百一十四歲. 云當滿足, 則上違乾鑿度, 元命苞, 中使獲麟不得在哀公十四年, 下不及命曆序獲麟至漢相去四蔀年數, 與奏記譜注不相應.

> The [apocryphal] *Yuan ming bao* and *Qian zao du* both make it 2,760,000 years from the beginning of the cosmos to the capture of the Lin [in 481 BC]. When we come to the [apocryphal] *Ming li xu*'s total from the capture of the Lin to the Han, it starts in the 23rd year of the *gengwu*.7 Obscuration, and extends through the *jiyou*.46 and *wuzi*.25 Obscurations and the 69th sui of the *dingmao*.4 Obscuration, making 275 years in all. In the first year of Han [206 BCE] the year was *yiwei*.32; if one goes back to the capture of the Lin, then the year was at *gengshen*. 57. Extending on backwards from this until one comes to the start of the cosmos, then the origin is at *gengshen*.57. Even though the Apocrypha have no [relevant] text, the reckonings are there before us. But Guang and Huang take it that from the start of the cosmos to the capture of the Lin is 2,759,886 years and that it is 161 years from the capture of the Lin to the Han. They have made a

[33] See *Chun Qiu Zuo Zhuan zhu shu* 春秋左傳註疏, Duke Ai, 14th year, chapter 59, 1030 in edition of *Shi san jing zhu shu*.

[34] Endymion Wilkinson and Asia Center Harvard University (1998) *Chinese history: a manual*. Cambridge, Mass., London: Harvard University Asia Center; Harvard University Press [distributor], 176.

> difference of 114 years less [in both cases]. If one says that this is to be made up, then far in the past it will contradict the *Qian zao du* and *Yuan ming bao*, in the intervening period it will mean that the capture of the Lin did not fall in the 14th year of Duke Ai [of Lu] [i.e. 481 BCE] and finally it will not reach the number of years [specified in] the *Ming li xu* for the four Obscurations from the capture of the Lin to the Han, so that this will not be answerable to the data noted in the memorial. (*Hou Han shu, zhi* 2, 3038–9; Cullen 2017, 406–7).

The technical details here are not particularly hard to follow if we take them step by step. Let us recall that an Obscuration is 76 years long, and that there are 20 of these in an Era cycle of 1,520 years. Each Obscuration is named after the sexagenary number of the day on which it begins, and this shifts by 39 from one Obscuration to the next. It is clear that the *Ming li xu* is using the Yin system with an Origin Head in 1567 BC, since this positions Obscuration Heads precisely as specified: see Box 7.2.

Box 7.2: Cai Yong's calculation of Obscuration Heads using the Yin system

1567 BCE is an Origin head for the Yin system, and hence is the first year of an Obscuration whose first day is *jiazi*.1.

The usual date for the capture of the Lin is the 14th year of Duke Ai 哀 of Lu, 481 BCE. To find the sexagenary number of the year, we note that:

$$1{,}567 - 481 = 1{,}086 = 18 \times 60 + 6$$

So since the first year was *jiayin*.51, this year is *gengshen*.57, as Cai Yong states.

For the Obscuration, we note that:

$$1{,}567 - 481 = 1{,}086 = 14 \times 76 + 22$$

So 481 BCE is the 23rd year of the 15th Obscuration since the Origin of the Yin system.

Now in a system of the quarter remainder type, the number of days in an Obscuration is:

$$76 \times 365\ \tfrac{1}{4} = 27{,}759 = 462 \times 60 + 39$$

continued

Box 7.2: *Continued*

So each Obscuration shifts the sexagenary day number of the first day by 39. Since the day number on which the first Obscuration begins is that of the Origin, *jiazi*.1, the 15th Obscuration will begin on day:

$$1 + 14 \times 39 = 547,$$

which on casting out whole multiples of 60 takes us to sexagenary day *gengwu*.7.

So the Lin was indeed captured in the 23rd year of the *gengwu*.7 Obscuration of the Yin system, a *gengshen*.57 year, as stated by Cai Yong.

As we move on to the start of the Han, we shift from the 14th Obscuration through the 15th Obscuration (first day *jiyou*.46), the 16th Obscuration (first day *wuzi*.25) and into the 17th Obscuration (first day *dingmao*.4). After 69 full years of that Obscuration, a total of 275 years since the capture of the Lin, the sexagenary year number will have increased by:

$275 = 4 \times 60 + 35$ years, which takes us to from a *gengshen*.57 year to *yiwei*.32, as Cai Yong states.

Since we know that the first year of the Han dynasty was 206 BCE, a *yiwei*.32 year according to the system used in Cai Yong's day, it is easy to check that the capture of the Lin was in a *gengshen*.57 year, and since 2,760,000 is a multiple of 60, the year of the beginning of the cosmos was also a *gengshen*.57 year.

Cai Yong then analyses his opponents' calculation and reveals its internal contradictions: it seems that they may have been initially unwilling to discard the tradition that the Lin had been captured in a *gengshen*.57 year. Thus the only way for them to make their favoured *jiayin*.51 year the start of the cosmos was to use a different (and as Cai Yong points out non-scriptural) number of years from that point to the capture year. A decrease of 114 years from 2,760,000 to 2,759,886 does this, since $114 = 60 + 54$, and $54 + 57 = 60 + 51$. However, it appears that for some unknown reason they also decreased the time from the capture to the beginning of the Han by that same amount: if we stick to the known historical fact that the dynasty began in a *yiwei*.32 year according to the Eastern Han reckoning, then the capture of the Lin would have been in a *jiayin*.51 year after all. Further, the historical record over the intervening years is far too solid to allow for the loss of over a century. As Cai Yong indicates, their inexpert attempt to give prominence to their favoured sexagenary year number is a complete failure.

There may have been many in the audience who found it difficult to check such reckoning for themselves, and perhaps for that reason Cai Yong now shifts to an astronomical point that was simple enough to be obvious to all.

> 當今曆正月癸亥朔, 光, 晃以為乙丑朔. 乙丑之與癸亥, 無題勒款識可與眾共別者, 須以弦望晦朔光魄虧滿可得而見者, 考其符驗.
>
> According to the present system, the first day of the first month [of the coming year] will be *guihai*.60, but Guang and Huang make it *yichou*.2. If it is a matter of [choosing between] *yichou*.2 and *guihai*.60, there is no way that one can simply carve out a way different from everybody else—the matter is something that can be made manifest through the crescents and full moons, the last and first days of the month, and waning and waxing of the dark and brightness of the moon, examining which fits in with a check [through observation]. (*Hou Han shu, zhi* 2, 3039; Cullen 2017, 407–8).

Cai Yong is correct in his statement that the official system of his day (that is, the Han Quarter Remainder) predicts the forthcoming New Year's Day of the civil year commencing in spring of 176 CE as stated, and the Yin system apparently used by his opponents does indeed place it two days later. Observation of the sun and moon would have supported the official system; the true conjunction was close to 21:30 Luoyang local time on 29 January 176 CE, a *guihai*.60 day, whereas at sunset on the *yichou*.2 day following (31 January) a thin first crescent would already have been visible about 24 degrees from the sun, making it clear that conjunction had occurred about two days previously, since the moon moves about 13 degrees in a day.

Cai Yong's final attack on astronomical grounds takes the form of a challenge. He points out that his opponents are relying on values for such things as the position of the winter solstice sun amongst the stars as set out in the 'Weft' text known as the [尚書]考靈曜 [*Shang shu*] *kao ling yao* '[The Book of Documents:] examining the numinous brightnesses'. These differ from those currently in use by astronomical officials, and observation confirms that the latter are correct:[35]

> 以今渾天圖儀檢天文, 亦不合於考靈曜. 光, 晃誠能自依其術, 更造望儀, 以追天度, 遠有驗於圖書, 近有效於三光, 可以易奪甘, 石, 窮服諸術者, 實宜用之. 難問光, 晃, 但言圖讖, 所言不服.
>
> Examining the celestial patterns with today's celestial sphere plotting instrument [i.e. an armillary sphere], [the result] is likewise not in accord with the *Kao ling yao*. If Guang and Huang are really able to depend on their methods, [let

[35] See the discussion of this point by Jia Kui, chapter 6, section 6.3.

them] go on to make an observational instrument, in order to pursue the celestial degrees, so that far off they may have verification from the writings of the Charts, and near at hand they may find a check in the Three Luminaries. Then they could change and supplant [the ancient astronomical authorities] Gan and Shi, and the matter having been submitted to the experts, it would really be fitting to make use of [their proposal]. [But] when I raised objections to Guang and Huang, all they did was to quote the Charts and Oracles, and their answers were not satisfactory. (*Hou Han shu, zhi* 2, 3039; Cullen 2017, 408)

For Cai Yong, only those able to appeal to instrumentation and observation are entitled to a view on celestial co-ordinates: here he invokes the role of the armillary instrument as a 'black box', an unchallenged source of data that his opponents can only answer by reference to scripture. However, fully aware of the dangers that an attack on the authority of scripture might entail, Cai Yong immediately covers himself by the claim that the original edict of 85 in which the Han system with its *gengshen*.57 origin was promulgated was itself firmly based in scripture, and drew copiously on apocrypha of various kinds. There remains the unfortunate fact that the social disorders referred to in the original accusation cannot simply be denied. However, Cai states, even though the sage rulers of remote antiquity payed close attention to astronomical systems and the calendar, they still suffered natural disasters and attacks from within and without—these cannot be attributed to incorrect system origins. In any case, the Han dynasty has already changed its system origin twice since it took over from the preceding Qin dynasty. Finally, as he points out, there have already been two major instances of proposals similar to those of his opponents in recent centuries, both of which have been carefully investigated and rejected. The edict establishing the Han system in 85 CE was soundly based in every aspect. With this final invocation of imperial authority and precedent his presentation ends.

Unlike his opponents, Cai Yong is not recorded as calling for his adversaries to suffer any judicial penalty. Nevertheless, we are told that as a result of the debate, the presiding Three Excellencies indicted them for *lèse-majesté* (literally *bu jing* 不敬 'disrespect') presumably for having attempted to mislead the emperor on an important matter. They were sentenced to penal banishment, but an imperial edict suspended the judicial action.[36] The remark on this conclusion by the early sixth century CE commentator Liu Zhao 劉昭 is highly significant:

[36] It was by no means uncommon in imperial China for a severe sentence imposed by an official to be commuted to a lighter penalty by the emperor; one might almost say that severe sentences were imposed in some cases in the knowledge that the emperor might display his humanity by commuting them.

不有君子, 其能國乎？觀蔡邕之議, 可以言天機矣. 賢明在朝, 弘益遠哉！公卿結正, 足懲淺妄之徒, 詔書勿治, 亦深「盍各」之致.

> Without gentlemen, how could one have a state at all? Looking at Cai Yong's *yi*, one can say it is a matter of vital importance [to the state]. If the worthy and brilliant are at court, then their vast benefits reach afar. If the ducal ministers bind themselves to rectitude, that is enough to settle mean and irresponsible people. The edict ordering that the prosecutions [of Guang and Huang] should not proceed is surely a full expression of [the spirit of] 'Why not all [speak your mind]' [as in the *Analects* of Confucius]. (*Hou Han shu, zhi* 2, 3040, commentary; Cullen 2017, 410)

We should note carefully the claim made here that issues so complex as those on the table in 175 CE could only be settled by a 'gentleman' 君子 *jun zi*. However, the situation differs markedly from more familiar discussions involving that term in the context of science in 17th and 18th century England.[37] Truthful reporting of observation is not the issue here: the truth of the Grand Clerk's records is not disputed, even though they embody the results of the silent and frequently patronized 史官 *shi guan*, mere functionaries or operants. The point about the *jun zi* is not that he is reliable on one critical point or other, but that he alone can make judgements embracing the full variety of relevant factors, ranging from the scriptures on which Huang and Guang placed almost all their weight, to the need for technical change and the importance of instrumentation as a means to reliable knowledge. Finally, we may note Liu Zhao's approval of what he sees as the emperor's support for free expression in the context of the institution of the *yi*. One can be in the wrong without deserving to be treated as a criminal—even when, apparently, one has demanded such treatment for others.

7.2.4 The heavens as a locus of controversy: issues and resources for debate

What kinds of resources were open to controversialists discussing the heavens? Let us look back at two examples from chapter 6. In the first place we may consider the tactics used in the dispute over lunar eclipse prediction in 62–69 CE (chapter 6, section 6.1). The dispute opened with Yang Cen performing the manoeuvre that we may call (following Latour) 'modalizing' the official system, that is, attacking its claim to be treated as an unquestioned 'black box'.

[37] Steven Shapin and Simon Schaffer (1985) *Leviathan and the air-pump: Hobbes, Boyle, and the experimental life*. Princeton; Oxford, Princeton University Press

This he achieved easily enough by making one criticism of its predictions of a class of events that is hard to miss—a lunar eclipse. He then continued his success with a run of more routine predictions—only to see Zhang Cheng and his colleagues walk in through the door he had opened, and take over the task of introducing new methods of calculation. One factor that may have allowed Zhang Cheng et al. to enter the competition may have been that they could present themselves as operating *systematically* by using 'quarter remainder' methods as a coherent alternative methodology, in contrast with Yang Cen's one-day shift. The aim is not simply to predict the celestial phenomena somehow or other, but to demonstrate that one can do so through what we shall see in the next chapter (section 8.1.1) was known as a *shi fa* 師法 'master method'—a coherent and methodical scheme, possibly validated by authoritative transmission.

Zhang Cheng and his colleagues did manage to present their work as having a more systematic basis than that of Yang Cen—despite the fact they did not succeed completely, as is shown by the statement that they 'were not yet able to distinguish clearly what the system origin should be, or to comprehensively compare the fractions and degrees'. An incomplete system that produced good predictions might be allowed to replace one that was coherent (as the official system was), but made predictions that were obviously in error.

The other example already considered was the campaign of Jia Kui to persuade the Clerk's Officials to change the accepted position of the winter solstice, and to use the ecliptic, rather than the equator, as their basic reference frame for the motions of the sun and moon. One significant point is that in running his argument about the shifting position of the winter solstice, Jia Kui is able make checks of predictions against records of 70 solar eclipses during the Han dynasty, and 24 before that time. One great advantage for the Chinese imperial astronomer clearly lies in his direct access to such a rich databank, an equivalent to which astronomers of the Hellenic world such as Ptolemy of Alexandria had to seek from Mesopotamia, across significant boundaries of language and culture. The ability to settle disputes by reference to the record is no doubt a formative factor in the way debates on astronomical issues were structured and conducted in imperial China.

CHAPTER 8

Liu Hong and the conquest of the moon

In this chapter, we look at the work of Liu Hong, who emerges as the principal technical consultant involved in the later phases of the debates discussed in the last chapter, although it seems that he only had an official post concerned with celestial observation or calculation at the beginning of his career. He created the last great astronomical system that we shall discuss—the Qian xiang li *'Uranic Manifestation' system. This was the first system to give a complete account of the main irregularities of lunar motion, based on a subtle analysis of the mass of data gathered by the routine observations of Han sky-watchers in preceding centuries. The methods used by Liu Hong set the pattern for the handling of such questions in later centuries, and show the full power of the style of algorithmic modelling that was characteristic of the East Asian tradition.*

8.1 The work of Liu Hong

Liu Hong 劉洪, who worked with Cai Yong to compile the major collection of documents on which we have frequently drawn in this and previous chapters, is the last major figure whose work will be discussed in this book. He was also the creator of the Uranic Manifestation system *Qian xiang li* 乾象曆, the last astronomical system that we shall discuss at length. Our only substantive biographical account of Liu Hong comes from the sixth century commentary to the *Hou Han shu* monograph on astronomical systems, a monograph which consists largely of the material that he prepared with Cai Yong (see chapter 6, section 6.1):

> 袁山松書曰:「劉洪字元卓, 泰山蒙陰人也. 魯王之宗室也. 延熹中, 以校尉應太史徵, 拜郎中, 遷常山長史, 以父憂去官. 後為上計掾, 拜郎中, 檢東觀著作律曆記, 遷謁者, 穀城門候, 會稽東部都尉. 徵還, 未至, 領山陽太守, 卒官. 洪善筭, 當世無偶, 作七曜術. 及在東觀, 與蔡邕共述律曆記, 考驗天官. 及造乾象術, 十餘年, 考驗日月, 與象相應, 皆傳于世.」
>
> Yuan Shansong[1] wrote: 'Liu Hong's literary name was Yuanzhuo. He was from Mengyin near Mount Tai [in modern Shandong]. He was of the ancestral house of the Kings of Lu. In the Yanxi reign period [158–166 CE] he was appointed to office when the commandant [of his commandery?] responded to [a request for recommendations from] the Grand Clerk, and he was made a Gentleman of the Palace. He was transferred to become Senior Scribe of Changshan. He left office to mourn for his father. Later he was made Senior Reckoning Clerk, and was [once more] made a Palace Gentleman. He made selections of documents in the Dong guan [archives] to make a record of harmonics and astronomical systems; then he became [in turn] a Receptionist, the Watcher at the Gucheng Gate, and commandant of the eastern section of Kuaiji [in modern Zhejiang]. He was summoned to return [to the capital], but had not arrived [when he was] given the additional post of Governor of Shanyang [in modern Shandong]; he died in office. Hong excelled in calculation, and in his time he had no equal; he created the method of the Seven Brilliances [i.e. the sun, moon and five planets]. When he was in the Dong guan [archives], he worked on his record on harmonics and astronomical systems with Cai Yong, and made checks on the asterisms. When he came to create the methods of his Uranic Manifestation system, he checked it against the sun and moon for more than ten years, and found it to correspond to the phenomena; it has all been passed down to our time.' (*Hou Han shu, zhi* 2, 3043, commentary)

Liu Hong's lifetime is commonly said to have extended from about 130 CE to around 200 or 210 CE, but these two end-dates are somewhat conjectural. His work in the Dong guan archives with Cai Yong took place in 178 CE, as we saw in the last chapter (*Hou Han shu, zhi* 3, 3082; Cullen 2017, 234), and we shall see later (in section 8.1.1) that he was consulted on astronomical matters by the Han court in 174 and 179 CE. If we estimate him as being around 30 in 160 CE, about the time he took up his first post, he would have been about 50 in 180 CE.

[1] *Jin shu* 10, 253 records his death in the summer of 401 CE.

His Uranic Manifestation system is preserved in the *Jin shu* 晉書, which was compiled in the seventh century.[2] The editor responsible for the monograph on harmonics and astronomical systems *Lü li zhi* 律曆志 in the *Jin shu* was Li Chunfeng 李淳風 (602–670), who tells us explicitly that he aimed to continue his account of such matters from where the *Hou Han shu* broke off (*Jin shu* 17, 498). At the very beginning of the account of the Uranic Manifestation system, there is a date that may indicate when Liu Hong last worked on the text as we have it today. It occurs when he tells us, in effect, the 'system origin' he is using as the basis for his calculations:

> 上元己丑以來, 至建安十一年丙戌, 歲積七千三百七十八年.
>
> From the High Origin *jichou*.26 to the 11th year of the Jian'an period [206–207 CE] *bingxu*.23, the year accumulation is 7,378 years. (*Jin shu* 17, 504; Cullen 2017, 241)[3]

Liu Hong is here taking it for granted that his readers will be using his system after the year in question, so that all they have to do is to add the years passed since that year to 7,378. At the time he wrote that note, he might have been in his seventies. Li Chunfeng's account enables us to add another date to Liu Hong's biography, telling us how the chain of transmission of his work began—apparently about ten years before the text as we have it today was finalized:

> 獻帝建安元年, 鄭玄受其法, 以為窮幽極微, 又加注釋焉.
>
> In the first year of the Jian'an reign period of Xiandi [196 CE], Zheng Xuan received his [i.e. Liu Hong's] methods; he judged them to be extremely profound and subtle, and added an explanatory commentary to them. (*Jin shu* 17, 498)

Zheng Xuan 鄭玄 (127–200 CE), who would have been aged 69 at the time of the meeting with Liu Hong, was by that time a highly respected classical scholar. Zheng's biography confirms that the Uranic Manifestation system was amongst the texts on which he wrote commentaries; given that the other works on which he wrote were revered classical texts such as the *Book of Change*, the *Book of Documents* and the *Analects of Confucius*, it is evident that he must have seen the Uranic Manifestation system as having some importance.[4]

[2] *Jin shu* 晉書 *(History of the Jin dynasty, 265–419 CE)*. (c. 648 CE. Punctuated edition of 1974). Fang Xuanling 房玄齡 (579–648 CE), Beijing, Zhonghua Press, 17, 504–31; translated in full with an introduction and explanatory commentary in Cullen 2017, 235–355.

[3] This specification is repeated in identical terms at the beginning of the section of the system giving details of planetary calculations, at *Jin shu* 17, 519; Cullen 2017, 311.

[4] *Hou Han shu* 35, 1212. Zheng Xuan's commentary on the Uranic Manifestation system is now lost.

He would have been well equipped to understand such a complex piece of technical writing: in his youth he is said to have studied both the Triple Concordance system *San tong li* 三統歷, and the 'Mathematical procedures in nine sections' *Jiu zhang suan shu* 九章算術 (the 'Nine chapters'). It was his skill in calculation that had first won him admission to the group allowed personal access to Ma Rong 馬融 (79–166 CE), the greatest classical scholar of the preceding generation.[5]

Zheng Xuan was from the same part of the empire as Liu Hong: his biography tells us that his family was from Gaomi 高密, which is also in modern Shandong.[6] What must have been (given the complexity of the Uranic Manifestation) a fairly prolonged series of exchanges between the two men could well have taken place when Liu Hong was in his final posting at Shanyang. What we may call anachronistically the 'Shandong connection' is also evident in the later network of transmission of his work. This takes us from the Han into the period whose name comes from the 'Three Kingdoms' *San guo* 三國 which followed Han; these were Wei 魏 in the north (220–265 CE), Shu 蜀 in the west (221–263 CE), and Wu 吳 (222–280 CE) in the south In a memorial prepared in 223 CE as part of a discussion on calendrical matters in the state of Wei,[7] a certain Xu Yue 徐岳 said of him that:

> 劉洪以曆後天, 潛精內思二十餘載 [...] 劉歆以來, 未有洪比.
>
> Because the [contemporary] astronomical system was falling behind the heavens, Liu Hong immersed himself in subtle thought for more than twenty years [...] Since the time of Liu Xin, there has been nobody comparable with him. (*Jin shu* 17, 499–500)

We shall discuss Xu Yue's memorial in more detail later on. Li Chunfeng's editorial writing in the *Jin shu* tells us that Liu Hong's system excited interest in another of the successor states, Wu. In his account Li Chunfeng reveals that Xu Yue, like Zheng Xuan, came from the same area as Liu Hong, and that it was through Xu Yue that the Uranic Manifestation was passed to Wu, where it was adopted as the official astronomical system:

> 其劉氏在蜀, 仍漢四分曆. 吳中書令闞澤受劉洪乾象法於東萊徐岳, 又加解注. 中常 侍王蕃以洪術精妙, 用推渾天之理, 以制儀象及論, 故孫氏用乾象曆, 至吳亡.

[5] *Hou Han shu* 35, 1207.

[6] *Hou Han shu* 35, 1207.

[7] For this date, see section 8.1.2.

> The Liu family [the imperial clan of Han] was [in power] in Shu, and so they continued the use of the Han Quarter Remainder system. In Wu, the Director of the Palace Secretariat, Kan Ze received the methods of Liu Hong's Uranic Manifestation [system] from Xu Yue of Donglai [in modern Shandong], and added explanations and a commentary. The Regular Palace Attendant Wang Fan thought that Hong's procedures were subtly marvellous, and used them as the principles for his calculations about the celestial sphere; in order to make an instrument and its associated discussions. So the Sun clan [the rulers of Wu] used the Uranic Manifestation system until the fall of Wu. (*Jin shu* 17, 503)

A short work attributed to Xu Yue, *Shu shu ji yi* 數術記遺 'Memoirs on numerical procedures' says that he knew 'Liu Kuaiji 劉會稽' of Taishan 泰山, who was an expert in the matters with which his book was concerned; as we know from Yuan Shansong's account, Liu Hong had been a commandant in Kuaiji.[8] And to underline once more the importance of Liu Hong's personal itinerary, we may note that Kan Ze himself was from Kuaiji.[9] Kan Ze's commentary is mentioned in the catalogue of the imperial library of the Sui 隋 dynasty (581–618 CE), but is not subsequently referred to. In all likelihood this work was lost with 80–90% of the imperial collection when it was being transported across the Yellow River to the capital of the victorious Tang 唐 dynasty in 622 CE.[10] Material by Xu Yue discussing the technical details of the Uranic Manifestation system and its merits in comparison with other systems appear several times in the material assembled by Li Chunfeng in the *Jin shu*, and it is evident from what Xu Yue says that he has a good understanding of the methods of Liu Hong's system.[11]

Xu Yue's comparison of Liu Hong with Liu Xin showed a very high estimate of the abilities of both men. The most significant evidence of Liu Hong's achievements in the field of celestial calculation is found in the sections of his Uranic Manifestation system dealing with the motions of the moon. It may therefore be significant that all three references in contemporary sources to his having been

[8] See *Shu shu ji yi* in Guo Shuchun 郭書春 and Liu Dun 劉鈍, Eds. (2001). *Suan jing shi shu* 算經十書 *(The ten mathematical classics)*. Taibei, Jiu zhang chu ban she 九章出版社, 445. It is by no means certain that this rather strange little work is actually by Xu Yue, rather than by a later hand: for discussion, see Guo Shuchun 郭書春 and Liu Dun 劉鈍, Eds. (2001), 23–4.

[9] *San guo zhi* 53, 1249.

[10] *Sui shu* 34, 1022; Denis Twitchett (1979) *The Cambridge history of China. Vol 3, Sui and T'ang China, 589–906, Part I*. Cambridge, Cambridge University Press, 217.

[11] Thus, for instance, he gives calculated results for the times of appearance of planets that correspond to the predictions of the Uranic Manifestation system: see section 8.1.2. He also makes statements about the interval between the times of three observed solar eclipses and the predictions of the Uranic Manifestation (in effect its predictions for the true conjunction) that correspond quite well with the results of calculation: see *Jin shu* 17, 500.

called upon to give expert advice relate to the moon, and more particularly to lunar eclipses.

The importance laid on lunar eclipse predictions at this period was technical rather than prognosticatory. The long list of solar phenomena in the omen records of *Hou Han shu, zhi* 18, which includes 72 solar eclipses, is followed by notes of just two lunar eclipses, which are characterized as *yue shi fei qi yue* 月蝕非其月 'the moon being eclipsed in the wrong month'.[12] The main reason for attending to lunar eclipses was that they provided an unambiguous test of the accuracy of an astronomical system in predicting positions of the sun and moon, since they can only occur when the moon is very close to being opposite the sun in the sky, so that (in modern terms) it falls into the relatively narrow cone of shadow cast by the earth (section 4.4.5.4). What is more, lunar eclipses are relatively easily to observe, since if the moon is in shadow, the fact will be evident to anybody on the approximately half of the earth's surface from which the moon is visible during its relatively slow passage through the shadow. The same is by no means true of a solar eclipse, as we shall see later on.

Predicting lunar eclipses—or at least predicting when a lunar eclipse was likely—was by the time of Liu Hong not a complex procedure in principle. As we have seen, all systems operated in early imperial China used a simple cycle that assumed that the first opposition after system origin had a lunar eclipse, and that thereafter in a certain number of lunations a certain number of lunar eclipses would occur. Both the Triple Concordance and the Han Quarter Remainder systems used cycles with 23 lunar eclipses in 235 lunations; the Uranic Manifestation system modified this slightly to 1,882 lunar eclipses in 11,045 lunations, amounting to a decrease of about 0.01% in the mean interval between eclipses. Since eclipses can occur only at the middle of the moon's cycle of phases, and hence will be separated from one another by a whole number of lunar cycles, the practical effect will be to slightly increase the number of eclipses separated by five lunations rather than six.[13]

[12] The first eclipse actually occurred in the 11th month of the year 157–158 CE, when it was predicted for the 12th month, and the second occurred in the first month of 165–167, whereas it had been predicted for the second month. See *Hou Han shu, zhi* 18, 3374.

[13] On lunar eclipse prediction methods in the Triple Concordance, Han Quarter Remainder and Uranic Manifestation systems, see Cullen 2017, 99–100, 157–9, 180–5 and 261–3. See also Shi Yunli 石云里 and Xing Gang 邢钢 (2006) 'Zhong guo Han dai de ri yue shi ji suan ji qi dui xing zhan guan de ying xiang 中国汉代的日月食计算及其对星占观的影响 (The Calculation and Divination of Luni-solar Eclipses of the Han Dynasty and Its Impact on the Astrological Thoughts).' *Zi ran bian zheng fa tong xun* 自然辩证法通讯 *(Journal of the Dialectics of Nature)* 28, (2): 79–85.

8.1.1 Liu Hong as trusted consultant—and the case of an expert commoner

The references to Liu Hong's involvement with such matters occur in the collection of documents that he formed in collaboration with Cai Yong, and are therefore likely to give what he would have regarded as a representative account of his work.

The first reference relates to enquiries set in motion in 174:

> 到熹平三年, 二十九年之中, 先曆食者十六事. 常山長史劉洪上作七曜術.
>
> Up to the third year of the Xiping period [AD 174–5], an interval of 29 years, there were 16 instances of [lunar] eclipses in advance of [the predictions of] the system. Liu Hong, Senior Scribe of Changshan, submitted the worked calculations he had done [for the] Seven Brilliances [i.e. the sun, moon and five visible planets]. (*Hou Han shu, zhi* 2, 3040; Cullen 2017, 411)

One example of the phenomenon referred to here—lunar eclipses being seen earlier than predicted by the current system—may be seen in early 174 CE. The Han Quarter Remainder system predicted a lunar eclipse on the day of opposition in the second civil month, 5 April, but in fact there was no eclipse at that time. One did, however, occur at the opposition of the preceding month, on 6 March, and it would have been clearly visible from Luoyang, with nearly ¾ of the moon's disc in full shadow at about 6:30 p.m. with the moon then being 12° above the eastern horizon.

We are told that an edict was issued instructing the Grand Clerk and others to make detailed checks on Liu Hong's lunar calculations using *ba yuan shu* 八元術 'The method(s) of the Eight Origins', and that their calculations agreed with those of Liu Hong.[14] Given that the edict was issued after Liu Hong made his submission, it would seem that it was Liu who took the initiative of raising the matter. Perhaps we can see here some evidence of the 20 years' work leading to the Uranic Manifestation system to which Xu Yue referred?

In the years that followed, Liu Hong was asked to participate in two further enquiries involving proposals relating to lunar eclipse prediction, both of which of involved wide consultation over an extended period (Cullen 2017, 410–19). The first such enquiry had its origins a century before, when in 90 CE the

[14] The 'Eight Origins', otherwise unknown, may be related to the idea that there had anciently been six different *li* 'systems', each with their own system origins (3.4.1), to which could be added the Triple Concordance and Han Quarter Remainder systems, thus making eight in all.

commoner Zong Gan 宗紺, a holder of the Eight Rank,[15] criticized the prediction of a lunar eclipse by the Han Quarter Remainder system then in use. On the 12th day of the first civil month of that year (1 March) he pointed out that the official system predicted a lunar eclipse for the opposition day of the next month, which would have been on 3 April. He claimed, however, that the eclipse would instead take place on the opposition day of the current month, corresponding to 5 March, only four days after his prediction. A good total lunar eclipse was in fact visible from Luoyang at about 00:30 local time on that day, with the moon about 60° above the southern horizon. This success won Zong Gan appointment as an Expectant Official, and his methods of calculation were officially adopted. They are said to have continued in use for the next 56 years (*Hou Han shu, zhi* 2, 3041). Then in 175 CE:

> 紺孫誠上書言:「受紺法術,當復改,今年十二月當食,而官曆以後年正月.」
>
> [Zong] Gan's grandson [Zong] Cheng submitted a memorial saying 'I have received the model methods of Gan. They require to be revised. In the present year there should be an eclipse in the twelfth month [31 December 175–28 January 176], while the official system has it in the first month of the next year.' (*Hou Han shu, zhi* 2, 3041; Cullen 2017, 412)

Zong Cheng's prediction was in fact correct.[16] This was apparently the starting point for a somewhat rancorous dispute that came to the surface four years later, when in 179 Zong Cheng predicted a lunar eclipse in the fourth month, and the Grand Clerk and one of his staff, Zhang Xun 張恂, predicted a likely eclipse in the third month. Had it been possible to conduct observations, the issue might have been settled—since there was indeed an eclipse on the opposition day of the third month, 9 May. This was a penumbral eclipse that reached its maximum about 00:20 local Luoyang time on that day, when the moon was 38° above the southern horizon. However, the weather was cloudy from the third to the fifth months—at least on the days when eclipses might have been observed. Xun and his colleagues insisted, however, that their calculation was 'close', and submitted to the throne that their method should be adopted in preference to Zong Cheng's.

[15] The male population of Han were allocated positions in a system of twenty ranks. The eighth rank was the highest grade awarded to those not holding an official post, and entitled the holder to certain privileges such as exemption from some labour services, and lighter punishment for offences. See Twitchett, Loewe and Fairbank (1986), 552–3, and in more detail M. Loewe (1960) 'The Orders of Aristocratic Rank of Han China.' *T'oung Pao* 48, (1/3): 97–174.

[16] The opposition day of the 12th month was 14 January 176 CE. A clear penumbral lunar eclipse was in fact visible on that day, reaching its maximum at 06:00, when the moon was still 12° above the western horizon. It had begun about 2 hours earlier, when the moon's altitude was around 35°.

The next year, however, repeated memorials of protest from Zong Cheng's elder brother led to an edict being issued to the Chamberlain for Ceremonials (*Tai chang* 太常), the high official who was the Grand Clerk's immediate superior, commanding him to enquire into the matter. The Chamberlain assembled a group of four persons, of whom Liu Hong (then the Watcher at the Gucheng Gate, *Gu cheng men hou* 穀城門候) was one, and charged them to 'comprehensively review the observation records, and deliberate equitably on disputed questions' (*fu jiao zhu ji, ping yi nan wen* 覆校注記, 平議難問). Zhang Xun and Zong Cheng both made further submissions.

The resultant report was lengthy, and on the whole seems to have done its best not to offend either side. It noted that the key technical difference between the two approaches was that Zong Cheng used the well-known eclipse cycle that provided for 23 eclipses in 135 lunations, whereas Zhang Xun used his own cycle of 961 eclipses in 5,640 months (we may note that since 5,640/961 = 134.98/23, this change was only a minor adjustment). However, there had so far been no clear observational test to settle who was in the right:

> 未驗無以知其是, 未差無以知其失.
>
> If there has not yet been a verification, there is no way to know something is right. If there has not yet been an error, there is no way to know something is at fault. (*Hou Han shu, zhi* 2, 3041; Cullen 2017, 415)

Zhang Xun was complimented for his skill as both an observer and a calculator, but the conclusion was that the methods used by Zong Cheng were well supported by previous practice and repeatedly stated official policy, and should be adopted. On the other hand, the Clerk's officials should continue to keep Zhang Xun's method under review, and if at some time in the future it was found to be verified, it might then be adopted. Neither side was satisfied with this, and both sent in further protests. Zong Cheng in particular complained that what he called '[Liu] Hong's report' had treated him presumptuously—perhaps because Liu had refrained from stating that his method had been decisively supported by the evidence. At this point higher authority intervened with evident exasperation, and the two complainants were convicted of deception: Zong Cheng and Zhang Xun were allowed to redeem their offences at the cost of two months' salary, while Zong Cheng's brother was made to serve two months as a construction foreman. Hong's recommendation in favour of Zong Cheng's methods was subsequently followed.

Although Zong Gan had been a commoner when in the first century CE he first approached the court with his proposals, the fact that his grandson Zong Cheng was punished by loss of salary suggests that his descendants had remained in official posts after his success. Nevertheless, the expertise on which Zong Cheng relied seems to have originated in private family circles.

The next issue relating to lunar eclipses in which Liu Hong became involved arose from the claims of another commoner expert. Like Zong Gan, Wang Han 王漢 was a holder of the Eighth Rank. He arrived at court in 179 CE, the same year as the disputes outlined earlier in this section, bringing with him observation notes of 96 lunar eclipses, extending back 93 years—presumably the result of continuous interest in such matters by his family. He also appears to have had his own method for predicting eclipses, involving the use of his own 'system origin'. The Grand Clerk reported that Wang's observations were not entirely consistent with his department's records; there were two cases in which no record of a visible eclipse corresponded to what was on Wang's list, and 29 cases where records of what seemed to be the same event differed significantly. The Secretariat called on Liu Hong to look into Wang's claims. Liu Hong's subsequent report concentrates on a major issue: the 'origin' used by Wang Han for his eclipse calculations, that is, the point in the past from which he counted off his eclipse cycles. The instruction issued to Liu Hong included these words:

> 推元課分, 考校月食. 審己巳元密近, 有師法, 洪便從漢受; 不能, 對.
>
> [Now Liu Hong is to] calculate back to the origin, and check the fractions, examining and verifying lunar eclipses, and to look into the accuracy of the *jisi*.6 origin [used by Wang Han]. If [Wang] has a 'master method', let Hong receive it from Han; if he is unable to do so, reply [accordingly]. (*Hou Han shu, zhi* 2, 3042; Cullen 2017, 417)

To ask whether Wang Han possessed a 'master method' *shi fa* 師法 amounted to asking whether he based his predictions on some systematic and distinctive method which had the authority that came from master–pupil transmission.[17] A *shi fa* had to be consistent, rather than ad hoc: when Liu Hong said of Zhang Xun and Zong Cheng that they both made adjustments as suited them, and that their writings had no guiding principle, but just adopted whatever 'kept up with the heavens' (*qu zhui tian er yi* 取追天而已) he was implying that they certainly did not have a 'master method'.

[17] It was said of one of Wang Han's contemporaries, Liu Kuan 劉寬 (120–185 CE): 星官, 風角, 筭歷, 皆究極師法, 稱為通儒 'In [divination by] the stellar offices [i.e. the constellations correlated with human activity] and wind directions, and in calculating [astronomical] sequences, in every case he thoroughly investigated the *shi fa*, and had the reputation of a versatile scholar' (*Hou Han shu* 25, 891 comm.).

The use of a distinctive 'system origin' would certainly be a valid way of laying claim to methodological originality, and also raised wider and potentially more politically sensitive issues. Indeed, the instructions from the Secretariat specifically referred to the events of a few years previously, discussed in the previous chapter, when Cai Yong had publicly rebutted the claims of Feng Guang and Chen Huang, who had attributed current social disturbances and frontier problems to the use of the wrong origin. Fortunately for Liu Hong, these were matters in which he was an expert. He points out that Wang's '*jisi*.6 origin' was in fact the epochal day of the system mentioned in a well-known apocryphal work the *[Shang shu] kao ling yao* [尚書]考靈曜 'Examination of the Mysterious Brilliances (i.e. the celestial bodies) [in the *Book of Documents*]', whose epochal year was *yimao*.52. This, he notes, is simply the Zhuan Xu system, which he claims (in accordance with the view current in the Eastern Han) was used in the Qin and early Han (see chapter 3, section 3.4.2). Feng Guang and Chen Huang had similarly used another obsolete system, but with its origin in a *jiayin*.51 year: although Liu Hong does not say so, this is the system we have already met in the Western Han as the Yin system (chapter 3, section 3.4). That system was accurate in the time of Confucius, he says, while the Zhuan Xu system was also accurate in its time, but required change by the time of the Grand Inception reform of 104 BCE. The fact that both systems are mentioned in apocryphal works does not mean that they can be expected to fit current circumstances. But when asked to justify his choice of system, all Wang Han could reply was that 'his forefathers had their books' *xian ren you shu er yi* 先人有書而已.

Liu Hong goes out of his way to emphasize how well known to him both systems are: he gives details of precisely how far apart are their predictions: conjunction predictions differ by 304 parts (out of 940) and *qi* predictions by 29 parts (out of 32). Wang Han's use of an epoch on a *jisi*.6 day simply came, Liu Hong says, from his having read about it somewhere, and his mistaken assumption that this was something that the court would not know about.

But:

> 不知聖人獨有興廢之義, 史官有附天密術.
>
> [He was] unaware that the Sage alone possesses the true righteousness for raising up and casting down, and that the Clerks had accurate methods for staying close to the Heavens. (*Hou Han shu, zhi* 2, 3042; Cullen 2017, 419)

To add to the impression that Wang Han was quite out of touch with current practice, Liu Hong notes that he had criticized the use of the first year of the Heping 河平 reign period (28 BCE) as an origin for eclipse cycles: but this had long ago been abandoned.[18] Wang was 'making trouble over matters that are done with'. All in all, 'Even if he had a "master method", it would be just the same as if he had none' *sui you shi fa, yu wu tong* 雖有師法, 與無同. And thereupon Wang was sent back home. Liu Hong's approach to evaluating Wang Han seems quite different from the more tactful approach he had taken to Zong Cheng and Zhang Xun. It may be that he had already had enough of evaluating other peoples' ideas for one year, or it may be that he was irritated by the Secretariat's suggestion that he might learn something from Wang. As we shall see, he may already have had good reason to feel that was unlikely.

And at this point the documents collected by Cai Yong and Liu Hong come to an end. A decade after the events recorded here, the court at Luoyang had ceased to function: on 4 April 190 CE, Dong Zhuo forced the emperor to shift the capital to Chang'an, and then burned Luoyang.[19] Vast losses of documents appear to have occurred, and presumably discussions of calendrical matters were not a practical possibility for some time. By that time, it seems likely that Liu Hong had left the court to take up the provincial posts he occupied in the closing years of his life.

8.1.2 The Uranic Manifestation system

The materials on Liu Hong's activities in the main text of the *Hou Han shu* make no reference to the creation of his *Qian xiang li* 乾象曆 'Uranic Manifestation' system. There are no signs that it was ever proposed as a replacement for the Han Quarter Remainder system during the period when dynastic institutions continued to operate normally. However, we do know that it must have been completed in some form by 196 CE, over two decades before the last nominal Han emperor abdicated, since it was in that year that, as we have seen, the great scholar Zheng Xuan 'received' its methods; the last date mentioned in the system as we have it today corresponds to 206 CE, ten years later.[20]

Fortunately, the *Jin shu* gives us a complete specification of Liu Hong's system (though alas without any commentary from either Zheng Xuan or Kan Ze).[21]

What were seen as the distinctive features of Liu Hong's new system? According to Li Chunfeng:

[18] In fact it had only been used for five years from 85 CE when the Han Quarter Remainder system had first been introduced: see *Hou Han shu, zhi* 2, 3041; Cullen 2017, 410–11.

[19] Twitchett, Loewe and Fairbank (1986), 348.

[20] See above in section 8.1.

[21] See *Jin shu* 17, 504–31; for a full translation with introduction and explanatory commentary, see Cullen 2017, 235–355.

> 漢靈帝時, 會稽東部尉 劉洪, 考史官自古迄今曆注, 原其進退之行, 察其出入之驗, 視 其往來, 度其終始, 始悟四分於天疏闊, 皆斗分太多故也. 更以五百八十九為紀法, 百四十 五為斗分, 作乾象法, 冬至日日在斗 [二十一度] (二十二)度,[22] 以術追日, 月, 五星之行, 推而上則合於 古, 引而下則應於今.
>
> In the time of Lingdi [r. 168–188 CE] Liu Hong, commandant of the eastern district of Kuaiji [near Shaoxing in modern Zhejiang] examined the astronomical systems and their annotations by the Clerk's officials from antiquity to the present, traced the reasons for the advances and retardations of their motions, looking into the verification of their discrepancies, observed their coming and going, and measured out their ends and beginnings. Thus he first realized that the divergence of the [Han] Quarter Remainder [system] from the heavens was because its Dipper Parts were too great. Thus he took 589 as his Era Factor, and 145 as the Dipper Parts. He made the Uranic Manifestation system; on the day of the winter solstice the sun was at 21 *du* of Dipper. His methods pursued the motions of the sun, moon and five planets; he extended it upwards to fit in with antiquity, and extended it down to answer to the present. (*Jin shu* 17, 498)

The note above confirms the impression given in the *Hou Han shu* that Liu Hong had been interested in such matters as lunar motion at least a decade before the meeting with Zheng Xuan. Now according to systems of the quarter remainder type, the interval between winter solstices was 365 ¼ days, 365.25 days. The 'Dipper Parts' *dou fen* 斗分 represents the fractional part of this quantity; its denominator in the Uranic Manifestation system is Era Factor [589]. Since $\frac{1}{4} = {}^{147.25}\!/_{589}$, Liu Hong's choice of constants amounts to a slight reduction in the Dipper Parts. A value based on modern calculations would be closer to 365.243 days.[23] Liu Hong's figures would yield $(365 + {}^{145}\!/_{589}) = 365.246$ days.

Once the year length was fixed, the length of the lunation was determined by the fact that Liu Hong had decided to retain the well-known equivalence of 19 solar cycles and 235 lunations, so that one mean lunation lasted $(365 + {}^{145}\!/_{589}) \times {}^{19}\!/_{235} = 23.53054$ days, as opposed to a modern value of 23.53059 days for this period (the mean synodic month).

So much for the motions of the sun and moon—at least the mean motion of the moon. But we are also told that Liu Hong 'pursued the motions of the … five planets'. Did he innovate in any way when he laid down the basic constants for these bodies? Table 8.1 sets out a comparison of the Uranic Manifestation

[22] Correcting in accordance with the note in the Zhonghua edition.

[23] See Meeus and Savoie (1992), 42. The effect of Liu Hong's change in Dipper Parts would have mounted up to a whole day in 262 years.

Table 8.1 ***Planetary constants in three systems***

	Triple Concordance			Triple Concordance Appearance number changed to Han Quarter Remainder denominator	Han Quarter Remainder			Han Quarter Remainder cycle rate changed to supernal manifestation solar rate denominator	Supernal manifestation			Modern values for synodic periodic, years
	Year Number	Appearance/ Return Number	Synodic period = Year Number/ Appearance Number		Cycle Rate	Solar Rate	Synodic period = Solar rate/ Cycle Rate		Cycle Rate	Solar Rate	Synodic period= Solar rate/ Cycle Rate	
Jupiter	1,728	1,583	1.0915982	4,328.52	4,327	4,725	1.0919806	6,722.65	6,722	7,341	1.0920857	1.092
Mars	13,824	6,469	2.1369609	877.88	879	1,876	2.1342435	3,406.83	3,407	7,271	2.1341356	2.135
Saturn	4,320	4,175	1.0347305	9,098.99	9,096	9,415	1.0350704	3,529.23	3,529	3,653	1.0351374	1.035
Venus	3,456	2,161	1.5992596	5,828.95	5,830	4,661	1.5989708	9,022.05	9,022	7,213	1.5989803	1.599
Mercury	9,216	29,041	0.3173444	11,905.05	11,908	1,889	0.3172657	11,561.29	11,561	1,834	0.3172736	0.317

system with the constants of the two preceding systems, the Triple Concordance system and the Han Quarter Remainder system.

For comparison between the systems, the intermediate columns between each group show the number of planetary cycles given, revised so as to fit into the number of years specified in the following system. It is clear that hardly any change in the synodic period has actually been made in the Uranic Manifestation in comparison with its predecessor; the changes between the first and second systems were a little more marked. However, one special point of interest in Liu Hong's planetary system is that it is the only one of the three for which we have any record of a detailed comparison of its predictions with observation.

This comes from 223 CE, shortly after the end of the Han dynasty, when a discussion of the merits of competing systems took place at the court of the kingdom of Wu. Xu Yue, who as we know had expert knowledge of the Uranic Manifestation system, listed fourteen observations of planetary phases—one for Jupiter, three for Saturn, two for Venus and eight for Mercury, including first and last dawn and dusk visibilities—and compared them not only with the predictions of the Uranic Manifestation system, but also with those of the *Huang chu* 黃初 'Yellow Inception' system advocated by his opponent Han Yi 韓翊.[24] In each case he gives the precise date on which the phase was recorded as having been observed, followed by the dates predicted by each system, and the differences in days between observation and prediction. In the case of Liu Hong's system, it is possible to carry out the calculations following the Uranic Manifestation specifications that must have lain behind the results given by Xu Yue, and, reassuringly enough, it turns out that the results produced are in agreement with his.[25] The discrepancies between observation and prediction

[24] See *Jin shu* 17, 500–2. The exchanges in which Xu Yue takes part are said to have taken place 'in the Huangchu reign period (220–226 CE) of Wendi of the Wei' *Wei wen di huang chu zhong* 魏文帝黃初中 (*Jin shu* 17, 498). The earliest planetary phenomenon he lists is said to have been observed on the 26th day, *renchen*.29, of the 11th month of the second year, corresponding to 21 December 221, and the latest fell on the 22nd day, *renzi*.49, of the 11th month of the third year, corresponding to 11 January 223. Thus a date for this discussion in 223 seems strongly indicated. Compare the discussion of these phenomena in Morgan (2013) 334–6.

[25] In one case, a phase of Venus, we need to correct a stated discrepancy of 23 days between observation and prediction to 22 days, so as to make it consistent with the two sexagenary dates given; this is probably a simple copyist's confusion of the characters 三 and 二. In one other case, a dusk appearance of Mercury in early 223 CE, modern recalculation using Uranic Manifestation constants gives a discrepancy between prediction and observation of 15 days rather than 16 days, but the 16 days figure is consistent with the sexagenary days given in the *Jin shu*; the difference could have been produced by a calculator who rounded down the interval from conjunction to dusk appearance to whole days. For the other seven Mercury phases, modern calculations using the Uranic Manifestation constants are identical with those in the *Jin shu*.

are carefully noted, and it is clear that for Xu Yue and his colleagues they are an important factor in evaluating the merits of different systems. Xu Yue notes that the Uranic Manifestation system scored 'seven close [to observation] and two on target' 七近二中 *qi jin er zhong*, while the *Huang chu* system only scored 'five close and one on target' 五近一中 *wu jin yi zhong*. Judging from the results, it seems that 'close' for Xu Yue means a prediction within five days or less of actual observation.

We may note that in any prediction of a planetary phase, such as a dusk or dawn appearance (first visibility) or disappearance (last visibility) of a planet, there are two factors in play:

(a) The accuracy of prediction of the moment when the planet is in conjunction with the sun, which is the key point in predicting the dates of visible planetary phases.
(b) The way the system in question sets conditions for the visibility of a given phase.

The latter is a highly complex question, which I have already discussed elsewhere in connection with the *Wu xing zhan* text.[26] As pointed out there, in the ancient Mediterranean world, Ptolemy attempted to predict planetary visibility on the basis of a 'normal arc' (the angular distance of the sun below the horizon as the planet crosses the horizon) which if exceeded should lead to a report of the planet being seen.[27] These arcs were given by Ptolemy as:

Saturn 11°
Jupiter 10°
Mars 11 ½°
Venus 5°
Mercury 10°

These universally applicable quantities he derives from a set of observations said to have been made in summer, with the Sun near the beginning of Cancer,[28] when the distances of the planets from the sun along the ecliptic at their first visibility were found as:

Saturn 14°
Jupiter 12 ¾°

[26] See chapter 4, section 4.4.3, and Cullen (2011b), 234–6, and further in Robinson (2009).
[27] Toomer (1998), 636–47.
[28] Toomer notes that this specification, with other references, suggests that Ptolemy took these figures from a Babylonian source: Toomer (1998), 637.

Mars 14 ½°
Venus 5 ⅔°
Mercury 11 ½°

The Uranic Manifestation system simply states how many days after conjunction must pass before a planet first becomes visible. However, since daily motion during this period is also stated, it is possible to deduce the critical distances from the sun that are implied by this. The results are given below, and are stated, for convenience, in the same order and to the same precision as Ptolemy; the differences between degrees and *du* are not significant at this scale:

Saturn 14 ½ *du*
Jupiter 13 ⅓ *du*
Mars 16 *du*
Venus 9 *du*
Mercury 16 *du*

For the first three planets, the figures given are quite close to Ptolemy's. The Uranic Manifestation does, however, seem much more pessimistic than Ptolemy about the chances of seeing Venus, and a little more pessimistic for Mercury. If, however, we look at the elongations from the sun of the planets given by modern calculations at the actually observed dates of first and last visibility reported in the *Jin shu*, we find data very different from this theoretical scheme:

Saturn 20°, 12 ½°, 23°
Jupiter 15°
Mars—no data
Venus 14 ½°, 18°
Mercury 22°, 19 ½°, 18°, 27°, 18°, 11 ½°, 14°, 18°

It seems, therefore, that a considerable part of the differences between observation and prediction must be ascribed not just to discrepancies in calculating the moments of conjunction, but also to the difficulties of any attempt to make any simple mathematical model of the complexities of horizon visibility phenomena, which will depend on:

(a) The brightness of the planet.
(b) The proximity and direction of the sun.
(c) Current atmospheric conditions near the horizon.
(d) The acuity of the observer's eyesight.
(e) The degree of conviction of having seen the planet that the observer demands before he feels confident enough to make a report.

Of these, only (b) can easily be modelled mathematically, and (d) and (e) are not susceptible to modelling at all, since they depend on the observer.

The planets, like the sun and moon, follow the ecliptic in their motion round the celestial sphere. As we have seen, it was established by Jia Kui that the ecliptic, not the equator, was to be the basic frame of reference for solar, lunar and planetary motion. In introducing Liu Hong's new system, the editors of the *Jin shu* note that he was the first to design a *li* on this basis:

> ... 日行黃道, 於赤道宿度復有進退. 方於前法, 轉為精密矣.
>
> ... as the sun travelled the Yellow Road [the ecliptic], there would be advance and retardation of the *du* of lodges on the Red Road [the equator]. By this, he made [his methods] more precise than their predecessors. (*Jin shu* 17, 498)

The description of these 'advances and retardations' in Liu Hong's text is brief and somewhat oddly placed, but it says enough to make it plain that he is following the method set out by Zhang Heng (chapter 6, section 6.5.3):

> 推有進退, 進加退減所得也. 進退有差, 起二分度後, 率四度轉增少, 少每半者, 三而轉之, 差滿三止, 歷五度而減如初.
>
> To predict when there will be advance or retardation: this is obtained from adding advance and subtracting retardation. There are differences in advance and retardation: beginning after the degrees of the two equinoxes,[29] the rate is that with every cycle of four *du* one increases [the advance or retardation] by a quarter, and each time the quarters make up to a half, one takes a cycle of three. When the difference mounts up to three *du* one stops, and goes through five *du*, and then subtract as before. (Jin shu 17, 509; Cullen 2017, 267–8)

Liu Hong is describing the same pattern as Zhang Heng, but with extreme brevity. Each *qi* begins with two four-day periods, during each of which the advance/retardation increases by ¼ *du*. By now the advance/retardation has reached ½ *du*, and the next ¼ *du* is allocated to only three days. We are not told what happens to complete the *qi*, but obviously there is another four-day period with ¼ *du* difference added, as in Zhang Heng's scheme. Finally (at the end of three *qi*, though Liu Hong does not say this explicitly) the total

[29] It is interesting to note that Zhang Heng began his reckoning from the winter solstice—but the pattern followed is exactly the same.

difference amounts to three *du*, at which point the sun moves through five *du* (which takes five days) without any change in the advance/retardation, just as Zhang Heng tells us.[30]

8.2 Modelling the motions of the moon

But the significance of Liu Hong's innovations did not lie in the lengths of years or months, or in his treatment of the planets.[31] Li Chunfeng's discussion of his work continues:

> 創制[月] (日)行 遲速, 兼考月行, 陰陽交錯於黃道表裏, 日行黃道, 於赤道宿度復有進退. 方於前法, 轉為精密矣.
>
> He also created the innovation of the slowing and acceleration of the moon,[32] with a comprehensive examination of the yin and yang sections of the moon's motion, as it crosses to the outside and inside of the Yellow Road [the ecliptic], [and also provided for the fact that] as the sun travelled the Yellow Road, there would be advance and retardation of the *du* of lodges on the Red Road [the equator]. By this, he made [his methods] more precise than their predecessors. (*Jin shu* 17, 498)

We have already seen that the topic of the moon's departures from steady motion along the ecliptic was a centre of attention for a number of people before Liu Hong, certainly as far back as Jia Kui around 100 CE, if not earlier (see section 6.3.5).

[30] Yan Dunjie 嚴敦傑 (1958) 'Zhong guo gu dai de huang chi dao cha ji suan fa 中國古代的黃赤道差計算法 (Method for Reduction of the Ecliptic as used by the Ancient Chinese Astronomer).' *Kexueshi jikan* 科學史集刊 1: 47–58, 47–8, quotes a late Qing scholar's explanation of Liu Hong's passage close to that given here. Yan's description of what happens near the maximum value is in my view mistaken, however. Whereas (like Zhang Heng) Liu Hong is clearly telling us that the last five days of each three-qi 'half-season' have the same advance/retardation value, Yan's explanation suggests that the stationary five days occur only at the end of a whole six-qi season, i.e. they first occur at summer solstice. Moreover, he does not seem to have noticed that Liu Hong agrees with Zhang Heng in precise detail, not simply in the fact that their advance/retardation values both vary between ± 3 du. See the somewhat different interpretation of these fractions in Lien (2012).

[31] The discussion that follows builds on and expands the content of Cullen (2002). I would like to thank Tang Quan 唐泉 for his kindness in reading and commenting in detail on an earlier draft of this section of the book; as ever, I am wholly responsible for any remaining errors or omissions.

[32] As the modern editors note, in the light of what follows, 日 is clearly a copyist's error for 月, since it was precisely the slowing and acceleration of the moon that Liu Hong was the first to analyse in detail. Without this emendation we would have to accept that the *Jin shu* was making the highly implausible suggestion that Liu Hong had anticipated the accepted date of the Chinese discovery of the variation in speed of solar motion by four centuries. This view is strengthened by the fact that not long after this editorial remark, Li Chunfeng quotes Xu Yue as saying that Liu Hong *wei yue xing chi ji* 為月行遲疾 'made the slowing and acceleration of the moon's motion': *Jin shu* 17, 499.

If, however, Jia Kui's 'Nine Roads' method really did provide a complete account of the aspects of lunar motion covered by Liu Hong's system, it seems that it may have been forgotten by the time that Cai Yong and Liu Hong worked. As in other *li*, the specifications of the Uranic Manifestation system in the *Jin shu* give instructions for operating the system, but do not indicate any sources from which its methods may have been drawn.

Li Chunfeng credits Liu Hong with two innovations in understanding the behaviour of the moon—first his analysis of the variation in its speed of displacement against the background of the stars, and secondly his method for predicting its changing distance from the ecliptic; in modern terms, its latitude. Taken separately, each of these innovations certainly increased the accuracy with which the motion of the moon could be predicted. But it was their combination that opened a major new area for experts in *li*, since it became possible to:

(a) Find the true moment of conjunction or opposition of sun and moon, rather than relying on estimates of the mean phases based on the notion that the moon moved at constant speed, which is all that had been possible before.

(b) Find where the sun and moon at true conjunction are located in relation to the two nodes—the points where the path of the moon crosses the ecliptic, the path of the sun. If a conjunction takes place close enough to a node, there is a possibility that the moon may partly obscure the sun, leading to a solar eclipse. Thus, for the first time, specialists in *li* were able to give advance warning that a solar eclipse might take place on a given date.[33]

We shall now examine each innovation in turn, after which we shall consider the question of their possible use in eclipse prediction.

8.2.1 Liu Hong's speed sequence

Liu Hong addressed the question of the moon's varying speed—and hence the difference between its true position and the position predicted on the basis of constant mean motion—in his *chi ji li* 遲疾歷 'speed sequence' table.[34] Table 8.2 is a translation (both in language and, to some extent, in structure) of the

[33] Similarly, and independently of the cycle-based methods previously employed, it was possible to use the closeness of the moon to a node at opposition to estimate the likelihood of a lunar eclipse.

[34] Literally 'slow and fast sequence'; see the fully commented translation of relevant material in Cullen 2017, 268–93 for further details.

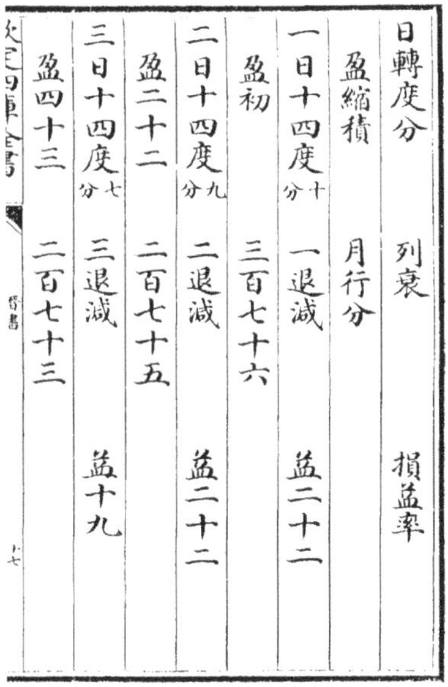

日轉度分	列衰	損益率
盈縮積	月行分	
一日十四度十分	一退減	益二十二
盈初	三百七十六	
二日十四度九分	二退減	益二十二
盈二十二	二百七十五	
三日十四度七分	三退減	益十九
盈四十三	二百七十三	

欽定四庫全書 晉書 十七

Figure 8.1 Beginning of Liu Hong's speed sequence table in *Jin shu* 17, 17a, *Si ku quan shu* edition, c. 1782.

tabulation given by Liu Hong in *Jin shu*, 17, 516 (Cullen 2017, 271–4). In its pre-modern form, the original table begins as shown in Figure 8.1.

The two columns on the right give the data headings: in the first column we have 'Day count, *du* and parts', 'Successive reduction' and 'Rate of decrease or increase', while in the second column we have 'Excess and deficit accumulation' and 'Parts of lunar motion'. The third column gives us the first three of these quantities for the first day, and the fourth gives the last two for the same day. Columns 5 and 6 give data for the second day, and so on.

For clarity, in the full rendering of the table given here, some of the data have been split between columns, so that, for instance, 'Day count, *du* and parts' is split into columns 1, 2 and 3.

As to the significance of the data given, column 1 gives the number of the day of the speed sequence, that is, the day in the cycle over which the variation in lunar speed repeats. Apart from the fractional last day, these days are the same length as a solar day (presumed constant at this period), but do not in general begin at midnight. The length of the speed sequence is equivalent to the anomalistic month, which is the interval between the moon's passage past perigee (the point nearest to the earth in the moon's elliptical orbit), when the moon moves fastest: see chapter 6, section 6.3.5.1. The modern mean value is 27.55455 days; the length

Table 8.2 ***Uranic Manifestation system speed sequence table***

1	2	3	4	5	6	7	8	9
Day of sequence	*du*	Parts [1/19 *du*]	Rate [1/19 *du*]	Lead (+) lag (−)	Additive (+) subtractive (−)	Rate of Lessening (−) or Increase (+) [1/19 *du*]	Accumulated Excess or Deficit [1/19 *du*]	Parts of lunar motion [1/19 *du*]
1	14	10	1	−	−	22	0	276
2	14	9	2	−	−	21	22	275
3	14	7	3	−	−	19	43	273
4	14	4	4	−	−	16	62	270
5	14	0	4	−	−	12	78	266
6	13	15	4	−	−	8	90	262
7	13	11	4	−	−	4	98	258
8	13	7	4	−	−	0	102	254
9	13	3	4	−	+	−4	102	250
10	12	18	3	−	+	−8	98	246
11	12	15	4	−	+	−11	90	243
12	12	11	3	−	+	−15	79	239
13	12	8	2	−	+	−18	64	236
14	12	6	1	−	+	−20	46	234
15	12	5	1	+	−	−21	26	233

16	12	6	2	+	−	−20	5	234
17	12	8	3	+	−	18	−15	236
18	12	11	4	+	−	15	−33	239
19	12	15	3	+	−	11	−48	243
20	12	18	4	+	−	8	−59	246
21	13	3	4	+	−	4	−67	250
22	13	7	4	+	+	0	−71	254
23	13	11	4	+	+	−4	−71	258
24	13	15	4	+	+	−8	−67	262
25	14	0	4	+	+	−12	−59	266
26	14	4	3	+	+	−16	−47	270
27	14	7	3	+	+	−19	−31	273
28 ($^{3,303}/_{5,969}$ days)	14	$9 + {}^{755}/_{1,101}$	$^{1}/_{3}$	+	+	$-(21 + {}^{755}/_{1,101})$	−12	$275 + {}^{755}/_{1,101}$

of the Uranic Manifestation speed sequence is $27 + {}^{3,303}/_{5,969}$ days[35] = 27.55336 days, just under two minutes shorter.

Columns 2 and 3 give the moon's daily motion in *du* and ($^{1}/_{19}$) *du*, while 9 gives it in ($^{1}/_{19}$) *du*—e.g. 14 *du* and 10 parts (at a scale of 19) yields a total of $14 \times 19 + 10 = 276$. As we saw in chapter 6, the moon's minimum daily displacement during a cycle is usually not far from 12 *du*. The maximum displacement is more variable: it can rise to more than 15 *du* on occasion, though the maximum in a given cycle can be as low as 14 ½ *du*. The minimum shown on the table is 12 $^{5}/_{19}$ *du*, which seems reasonable, but the maximum is set at 14 *du* $^{10}/_{19}$, which seems to represent rather less than the average maximum over a series of cycles.

Column 7 gives the difference between column 9 and the mean daily motion of $^{254}/_{19}$ *du* (13 $^{7}/_{19}$ *du*), in units of $^{1}/_{19}$ *du*. This column is thus the first difference of column 8, which gives the accumulated shift of the true moon from the position of the mean moon. Column 4 functions as the first difference of both column 3 and column 7, with the signs being given respectively by column 5, and column 6 (with the proviso that this gives the sign of change in the absolute size of column 7). In column 8 the entry for day 18 has been corrected from 23 in the original for mathematical consistency. The bracketed fractions in entries for day 28 are supplied following Li Rui.[36]

The essence of the table is thus in columns 1 and 8. For any given day, we can calculate the mean position of the moon using (in effect) the mean daily motion of $^{254}/_{19}$ *du* = 13 $^{7}/_{19}$ *du*. Then we look in the table to find the necessary correction to allow for the effect of the moon's varying speed. Supposing that we are in day 10 of the system, then looking in column 8 gives a lead of 98 parts, and $^{98}/_{19}$ *du* = 5 $^{3}/_{19}$ *du*. We add this onto the figure obtained by our mean calculation, and this gives us the true position of the moon.

One very important effect of this correction will, of course, be to change our estimate of exactly when the sun and moon will be in conjunction: in the case just considered, the time of the true conjunction will be about 10 hours earlier than the calculation for the mean conjunction would have suggested (since the distance between sun and moon changes at about 12 $^{7}/_{19}$ *du* per 24 hours).[37] If a conjunction calculated by mean motion fell early in a given day, the correction to the

[35] Not $28 + {}^{3,303}/_{5,969}$ days, since the sequence begins on day 1, not day 0.

[36] See Li Rui 李銳 (1768–1817) (1993). 'Li shi yi shu 李氏遺書 (Transmitted works of Mr Li)' in *Zhong guo ke xue ji shu dian ji tong hui, tian wen juan* 中國科學技術典籍通彙天文卷 *(Compendium of classic texts of Chinese science and technology, astronomical section)*, Zhengzhou, Henan educational press. **2**: 701–818. 783b–785a.

[37] To get a more accurate result, we must, of course, use the actual speed of the moon at that time, as given in the table: see the example in Box 8.1 for details of how this is done.

true conjunction could easily shift it to the preceding day, which could in principle change the day on which a lunar month begins. In fact, however, during the period when the Uranic Manifestation system was adopted in the state of Wu, comparisons of recorded actual dates with calculation suggests that the calendar in use for civil dating was calculated on the basis of mean lunar motion only.

So the speed table is in principle fairly easy to understand and to use. On the other hand, it is no use if we do not know on a given day what point in the table corresponds to the date for which we wish to make calculations. For a modern reader with means of automatic computing, the obvious course would be to ask how many days have passed since the first speed sequence began at High Origin, and then cast out whole sequences of 27 $^{3,303}/_{5,969}$ days to find out where we are in the current cycle. Liu Hong, who is working within the context of a luni-solar calendar, asks a slightly different question: where in the sequence was the first day of the current lunar month, which included the moment of mean conjunction? In Liu Hong's words, we need to find where the conjunction 'enters the sequence' *ru li* 入曆. To make that calculation, he begins with the number of lunations completed since High Origin, the 'Accumulated Lunations' *ji yue* 積月, and proceeds as follows:

> 推合朔入曆
> 以上元積月乘朔行大小分，小分滿通數三十一從大分，大分滿曆周去之,
> 餘滿周法得一日,不盡為日餘.日(餘)命算外,所求合朔入曆也.

To predict where conjunctions enter the sequence:

> Multiply the Greater [11,801] and Lesser [25] Parts of new moon motion by the accumulated lunations [since] High Origin. Where the lesser fractional part fills the Compatibility Number 31 let that go with the greater fractional part. From the greater fractional part cast out what fills the Sequence Circuits [164,466], and for the surplus obtain one day for each time it fills the Circuit [day] Divisor [5,969]. What is not exhausted is the Day Remainder. Counting off the days, what is outside the count is the entry into the sequence of the conjunction which is sought. (*Jin shu* 17,513; Cullen 2017, 276–7)

In modern terms, this amounts to an instruction to do the following:

1. Multiply Accumulated Lunations by (11,801 + $^{25}/_{31}$), and take the whole number part of the result.
2. Cast out multiples of 164,466, each of which represents the days in a completed speed sequence, at a scale of 5,969.
3. Divide the remainder by 5,969 to get the number of days completed in the current sequence.

From 2. and 3., the implication is clear that one complete speed sequence has this number of days:

$$164{,}466/5{,}969 \text{ days} = 27 + {}^{3{,}303}\!/_{5{,}969} \text{ days} = 27.55336 \text{ days}$$

This is precisely the same as our earlier result. The slightly mysterious quantity (11,801 + $^{25}/_{31}$) is in fact simply the difference between the Uranic Manifestation values for days in the synodic month and the anomalistic month, in units of ($^{1}/_{164{,}466}$) of an anomalistic month. The addition of a 'lesser' fractional part of $^{25}/_{31}$ to what is already a fractional part of 11,801 is an expedient introduced by Liu Hong to improve accuracy.

How well does all this work? It seems quite likely that Liu Hong will have attempted to fit in with observations around his own time. Using NASA 'Horizons' ephemeris data, we can easily locate a lunar speed maximum on 17 November 206 CE, JD 1,796,619.208. The Uranic Manifestation procedure predicts the start of a speed sequence (and hence a maximum of speed) three days later. A maximum that occurred on 2 April 206 CE, JD 1,796,755.208, is predicted a day too late. Looking at the speed sequence table, a three-day error will give us a position correction for the moon of $^{43}/_{19}$ *du* = 2 $^{5}/_{19}$ *du*, about a 4-hour timing error for the true conjunction. Given that the conjunction is only directly observable at the time of a solar eclipse, this order of magnitude discrepancy would not have been obviously fatal.

A fully worked example of how the time of a true conjunction may be derived from a predicted time for a mean conjunction is shown in Box 8.1.

Box 8.1: Finding the mean and true conjunctions for the 11th civil month of Guanghe.1, 178–179 CE

Suppose we want to find the time of true conjunction at the start of the 11th civil month of the first year of the Guanghe 光和 reign period, which ran from early 178 CE to early 179 CE. This is also the first celestial month corresponding to the next civil year, which begins in 179 CE. We proceed as follows:

continued

Box 8.1: *Continued*

Finding the mean conjunction:

System origin for the Uranic Manifestation system was the start of the 11th civil month preceding the civil year 7172 BCE. So, allowing for the fact that there was no year 0 CE, the total years elapsed to the start of the month in question are:

$$179 - 1 - (-7{,}172) \text{ years} = 7{,}350 \text{ years.}$$

19 years contain 235 months, so the total number of months elapsed will be:

$$\text{Integral part of } 7{,}350 \times 235/19 = 90{,}907$$

One mean lunation in the Uranic Manifestation system has 43,026/1,457 days: Cullen (2017), 243. So the total days elapsed since the conjunction at High Origin are: $90{,}907 \times 43{,}026/1{,}457 \text{ days} = 2{,}684{,}533 + \frac{1}{1{,}457}$ days

Since $2{,}684{,}533 = 44{,}742 \times 60 + 13$, the sexagenary day number on which the mean conjunction falls will be 13 days greater than the sexagenary day number of High Origin, which was defined as *jiazi*.1, i.e. *dingchou*.14.

(This corresponds to 28 November 178 in the Julian calendar.) Thus we have predicted:

Mean conjunction falls $\frac{1}{1{,}457}$ day after midnight on 28 November 178 CE, i.e. at about 00:01.

Finding entry of the mean conjunction into speed sequence table

To find where we enter the speed sequence, we calculate:

Accumulated months $90{,}907 \times (11{,}801 + \frac{25}{31}) = 1{,}072{,}866{,}819 + \frac{3}{31}$

Cast out whole multiples of 164,466, representing whole speed sequences:

$$1{,}072{,}866{,}819 + \tfrac{3}{31} - 164{,}466 \times 6{,}523 = 55{,}101 + \tfrac{3}{31}$$

Divide by 5,969 to find whole days and fractions in the speed sequence:

$$55{,}101/5{,}969 = 9 + \tfrac{1{,}380}{5{,}969} \text{ (neglecting the fraction } \tfrac{3}{31})$$

So since the first day in the sequence is day 1, we are $\frac{1{,}380}{5{,}969}$ day into the 10th day of the speed sequence, so we use the data for this day.

Correcting to find the true conjunction

On the 10th day of the speed sequence, we find in the speed sequence table:

Parts of Lunar Motion: 246

continued

Box 8.1: *Continued*

(This is the speed of the moon in (1/19) *du* per day.)

Accumulated Excess or Deficit: 98

(This is the amount in (1/19) *du* by which the true moon is ahead of its mean position at the start of the day.)

Rate of Decrease or Increase: −8

(This is the change in Excess or Deficit that will take place during the present day. It is negative because the moon is moving below its mean speed.)

So we know that the moon is not in fact in conjunction with the sun, but some way ahead of the sun. How far ahead exactly? In effect we interpolate between the values of 98 Excess at the start of this day, and 98 − 8 = 90 at the start of the next day. We are 1,380/5,969 into the present day of the sequence, so the Excess will be

$$98 - 8 \times (1{,}380/5{,}969) = 573{,}922/5{,}969 = 96 + 898/5{,}969$$

That represents the number of (1/19) *du* between the present positions of the sun and moon.

Now the moon is moving at 246/19 *du* per day, but the sun is also moving, and in the same direction, at 1 *du* per day. So the speed of the moon relative to the sun (taken as constant during the day) is:

$$(246 - 19)/19 \text{ du per day} = 227/19 \text{ du per day.}$$

So the time to cover the distance of (96 + 898/5,969)/19 *du* will be:

$$[(96 + 898/5{,}969)/19 \text{ du}]/ [227/19 \text{ du / day}]$$
$$[573{,}922/5{,}969]/227 \text{ days} = 617/1{,}457 \text{ days}$$
to the nearest (1/1,457) day.

The mean lunation was 1/1,457 day after midnight on the present day, so the true conjunction is evidently (1,457 + 1 − 617)/1,457 = 841/1,457 day after midnight on the preceding day.

So to sum up:

True conjunction is 617/1,457 days (10 hours 10 minutes) earlier than mean conjunction, on 27 November, day *bingzi*.13, at 841/1,457 days after midnight (13:51 Luoyang local time).

(A modern calculation places the true conjunction at 09:47 on 27 November.)

8.2.2 Lunar latitude

Eclipses of the sun and moon can only occur at certain moments in a lunation—at a conjunction of sun and moon for a solar eclipse, and at opposition for a lunar eclipse.[38] By the time of Liu Hong, this was commonplace knowledge. Over half a century earlier Jia Kui had taken it for granted that a valid test of an astronomical system was to see if it would have predicted a conjunction (which in his day normally meant a mean conjunction) at the date that an observed solar eclipse was recorded: see chapter 6, section 6.3. In that light, he reviewed the dates of no less than 70 recorded solar eclipses to compare the performance of different systems. But while Jia Kui was clear that a solar eclipse must coincide with a conjunction, he knew perfectly well that the implication did not work in reverse: most conjunctions are not accompanied by an eclipse, and he appears to have had no means of telling at which conjunctions an eclipse might be expected. In the tabulation he called the *yin yang li* 陰陽歷 'yin-yang sequence', and its associated computation procedures, Liu Hong provided new tools that changed this situation radically, by enabling the circumstances of any given conjunction to be examined in an unprecedentedly detailed way.

In modern terms, we might model the situation at a conjunction near a node as in Figure 8.2.[39]

Here we see the sun S and moon M in conjunction shortly after passing a node N, one of the two points at which their paths cross. The angle α between the two paths is in reality about 6°, smaller than shown here. N is the 'ascending node', at which the moon crossed the ecliptic going from south to north. The sun is moving eastwards at 1 *du* per day relative to the celestial sphere, but the moon is moving on average about 13 times faster. The distance d represents how far the moon has moved since it passed the node to catch up with the slowly moving

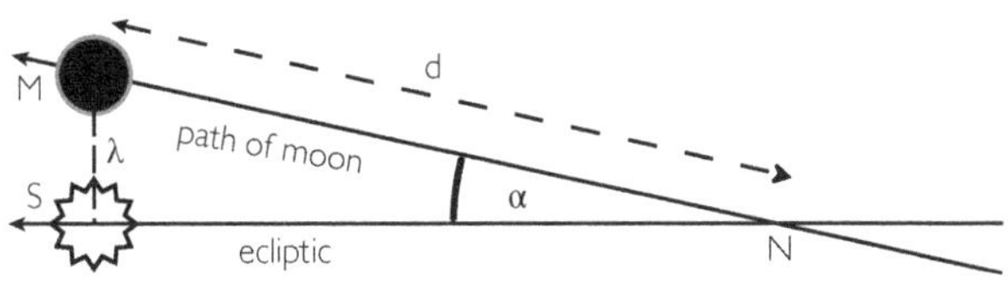

Figure 8.2 Sun and moon in conjunction near a node.

[38] See 6.3 and the fully commented translation of relevant material in Cullen 2017, 293–302 for further details.

[39] The diagram represents as a plane figure what is really the intersection of two circles on the celestial sphere. But so long as the region of the sphere's surface shown is small compared with its overall size, this is accurate enough for our purposes.

sun, while λ, the distance between the centres of the sun and moon, is in modern terms the moon's latitude. From this diagram, it seems that the critical point in deciding whether a solar eclipse can take place is whether d, and hence λ, are small enough so that M can overlap and hence obscure S to some extent at the moment of conjunction.

Liu Hong, however, does not refer explicitly to any diagrammatic representation such as Figure 8.2 as a basis for his analysis, and his vocabulary differs from that used here, not simply because it is in Chinese rather than in English, but rather because he seems to be using slightly different concepts. Thus, for instance, he does not distinguish the ascending and descending nodes by name, but instead divides the moon's path into two halves, *yin* 陰 when the moon is north of the ecliptic, and *yang* 陽 when it is to the south. He makes no direct reference to the angle α, consistently with the absence in China at this period of the concept of angle as an amount of rotation of a line. He quantifies d, the separation of the moon from the node, in what initially appear at first glance to be units of time—days—rather than distance on the celestial sphere measured in *du*. We are, however, told explicitly at a later stage that one of the quantities for which he gives values represents *yue qu huang dao du* 月去黃道度 'the *du* by which the moon is displaced from the ecliptic' (though at a scale of 12, so the numbers represent twelfths of a *du*), which is evidently equivalent to λ in the diagram.[40]

Liu Hong's analysis of the relevant factors is embodied in the tabulated data in his *yin yang li* 陰陽歷 'yin-yang sequence'. In a pre-modern layout, the table begins as shown in Figure 8.3.

The data headings are in the left-hand column of the right-hand page. They are, reading from top to bottom, 'Yin-yang sequence', 'Difference', 'Rate of decrease and increase' and 'Total number'. The first column of the left-hand page is headed 'Day 1', followed by '1, subtract', 'increase, 17' and 'start' (i.e. the 'Total number' is zero).

'Yin-yang sequence' gives the days of the sequence. 'Total number' gives the latitude of the moon (which may be north or south) in units of ($\frac{1}{12}$) *du*.[41]

[40] See Cullen (2017), 310. The significance of the terms *yin* and *yang* is made clear when we are told later on that in the *yin* section we must in effect subtract the given number of *du* from the sun's north polar distance to find the moon's north polar difference, but in the *yang* section we must add it.

[41] It is likely that Liu Hong may have been thinking in terms of so-called 'polar latitude' here, i.e. the distance of the moon from the ecliptic measured along a meridian of right ascension through the celestial poles, rather than along a meridian of solar longitude through the ecliptic poles, as would be normal modern practice. But to the order of accuracy of the date given here there is no significant difference between the two.

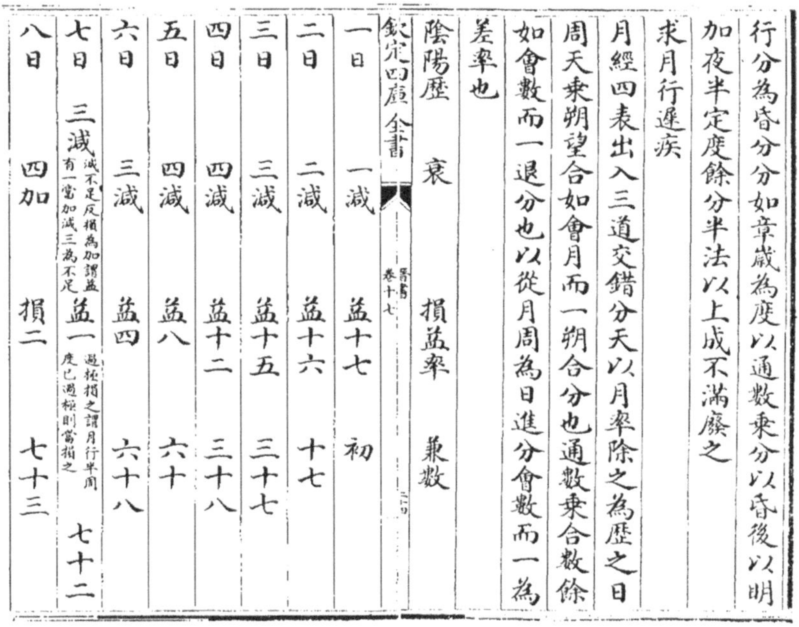

行分為昏分分如章歲為度以通數乘分以昏後以明
加夜半定度餘分半法以上成不滿廢之
求月行遲疾
月經四表出入三道交錯分天以月率除之為歷之日
周天乘朔望合如會月而一朔合分也通數乘合數餘
如會數而一退分也以從月周為日進分會數而一為
差率也

欽定四庫全書 晉書 卷十七

陰陽歷	衰	損益率	兼數
一日	一減	益十七	初
二日	二減	益十六	十七
三日	三減	益十五	三十七
四日	四減	益十二	三十八
五日	四減	益八	六十
六日	三減	益四	六十八
七日	三減 減不足反損為加謂益有一當加減三為不足	益一 過極損之謂月行半周度已過極則當損之	七十二
八日	四加	損二	七十三

Figure 8.3 Beginning of Liu Hong's yin-yang sequence table in *Jin shu* 17, 24a–b, *Si ku quan shu* edn., c. 1782.

The maximum latitude shown is $^{73}/_{12}$ *du*, which is 6.1 *du* or 6.0° to two significant figures. While the inclination of the moon's orbit to the ecliptic is in fact never more than about 5.3°, observation by a terrestrial observer may give different results, as we shall see below. 'Rate of decrease and increase' and 'Difference' are the first and second differences of 'Total number', i.e. of latitude.

Table 8.3 presents this data in a modern layout. The fractions in the last row are partly restored by the modern editors of the text used, following Li Rui, on the basis of mathematical consistency with the later procedures given by Liu Hong that make use of them, as are the data in the note to the first entry in column 1.[42] The basic procedure for using this table is straightforward. If we wish to find the position in the table corresponding to any given conjunction (i.e. where the conjunction 'enters the table'), so that we may find the latitude of the moon at that time, we take the number of completed months (and hence mean lunations) since the moon was at a node at system origin, and then carry out some simple

[42] See *Jin shu* 17, 515–516, Cullen 2017, 299–301, and Li Rui 李銳 (1768–1817) (1993), 788a–789a.

Table 8.3 *Uranic Manifestation system yin-yang sequence table*

1	2	3	4
Day of yin-yang sequence	Difference [1/12 *du*]	Rate of Decrease and Increase [1/12 *du*]	Total Number [1/12 *du*]
1	−1	17	0
2[1]	−1	16	17
3	−3	15	33
4	−4	12	48
5	−4	8	60
6	−3	4	68
7	−3[2]	1	72
8	4	−2[3]	73
9	4	−6	71
10	3	−10	65
11	2	−13	55
12	1	−15	42
13[4]	1[5]	−16	27
Fractional day 5,203/[/7,874 days][6]		−(16 [+ 306/473])	11

[1] Note in text: 'Limit Remainder 1,290, Fine Parts 457. This is the Earlier Limit.' – restored by Li Rui. See section 8.3.1 for the significance of this quantity.

[2] Note in text: '[Here] we are subtracting [from something] insufficient. So we turn it round, and make it an addition. This refers to the fact that we have an excess of 1, but must subtract 3, so there is insufficient [to subtract from].' The reference is to the fact that a difference of −3 applied to the value 1 produces the negative number −2.

[3] Note in text: '[At this point] we pass the extreme [latitude] and decrease it. This refers to the fact that the moon has passed the halfway point of the cycle [from one node to the next], so that the *du* [of latitude] have passed the extreme, and we must diminish them.'

[4] Note in text: 'Limit Remainder 3,912, Fine Parts 1,752. This is the Later Limit.' See section 8.3.1 for the significance of this quantity.

[5] Note in text: 'The sequence [here] begins the Long Fractional Day.' There is a similar note in the penultimate row of the speed sequence table; as there, this note seems to be referring to the prolongation of the last whole day by the part of a day with which the sequence ends.

[6] The numerator of this fraction is in the original, but the denominator is restored from the mathematical context.

multiplication and division, using the usual tactics to avoid very large numbers, such as casting out multiples of periods over which initial conditions repeat.[43] The first section of Box 8.2 shows an example of how we may do this for a given conjunction.

[43] For the details, see Cullen (2017), 303–4. Effectively, the moon is assumed to shift its position relative to the nodes by (18,328 + 914/2,209)/7,874 days of the yin-yang sequence from one mean conjunction to the next.

Box 8.2: Finding entry into the yin-yang sequence for mean and true conjunctions, and finding latitude of the moon

Finding entry of the mean conjunction into yin-yang sequence table

From the accumulated months since High Origin, 90,907 cast out Coincidence Months [11,045], to reduce size of numbers by eliminating cycles in which initial conditions repeat.

$$90{,}907 - 8 \times 11{,}045 = 2{,}547$$

Multiply by New Moon Conjunction Parts [18,328] and Fine Parts [914], the latter at a scale of Fine Parts Factor [2,209]:

$$2{,}547 \times (18{,}328 + {}^{914}\!/_{2{,}209}) = 46{,}682{,}469 + {}^{1{,}881}\!/_{2{,}209}$$

Cast out whole multiples of Circuits of Heaven [215,130], which represent whole circuits relative to the nodes:

$$46{,}682{,}469 - 216 \times 215{,}130 + {}^{1{,}881}\!/_{2{,}209} = 214{,}389 + {}^{1{,}881}\!/_{2{,}209}$$

As for the remainder, if it does not fill Sequence Circuits [107,565], that is entry into the Yang sequence.

Since 214,389 > 107,565 we are in the Yin half of the sequence, in which the moon is to the north of the ecliptic.

$$\text{Cast out } 107{,}565\text{: } 214{,}389 - 107{,}565 + {}^{1{,}881}\!/_{2{,}209} = 106{,}824 + {}^{1{,}881}\!/_{2{,}209}$$

This is parts and days of the sequence, at a scale of Lunar Circuits [7,874]
So we have: 13 days, 4,462 parts and 1,881 Fine Parts.
The sequence starts on day 1, so:

Mean conjunction falls 4,462 parts and 1,881 Fine Parts into the 14th day of the sequence.

Correcting to find entry of true conjunction into the sequence.

We found that the true conjunction was 617 day parts ahead of the mean conjunction, at a scale of Day Factor [1,457].

Liu Hong has told us earlier, in effect, that each day of time that elapses shifts the position of the conjunction by:

continued

Box 8.2: *Continued*

Daily Advance Parts [7,905 + $^{31}/_{47}$] relative to the nodes, at a scale of Lunar Circuits 7,874]: Cullen (2017), 298.

So we calculate:

(7,905 + $^{31}/_{47}$) × 617/1,457 = 3,347 + $^{1,839}/_{2,209}$ parts of days of the sequence.

Subtracting from the figures for the mean conjunction, we find:

True conjunction falls 1,115 parts and 42 Fine Parts into the 14th day of the sequence.

Finding the latitude of the moon at the true conjunction

For the 14th day of the sequence, we have:

Total number: 11

(This is the number of $^{1}/_{12}$ *du* of lunar latitude at the start of the day of the sequence.)

Rate of Decrease and Increase: −(16 + $^{306}/_{473}$)

(This is the change in Total Number per day of the sequence.)

So, ignoring Fine Parts, 1,115 parts into this day, the change in Total Number will be:

− (16 + $^{306}/_{473}$) × 1,115/7,874 = −2.357 (expressing the result in decimals to avoid a complex fraction)

So the Total Number is:

11 − 2.357 = 8.643

Which corresponds to a latitude of 8.643/12 *du* = 0.72 *du*

So at the true conjunction of the 11th civil month of 178 CE, which is predicted to fall at 13:51 Luoyang local time on 27 November, the predicted latitude of the moon is 0.72 *du*.

Since this is the result of a fairly complex multi-step calculation, it may be helpful to show that it is consistent with the results of a slightly simplified and more direct method.

We know that the High Origin of the Uranic Manifestation system was at Luoyang local midnight, 21 January 7172 BCE, which is JD −898,129.81250 (Cullen 2017, 241). The JD of the true conjunction of 27 November 178 CE is 1,786,402.76458, so the time elapsed since High Origin is:

continued

Box 8.2: *Continued*

$$(1{,}786{,}402.76458 - (-898{,}129.81250)) \text{ days} = 2{,}684{,}532.57708 \text{ days}$$

The time taken for the moon to make a circuit relative to the nodes (a draconitic month) is, according to the Uranic Manifestation (see note 47, or Cullen 2017, 304):

$$(365 + {}^{145}\!/\!_{489}) \times 235/19 \text{ days} \times 11{,}045/11{,}986 = 27.212150735 \text{ days}$$

Now 2,684,532.57708 = 98,652 × 27.212150735 − 0.51722922, so the moon is within 0.52 days of reaching a node, which we may approximate as 0.52 days of the yin-yang sequence. A whole day away from a node, the Total Number is 17, corresponding to a latitude of $^{17}\!/\!_{12}$ *du*. So 0.52 days corresponds to a lunar latitude of:

0.52 × $^{17}\!/\!_{12}$ *du* = 0.74 *du*, very close to the value obtained above.

One apparent problem with this table is that it does not seem to be long enough, since it only contains 13 whole days and one final fractional day. But the moon's cycle relative to one of the nodes, known as the draconitic or nodical month, has a mean value of a little over 27 days. The problem is resolved in great part by the observation that Liu Hong has saved space by giving us a table that applies equally to the time the moon spends to the north of the ecliptic (the yin sequence), and the time it spends to the south (the yang sequence). But if we look in more detail, a more serious problem becomes evident.

The total time taken by for the moon to perform a complete cycle relative to a node (in modern terms a 'draconitic' or 'nodical' month) will, one might suppose, be twice the number of days in the yin-yang sequence table, i.e.

$$2 \times (13 + {}^{5{,}203}\!/\!_{7{,}874}) \text{ days} = 27.32156 \text{ days (to 7 significant figures)}$$

There is, however, something a little disquieting about this value. In modern terms, it is much closer to the modern value for the length of the mean sidereal month, 27.32166 days, the period in which the moon makes a complete circuit of the heavens relative to the stars, than it is to the modern mean value of the draconitic month, 27.21222 days. If the nodes remained fixed relative to the stars, these two periods would be precisely equal. But as already explained (see chapter 6, section 6.3.5.1), the nodes perform a slow but steady motion westwards while the moon makes its much more rapid eastwards circuits. It should thus take less time for the moon to return to a given node than it does for it to return to a given position relative to the stars—so the draconitic month should be shorter than the synodic month.

If we calculate the precise value for the sidereal month implied by the Uranic Manifestation, the situation is even more striking: We are told that in an Era Cycle of 589 years, in which there are 215,130 days (the quantity called 'Circuits of Heaven' *zhou tian* 周天), the moon makes 7,874 complete circuits ('Lunar Circuits' *yue zhou* 月周) of the heavens.[44] So one complete circuit of the heavens by the moon (a sidereal month) takes:

$$215{,}130/7{,}874 \text{ days} = 27 + {}^{2{,}532}\!/_{7{,}874} \text{ days} = 2 \times (13 + {}^{5{,}203}\!/_{7{,}874}) \text{ days}$$

This is exactly equal to two yin-yang sequences as calculated above. Does this mean that Liu Hong really thinks, in effect, that the sidereal and draconitic months are equal, which could only be the case if he was ignorant of the fact that the nodes are in constant motion—something that had been recognized since the time of Jia Kui, whose writing on this topic he had edited with Cai Yong? That seems highly unlikely.

It is at such a moment that we might wish for a sight of the commentaries that Zheng Xuan and Kan Ze wrote on the Uranic Manifestation system. Unfortunately both are lost, so we must do our best with the text as it stands. Careful study of the procedures that Liu Hong specifies do, however, reveal how he is thinking, even though he does not explicitly address the problem in the terms in which we have stated it. Here is one approach to understanding what Liu Hong is doing:[45]

1. The division of the moon's orbit into the $27 + {}^{2{,}532}\!/_{7{,}874}$ days of the double yin-yang sequence is to be understood as a set of graduations of the moon's path into divisions of *space* on the celestial sphere rather than *time*.
2. Suppose we calculate the instant when the moon has crossed the ecliptic at a node. We then effectively adjust the graduated orbit until its zero point falls at that node.
3. For any later instant, we work out how many days we are after that starting point, and count off days of the yin-yang sequence round the

[44] See Cullen 2017, 239–40.
[45] For more detailed explanation, see Cullen 2017, 299–310.

circumference. If the moon's path remained fixed relative to the ecliptic, so that the nodes where the two paths cross were likewise fixed, then, using the table, we could simply read off the corresponding value for the lunar latitude, interpolating if necessary.

4. But now we have to allow for the fact that the zero point of our graduations, the node where we started counting days, will have shifted westwards since the moon passed the node. That means that the day graduations on the orbit will have shifted slightly westwards relative to the moon, and the moon will therefore be further eastwards relative to the yin-yang sequence.
5. We calculate this shift, and add it to the 'days of the sequence' we used in step 3, to obtain a new lunar position. We then read off the corrected value in the table, and that is the lunar latitude.

Figure 8.4 may help to make this process clear. The circle is imagined to be graduated in days of the yin-yang sequence, beginning at the node O and running anti-clockwise, taken as equivalent to eastwards. Let us suppose that about 3 days and 18 hours ago, the moon passed the node labelled O, and the yin-yang sequence thus began. We might therefore expect the moon to be at the position in the sequence marked M, and we would therefore use the values of latitude given for 3 and 4 days to interpolate, and hence find the latitude of the moon at this instant.

However, in the nearly four days it took the moon to move from the node to its present position, the node shifted westwards from O to O′ (in only 4 days the actual movement would be considerably smaller than shown in the diagram, about 0.04 day for each day elapsed, thus 0.16 day in this case). Hence, the graduations of the yin-yang sequence (which begin at the node) shift westwards too—which is equivalent to the position of the moon shifting *eastwards* relative to the yin-yang sequence by an equal amount, from M to M′. So we must use this new position, M′, to find the moon's latitude from the table.[46] The effect of this is that the moon will take less time to make a circuit relative

[46] In carrying out this procedure, it is tacitly assumed that movement of a node along the ecliptic results in an equal amount of displacement along the moon's orbit, which is not strictly true, since the moon's orbit is inclined to the ecliptic. However, given that the inclination is small, any resulting differences are negligible for our purposes. Ptolemy made the same simplifying assumption: *Almagest* IV.6, in Toomer (1998), 198. The maximum resulting error in longitude is only about 0.1°.

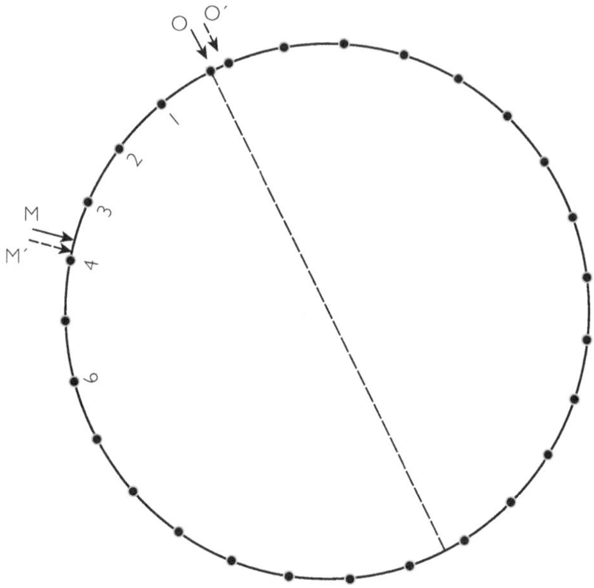

Figure 8.4 Effect of westwards shift of nodes on position of moon in yin-yang sequence

to the nodes than it takes to make a circuit relative to the heavens—in modern terms, the draconitic month is a little shorter than the sidereal month, as noted above.[47]

One further point remains. As we have seen, Liu Hong's 'speed sequence table' enables us to correct the moon's position to allow for its varying speed, and hence find the moment of true conjunction of sun and moon, as opposed to its 'mean conjunction' found on the assumption that the moon moves at constant speed round the heavens.[48] If we know the time difference between the mean and true conjunctions, we need to know how that will change the position of the

[47] The Uranic Manifestation system does not give an explicit value for a draconitic month. It is, however, fairly easy to calculate the value implied by the constants of the system. Coincidence Months [11,045] is the number of lunations in which Coincidence Rate [1882] lunar eclipses occur, i.e. the sun passes a given node $^{1,882}/_{2}$ = 941 times. So since in each lunation the moon makes one circuit relative to the sun, in 11,045 lunations the moon makes 11,045 + 941 = 11,986 circuits relative to the nodes, each circuit being a draconitic month. We know that a Uranic Manifestation mean synodic month is (365 + $^{145}/_{589}$) × 19/235 days (see section 8.1.2), so a mean draconitic month should be (365 + $^{145}/_{589}$) × 19/235 × 11,045/11,986 days = 27.2122 days, which is identical to the modern value to within the same precision. See Cullen 2017, 304.

[48] See section 8.2.1.

conjunction relative to the yin-yang sequence in order to find the latitude of the moon at the true conjunction. Liu Hong, who must have done such calculations many times, gives us an easy way to do that, by defining a quantity that represents the total eastwards shift of the sun (and hence of the moon, since we are at a conjunction) relative to the nodes in one day, measured in days of the sequence. This is Daily [or solar] Advance Parts *ri jin fen* 日進分: (7,905 + 31/47) days of the yin-yang sequence, at a scale of Lunar Circuits [7,874]. The second and third parts of Box 8.2 showed how we may correct the entry of the moon into the sequence at the mean conjunction into the entry for the true conjunction, and hence obtain the moon's latitude at that instant.

It is interesting to compare the predictions of the yin-yang sequence with modern calculations. Test calculations suggest that predictions of the dates of moon's passage across the ecliptic at a node (i.e. at the start of a yin-yang sequence) made using Liu Hong's methods match observation to within a few days. In Figure 8.5 we show the results of using the yin-yang sequence to calculate lunar latitude at a series of conjunctions from 205–206 CE, near the last date when we have evidence of Liu Hong having worked on his system, as implied by the definition of his system origin.[49] These are then compared with modern calculations of the moon's apparent latitude at those conjunctions,

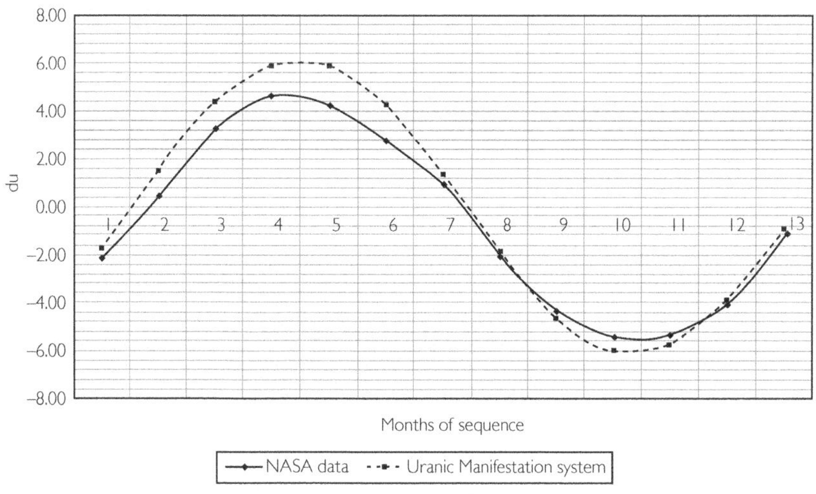

Figure 8.5 Latitude of moon at successive conjunctions from 28 November 205 CE to 17 December 206 CE.

[49] See section 8.1.

as seen by an observer at Luoyang, taken from the NASA online 'Horizons' ephemeris program. One point that emerges very clearly is that while Liu Hong's method predicts a symmetrical variation in lunar latitude on either side of the ecliptic, the moon's maximum observed south latitude of the moon for an observer at Luoyang is nearly a degree larger than its greatest north latitude.

In modern terms—or indeed in terms of any model of a cosmos based on a spherical earth with celestial bodies at varying distances from it[50]—the bias seen in the observed results shown in Figure 8.4 is only to be expected. This is a consequence of the fact that earth's radius of 6,400 km is not negligible in comparison with the distance of the moon from the earth, which varies between 356,500 km and 406,700 km, although it is small in comparison with the distance of the sun, some 150,000,000 km. Moving to different positions on the earth therefore changes the sight-line to the moon appreciably, while the sightline to the sun hardly changes at all. The apparent position of the moon therefore shifts relative to the sun, and hence relative to the ecliptic, on which the sun lies. This is one example of the effect known in modern terms as 'parallax'—the shift in the relative directions of objects at different distances from an observer as the observer or the objects move across the lines of sight.

For example, if we limit ourselves to considering movement of the observer's position along a north–south line, in Figure 8.6 (which is not to scale), the centre of the earth is C and the centre of the moon is at M; N and S are the north and south geographical poles of the earth, and EE′ shows the plane of the terrestrial equator. An observer at a position such as X on the line CM from the earth's centre to the moon's centre sees the moon directly overhead (at his 'zenith') along the sight-line CXM. An observer such as O, somewhere directly to the north of X, sees the moon along the sight-line OM, still on his meridian but some way away from his zenith, so that the direction of the moon relative to the sun (and hence to the ecliptic) is shifted towards the south in comparison with observations made from X. As a result, the moon's apparent latitude as measured by an observer at O is different from what the observer at X would have measured—decreasing in size if the moon seen from X was north of the ecliptic, and increasing if it was south of the ecliptic.

[50] That is, in any of the basic pictures of the cosmos used in the western tradition from the time of Ptolemy up to the present day.

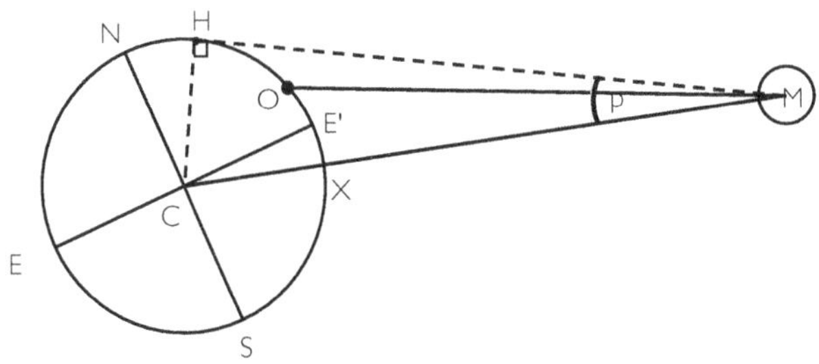

Figure 8.6 (Not to scale) Diagram showing the effect of lunar parallax for an observer in a northerly position on the earth.

This is precisely what we see in Figure 8.6. The angle OMC is the lunar parallax caused by moving from X to O.

The maximum possible value of lunar parallax caused by shifting the observer northwards from X is given by the angle p, as measured from a position such as H, from where the moon is only just seen on the horizon. Using the actual values for the diameter of the earth and distance of the moon, we find that this maximum possible parallax is about 1°.[51] This small change in the apparent latitude of the moon can, however, be critical in deciding whether or not a solar eclipse is visible to a given observer. Supposing, for instance, that an observer at X sees the moon directly aligned with the sun, and hence experiences a solar eclipse. Then an observer at H will see the moon about 1° south of the sun—and since the visible discs of both bodies are only about ½° across as seen from the earth, no part of the moon can overlap the sun, and an eclipse will not be seen. The effects of parallax in shifting the moon's apparent position relative to the sun will also be seen if the observer moves eastwards or westwards at any given instant—or indeed if we observe the moon at different times of day when its altitude above the horizon varies, though in the latter case we must also take account of the moon's steady change in latitude as it moves along the ecliptic during the day.

There was, however, nothing in the picture of the cosmos as Liu Hong is likely to have conceived of it that would have led him to expect the asymmetry in the

[51] In the figure, sin p = CH/CM. If we write CH = 6,400 km and CM = 356,500 km, we find p = 1.03°. Conversely, if we have a value of p derived from measurement, we can use that to estimate the distance of the moon in terms of the diameter of the earth. This was first attempted by Hipparchus in the second century BCE: see G. Toomer (1974b) 'Hipparchus on the Distances of the Sun and Moon.' *Archive for History of Exact Sciences* 14: 126, and *Almagest* V.11 in Toomer (1998), 243–4.

observed cycle of lunar latitude nowadays explained in terms of lunar parallax. It seems to have been a commonplace notion for his contemporaries that the sun, moon and planets were effectively attached to the inside of the celestial sphere. Their distances from it might vary slightly, but were insignificant compared with the distance of heaven from the earth. Thus, whereas in modern terms the moon is always much closer to the observer than is the sun, the Clerk's officials could argue in 255 CE that they were unable to predict whether, when the positions of sun and moon coincided, the moon would be in front of the sun, or behind it; in the first case there would be a solar eclipse, in the second case not.[52] For them, as for Liu Hong, the relative positions of sun and moon ought to appear identical for all terrestrial observers, and should not in theory have been affected by the observer's displacements on earth's surface, whether north–south or east–west.

In the centuries after Liu Hong, the asymmetry in lunar latitude variation was noted empirically, and a variety of systematic corrections were made to allow for it.[53] But how, if at all, might the effects of parallax have become evident to Liu Hong? We have no indication that he ever made direct and systematic measurements of lunar latitude that might have revealed the effect of parallax, although his figure of $7\frac{3}{12}$ *du* for the moon's maximum latitude must have had some observational basis. There were, however, other ways in which this phenomenon might have become evident to him, as we shall now see.

8.3 Solar eclipses as predictable portents

By the time of Liu Hong, lunar eclipses had, as already mentioned, long ceased to be seen as ominous. This was certainly not the case for solar eclipses. The standard histories included lists of solar eclipses in the sections reserved for

[52] See section 8.3.4.

[53] A good and clear summary of the mathematics of lunar parallax in relation to eclipse predictions in China is given in Nakayama (1969), 143–50. See the discussion of the general question of the treatment of the effect of parallax in pre-modern China in Tang Quan 唐泉 (2011) *Ri shi yu shi cha* 日食与视差 *(Solar eclipses and parallax)*. Beijing, Science Press, chapter 5, 161 ff. The first systematic attempts to take into account what we now know to be the effects of lunar parallax are found in astronomical systems of the Tang dynasty, such as the *Da yan li* 大衍曆 promulgated in 729 CE, and the *Xuan ming li* 宣明曆 promulgated in 822 CE. See Yabuuti Kiyoshi 藪内清 (1963). 'Astronomical tables in China from the Han to the T'ang dynasties' in *Chūgoku chūsei kagaku gijutsushi no kenyū* 中国中世科学技術史の研究 *(Research on the history of science and technology in medieval China)*, Tokyo: 445–92 (476–89), Yabuuti Kiyosi 藪内清 (1969), 331–7 and Tang Quan 唐泉 (2011), chapter 5, 195 ff. The corrections applied in the *Shou shi li* 授時曆 (1280 CE) are discussed in Sivin (2009), 499–500.

events seen as portents, frequently accompanied by discussions of their significance in terms of state policy or the emperor's personal conduct, such as the following example:

> 光和元年 [...] 十月丙子晦, 日有蝕之, 在箕四度. 箕為後宮口舌. 是月, 上聽讒廢宋皇后.
>
> In the first year of the Guanghe reign period, the last day of the tenth month, day *bingzi*.13 [27 November 178 CE], the sun was eclipsed, when it was at the fourth *du* of the lodge Winnower. Winnower relates to gossip in the private quarters of the palace. In this month, the emperor listened to slander, and dismissed Empress Song. (*Hou Han shu, zhi* 18, 3370)

The ritual system of the Han dynasty still provided for appropriate ceremonies to turn away any untoward consequences of such an event:

> 每月朔旦, 太史上其月曆, 有司, 侍郎, 尚書見讀其令, 奉行其政. 朔前後各二日, 皆牽羊酒至社下以祭日. 日有變, 割羊以祠社, 用救日(日)變.
>
> On the first day of every month, the Grand Clerk offers to the throne his calendar for the month; the officials, the attendant gentlemen, and the secretaries read the ordinances it prescribes, and carry out the [appropriate] administrative acts. On the two days before and after the first day of the month, they lead a sheep, with wine, to the shrine of the soil to sacrifice to the sun. If the sun suffers any alteration, they slaughter the sheep as an offering at the shrine, as a means of rescuing the sun from the alteration affecting it. (*Hou Han shu, zhi* 4, 3101)

The commentators on this passage all interpret the 'alteration' in question as connoting a possible solar eclipse, and cite records of ceremonies to assist the eclipsed sun, dating back to well before the imperial era. The fact that the rituals were prescribed for the days before and after the start of the lunar month is revealing. A new lunar month was, in principle, supposed to start on the day during which a new lunation began with conjunction of sun and moon; thus any solar eclipse might be expected to fall on the first day of a month. The motion of the moon in relation to the earth and sun is, however, complex, so that the true interval between conjunctions can vary from 29.18 days to about 29.93 days.[54] During the early imperial period, conjunctions were nevertheless predicted for calendrical purposes as following one another at a constant interval—in other words, the calendar was based on mean rather than true conjunctions. Thus, even if the calendar was being run so efficiently that the first day of a month always contained the 'mean conjunction', the true conjunction (and hence any

[54] See chapter 1, note 42.

solar eclipse) might occur on a slightly different day, such the last day of the preceding month, as in the above example.[55]

Such ceremonies continued to be seen as appropriate for centuries after the end of Han; in the 17th century Jesuit missionaries found that they were still practised, by which time solar eclipses were considered to be, in principle, completely predictable.[56] We shall see later what happened when, in the third century CE, Grand Clerks began to be able to warn in advance of the strong possibility (even if not yet the certainty) of a solar eclipse.

Solar eclipse prediction is a considerably more difficult task than predicting lunar eclipses, and neither the Triple Concordance nor the Han Quarter Remainder systems attempted it.[57] The difficulty lies in the different conditions that render lunar and solar eclipses visible by an observer on the earth. If the moon passes through the earth's shadow, any observer who can see the moon can see the eclipse happening. If, however, the moon's shadow falls on the earth, the small size of this shadow means that only some observers with a view of the sun will actually see the sun being obscured, when the shadow passes over their location.[58] An alternative but equivalent way of putting it would be to say that only those observers who are in just the right position will see the moon aligned with the sun, and thus obscuring it in whole or in part. In addition, the period when a lunar eclipse is obvious to an observer typically lasts much longer than in the case of solar eclipses—typically hours rather than minutes. Thus, although solar eclipses outnumber lunar eclipses by about 3 to 2 (if we exclude purely penumbral lunar eclipses),[59] few of those solar eclipses will be seen by a given observer, whereas about half of the lunar eclipses will be seen.

[55] Any systematic bias towards predicting conjunctions too early or too late would show up in the distribution of solar eclipses relative to the mean conjunction. As already noted, eclipses in a long sequence of records from Western Han fall on the last days of months much more frequently than on first days, thus suggesting that mean conjunctions were being placed too late: see chapter 2, section 2.1.1.

[56] Noël Golvers (1993) *The Astronomia Europaea of Ferdinand Verbiest, S.J. (Dillingen, 1687): Text, translation, notes and commentaries*. Nettetal, Steyler Verlag, 79–81.

[57] On the possibility that Liu Xin c. 10 CE may on one occasion have used the 135 lunation cycle to retrodict an earlier solar eclipse from a later recorded eclipse, see chapter 4, section 4.5. There are no other instances known to me of a possible attempt to predict solar eclipses using simple cycles in early imperial China; in any case, as Asger Aaboe (1972) 'Remarks on the Theoretical Treatment of Eclipses in Antiquity.' *Journal for the History of Astronomy* 3 (2): 105–118, 105 remarks, 'though there are excellent cycles for lunar eclipses ... there are no such cycles or periodic recurrences of solar eclipses for a given location on Earth.'

[58] The moon's umbral shadow on the earth's surface, from within which the sun appears completely obscured, is never much wider than about 270 km. Sometimes the long cone of the moon's umbra does not reach the earth at all, and no observer sees a total eclipse.

[59] Smart and Green (1979 (reprint of 6th edition 1977)), 398–400.

The earliest identifiable attempts to predict solar eclipses have been found in records compiled by late Babylonian astronomers working under the Seleucid empire in the fourth century BCE, and were based on the use of long-term cyclical patterns rather than on any physical model.[60] In the second century CE, Ptolemy set out detailed prescriptions for calculating eclipses, within the context of a geometrical model of the cosmos with a spherical earth at its centre.[61] It has usually been considered that the first Chinese attempts to predict solar eclipses are found in the description of the *Jing chu li* 景初曆 'Brilliant Inception [reign period]' system devised by Yang Wei 楊偉, and used from 237 CE, first in the kingdom of Wei 魏, and then elsewhere until 451 CE.[62] I believe, however, that there are strong indications that it was Liu Hong who first made a successful attempt on this problem. This conclusion is based in part on the fact that, as we have seen, he provided means for estimating the moment of true conjunction, and then for estimating the latitude of the moon at that moment, thus making it possible to tell whether an eclipse was likely. There are, however, other elements in his system whose presence is highly significant, and to these we now turn.

8.3.1 Eclipses and eclipse limits

What other evidence is there that Liu Hong may have been investigating the likelihood of solar eclipses? In answering that question, we may look for a comparison at the approach to eclipse prediction taken by Ptolemy of Alexandria, whose activities were roughly contemporary with Liu Hong. In addition to giving details of how the latitude of the moon may be calculated, he introduces the concept of the 'eclipse (or ecliptic) limit'[63]—a convenient means of telling

[60] See J. M. Steele (1997) 'Solar Eclipse Times Predicted by the Babylonians.' *Journal for the History of Astronomy* 28 (2): 133–9 and John M. Steele (2000a) 'Eclipse Prediction in Mesopotamia.' *Archive for History of Exact Sciences* 54 (5): 421–54.

[61] Toomer (1998), 310–20. As already mentioned, Ptolemy's methods also made it possible to account for the effect of an observer's geographical location—and hence the effect of lunar parallax.

[62] See for instance the summary in John M. Steele (2000b) *Observations and predictions of eclipse times by early astronomers*. Dordrecht; Boston, Kluwer Academic Publishers, 177. Steele's source is the account in Yabuuti Kiyoshi 藪内清 (1963), 476. The relevant sections of the *Jing chu* system are given in *Song shu* 12, 241–6. Steele has now (private communication, 2016) informed me that in the light of the evidence discussed here he considers that it is more probable that Liu Hong developed the method to predict solar eclipses that has previously been said to have first appeared in the *Jing chu* system.

[63] Ptolemy uses the adjective ἐκλειπτικος (ekleiptikos), which simply means 'related to eclipses': see *Claudii Ptolemaei opera quae exstant omnia. Vol.1, Pts I and II, Syntaxis mathematica/edidit J.L. Heiberg.* (1898). Claudius Ptolemy, Lipsiae, In aedibus B.G. Teubneri, 1898–1903., vol 1 part 1, 476. The great circle of the celestial sphere marked by the path of the sun is usually called 'the ecliptic' in modern English, since it is the circle on or near which the moon must be found in order for an eclipse to take place.

whether a given syzygy (a conjunction or opposition of sun and moon)—was likely to result in an eclipse:

> Now that we have explained the above methods [for predicting lunar latitudes], it would be appropriate to follow up with the considerations pertinent to the ecliptic limits for both solar and lunar eclipses. The purpose of this is that if we decide to compute, not all mean syzygies [in a given year] but just those which could fall into the category concerning eclipse prognostications, we may have a handy method of deciding which these are ... (Almagest VI.5, pages 282–3 in Toomer, G. J. (1998))

Thus he proceeds to calculate how far from a node the moon in opposition can be in longitude while still just undergoing a lunar eclipse (where the moon just touches the earth's shadow), and how far from the node in longitude the sun and moon may be at conjunction while a solar eclipse can still just take place (where the discs of the moon and the sun seen from the earth are just in contact). These are his 'eclipse limits'.[64]

An inspection of Liu Hong's lunar latitude table (Figure 8.5) reveals the presence of 'limits', which like those of Ptolemy mark positions just before and after a node. The second and thirteenth days of the sequence have the following annotations respectively:

> 限餘千二百九十, 微分四百五十七. 此為前限
> Limit remainder 1,290, Fine Parts 457. This is the Earlier Limit.[65]
>
> 限餘三千九百一十二, 微分一千七百五十二. 此為後限
> Limit remainder 3,912, Fine Parts 1,752. This is the Later Limit.

These Earlier and Later Limit Remainders, are, as Li Rui shows, eclipse limits, calculated on the traditional basis that if sun or moon are eclipsed exactly on a node, then a possible eclipse (of moon or sun respectively) half a synodic month away will just be missed.[66] Liu Hong states explicitly, when describing how to find the latitude of the moon at successive conjunctions with the sun:

> 入曆在前限蝕前, 後限蝕後者月行中道也.
>
> [If the conjunction] enters the [yin-yang] sequence before the earlier limit remainder, or after the later limit remainder, then this shows the moon is travelling the Middle Road [i.e. the ecliptic].[67] (*Jin shu* 17, 517; Cullen 2017, 304–6)

[64] See the discussion of Ptolemy's calculations and their results in Neugebauer (1975), 125–9.

[65] The figures in the table for the earlier limit, to which, as we shall see shortly, Liu Hong refers in his main text, are lost in transmitted versions of the text, but are restored by Li Rui 李銳 (1768–1817) (1993), 788a–b.

[66] See Li Rui 李銳 (1768–1817) (1993), 788b–779a.; see also Sivin (1969), 46.

[67] For the use of 'Middle Road' to refer to the ecliptic, see chapter 6, section 6.3.5.2.

Since we are specifically discussing the circumstances at conjunctions of the sun and moon, this is clearly the condition for a solar eclipse. But the eclipse limit can equally well be used to estimate the likelihood of a lunar eclipse. In the *Jing chu* system of Yang Wei, which was certainly based on Liu Hong's work, we are told explicitly that the limits there given can be used for both types of eclipse, and presumably the same applied to Liu Hong's limits (*Song shu* 12, 242–3).

The size of the two limits given by Liu Hong can be understood if we attempt to find the position of the moon relative to the 'days of the sequence' graduations that will represent an eclipse just missed, as described above. We start with the moon in conjunction with the sun precisely at a node, i.e. exactly at the start of a yin-yang sequence. Half a lunation (i.e. half a synodic month) later, the moon will be in opposition to the sun, but, while it will still be close to a node, it will (according to the traditional view) just miss a lunar eclipse. During the time elapsed since it was at a node in conjunction with the sun, the position of the moon relative to the 'days of the sequence' graduations will have increased from zero to:

$$(\tfrac{1}{2}) \times (18{,}328 + {}^{914}\!/_{2{,}209})/7{,}874 \text{ days of the sequence}^{68}$$
$$= 1 + (1{,}290 + {}^{457}\!/_{2{,}209})/7{,}874 \text{ days}$$

This is expressed by Liu Hong as 1 day with a remainder of 1,290 parts (at a scale of 7,874) and 457 Fine Parts *wei fen* 微分 (which are themselves at a scale of 2,209, and are added to the 1,290 parts). The use of additional 'Fine' parts enables Liu Hong to achieve higher precision without having, in effect, to write a very unwieldy fraction—in this case ${}^{2{,}850{,}067}\!/_{17{,}393{,}666}$.

Hence, in terms of days, the earlier eclipse limit is passed at a remainder of 1,290 into the second day of the sequence—and of course we still have the (insignificant) left over Fine Parts remainder. This is what is shown on the latitude table under the day 2 entry, as expected.

[68] See Cullen (2017), page 305. During Coincidence Months [11,045], the sun makes New and Full Moon Conjunction Number [941] circuits of heaven from west to east relative to a given node, while also making Coincidence Years [893] circuits of heaven. Hence in that time the node makes 941 − 893 = 48 circuits east to west relative to heaven, and in one month (in the sense of lunation) it moves 48/11,045 circuits. Meanwhile the sun will move 893/11,045 circuits relative to heaven west to east. Thus from one conjunction to the next the sun moves (893 + 48)/11,045 circuits relative to the node, and since each circuit has 21,530/7,874 days of the sequence, the shift in days of the sequence between successive conjunctions is (941/11,045) × (21,530/7,874) = (18,328 + ${}^{914}\!/_{2{,}209}$)/7,874 days of the sequence.

Working the other way round, if we want to find the conditions for the moon just to miss an eclipse half a synodic month before an eclipse occurs with the moon precisely at a node, then given that the total length of a sequence is 13 days $^{5,203}/_{7,874}$, if we subtract 1 day and $^{1,290}/_{7,874}$, with a 'Fine Parts' remainder of 457, we find as the result:

$$12 + (3{,}912 + {}^{1,752}/_{2,209})/7{,}874$$

So we have a remainder of 3,912 + 1,752 Fine Parts into the 13th day, under which the 'later limit' is shown, as is appropriate.

What do these limits imply in terms of angular displacement of the moon from the node? The first limit may be calculated as:

$$1 + (1{,}290 + {}^{457}/_{2,209})/7{,}874 \text{ days} = 1.16386 \text{ days of the sequence.}$$

Since the total number of days of the sequence is 2 × (13 + $^{5,203}/_{7,874}$) days = 27.32156 days, and the moon makes a complete circuit relative to a node while moving through the sequence, then the mean angular displacement of the moon relative to the node when it passes the first limit will be:

$$(215{,}130/589) \times 1.16386/27.32156 = 15.56\ du = 15.34°.$$[69]

Similar limits were used by Yang Wei in his *Jing chu* system:

> 去交度十五以上, 雖交不蝕也. 十以下是蝕, 十以上虧蝕微少 ...
>
> If [the moon] is 15 *du* or more from the node [at a conjunction], even at the node there will not be an eclipse.[70] If [the distance] is 10 [*du*] or less, this is a [definite] eclipse. If [the distance] is 10 or more, then the diminution [caused by] the eclipse will be rather less. (*Jin shu*, 18, 545)

The Uranic Manifestation system has one feature which the *Jing chu* system lacks: its yin-yang sequence enables actual values of lunar latitude to be obtained.

[69] Modern values for eclipse limits are of comparable size. A solar eclipse has a superior limit of 18.4° (an eclipse is impossible if the sun, and hence the moon, is further from a node than this at conjunction), and an inferior limit of 15.4° (an eclipse must take place if the sun, and hence the moon, is closer to a node than this at conjunction). For a lunar eclipse, limits may lie between the superior limit 12.3°, and the inferior limit 9.6°. Smart and Green (1979 (reprint of 6th edition 1977)), 383 & 389–90.

[70] This wording seems a little confused, but the significance is clear enough.

If we apply this to Liu Hong's Earlier and Later Limits, we can simply interpolate in the yin-yang sequence to find the corresponding latitudes. These are:

Earlier Limit: latitude of moon 1.64 *du*.
Later Limit: latitude of moon 1.59 *du*.

It seems unlikely that Liu Hong can have been unaware of the fact—obvious from actually viewing any eclipse of reasonable magnitude such as the large magnitude eclipse of 13 July 158 CE when the moon's apparent latitude at Luoyang was only 0.1°—that at a solar eclipse the disc of the moon overlaps the disc of the sun. How then can he have been satisfied with such wide limits, rather than choosing (say) 2,860 parts of a day of the sequence from a node, which would have given a value for latitude close to 0.5 *du*, the maximum possible before the discs of sun and moon (each about 0.5 *du* across[71]) begin to overlap? The answer to that will, I suggest, emerge when we consider what he might have seen if he had compared the predictions of his system with the record of eclipses actually observed—some, perhaps, observed by himself in addition to the observers of the Grand Clerk's staff.

To sum up: Liu Hong had constructed all the basic theoretical equipment needed to predict solar eclipses—a method of calculating what is in effect the distance of the moon from the node in longitude at the moment of true conjunction, and of finding its latitude at that instant, as well as eclipse limits which add a convenient means of estimating the likelihood of eclipses taking place. It is hard to see why he would go to the trouble of constructing such systems if eclipse prediction had not been his aim. Since he can calculate lunar latitudes, his methods are actually more sophisticated than those in the *Jing chu li*, which lacks this feature but has up to now generally been identified as the first system to attempt to predict solar eclipses.

There are three further questions we need to ask in order to form a judgement on whether Liu Hong ever used his system to predict solar eclipses:

(a) If he had tested his system against the records of eclipses available to him, would that have encouraged him to think that his system worked in practice?
(b) Are there any direct references to his actually predicting a solar eclipse?
(c) What signs are there that his activity encouraged others to attempt similar predictions?

We shall now investigate each of these points in turn.

[71] See for instance the account given by Zhang Heng: chapter 6, section 6.5.1.

8.3.2 Testing the prediction system

Testing the predictions of astronomical systems against observation was established Han practice. Thus we have seen that in 92 CE Jia Kui checked the luni-solar conjunction predictions of the Triple Concordance and Han Quarter Remainder systems against records of seventy solar eclipses from recent times and another twenty four from the remoter past—exploiting the fact that a solar eclipse can only happen at a conjunction.[72] We have seen that when Liu Hong worked on lunar eclipse prediction in 174 CE, he did so after examining the previous 29 years of records of lunar eclipses, which were frequently occurring earlier than predicted by the Han Quarter remainder system then in use.[73] Further, Yuan Shansong and Xu Yue told us that Liu Hong had spent ten or even twenty years elaborating and testing his Uranic Manifestation system, the former referring explicitly to his having 'checked it against the sun and moon'.[74] In the case of the parts of his system relating to solar eclipses it would have been obvious for him to test them against the records of such events in official sources.[75]

A reasonable selection of eclipses for such a test might have run from the beginning of the Yanxi reign period (158–166 CE), during which Liu Hong was sent to the capital in response to a summons from the Grand Clerk, to the end of the Xiping period (172–177 CE), during which he was consulted on lunar eclipses. That would also be consistent with the time periods given by Yuan Shansong and Xu Yue, and, as we shall see later, there are indications that by the end of the Xiping period Liu Hong may have arrived at a marked degree of confidence that his solar eclipse system made useful predictions.

The data shown in Table 8.4 are based on the listing of solar eclipses given in *Hou Han shu, zhi* 18, 3368–70. Where no remark is made, this source simply records that there was an observation of a solar eclipse on the date stated. In all the cases listed here, modern calculations confirm that an observable eclipse did take place. Reading the columns from left to right, an index number is followed by the date in the Julian calendar, followed by the Uranic Manifestation system prediction of the moment of true conjunction, the day and day fraction of the

[72] See chapter 6, section 6.3.

[73] See section 8.1.1.

[74] See section 8.1.

[75] Apart from the evidence of his access to records of lunar eclipses just mentioned, we know that he was given the privilege (previously refused to Zhang Heng) of access to confidential government archives: see chapter 6, section 6.1.

Table 8.4 ***Testing Liu Hong's eclipse predictions, 158–177 CE***

Number	Year	Month	Day	Uranic Manifestation prediction of time of true conjunction	Day of sequence	Day fraction	Within eclipse limit? <(2,1290) or >(13, 3912)	Uranic Manifestation latitude prediction / du	Modern estimate of time of conjunction seen at Luoyang	Latitude observed at Luoyang at actual conjunction / du	Remarks on recorded eclipse in *Hou Han shu*
1	158	July	13	21 : 05	1	2,430	YES	0.44	17:35	0.10	
2	165	February	28	12 : 20	1	486	YES	0.09	16:27	0.37	
3	166	February	18	06 : 17	2	440	YES	1.49	06:44	0.33	'Not seen by the Clerk's officials. Reported from commanderies and kingdoms.' A dawn eclipse.
4	167	July	4	19 : 00	13	6,410	YES	1.16	16:28	0.20	
5	168	June	23	09 : 07	14	1,500	YES	−0.02	05:45	−0.52	Eclipse was barely visible at Luoyang, but was visible as a partial eclipse further south; probably reported.[7]
6	168	December	17	14 : 59	1	2,785	YES	0.5	14:58	−0.11	
7	169	December	6	17 : 14	1	5,442	YES	0.98	16:21	0.52	'Reported by Youfufeng [commandery]' (34.3679° N 107.8816° E). Grazing contact about 15m before conjunction.

continued

Table 8.4 *Continued*

Number	Year	Month	Day	Uranic Manifestation prediction of time of true conjunction	Day of sequence	Day fraction	Within eclipse limit? <(2,1290) or >(13, 3912)	Uranic Manifestation latitude prediction / du	Modern estimate of time of conjunction seen at Luoyang	Latitude observed at Luoyang at actual conjunction / du	Remarks on recorded eclipse in *Hou Han shu*
8	170	May	4	01 : 15	14	105	YES	0.90	02:19	0.25	'Reported by the Chancellor of Liang' (34.4248° N 115.6428° E). Predicted conjunction was at 01:14 and actual eclipse at 02:05, when latitude of moon was 0.24 deg,—so no observation was possible.[8]
9	171	April	23	04 : 16	14	3,017	YES	0.39	05:15	−0.41	A dawn eclipse
10	174	February	19	11 : 17	13	4,809	YES	1.44	15:46	0.38	
11	177	December	8	10 : 13	13	5,881	YES	1.25	06:08	0.48	'Reported by the Chancellor of Zhao' A dawn eclipse, reported from east of Luoyang, where a small partial eclipse was visible. (Moon was just to east of sun at Luoyang.)

[7] See David W. Pankenier (2012) 'On the reliability of Han dynasty solar eclipse records.' *Journal of Astronomical History and Heritage* 15 (3): 200–212, 206.

[8] It is possible that this eclipse was extrapolated and reported on the basis that a total eclipse of the moon followed on the full moon of May 18, visible high in the sky around 01:00.

predicted true conjunction in the yin-yang sequence, a note of whether that falls within the eclipse limits, and then the value of lunar latitude predicted at true conjunction by the Uranic Manifestation. There then follow modern estimates of the time of true conjunction, and of the lunar latitude at those instants.

As for the remarks in the rightmost column, it is notable that the records of the eclipses of 166, 169, 170 and 177 CE explicitly state that they were observed some way from the capital at Luoyang.[76] This appears also to have been the case with the eclipse of 168 CE, which was not likely to have been visible at the capital, where at best the moon's disc would have been seen as barely grazing the southern edge of the sun. The eclipse would, however, have increased in magnitude as the observer's position moved to the south, causing the moon's apparent position relative to the sun to shift northwards.

What is Liu Hong likely to have looked for in testing his predictions against a list of recorded eclipses? As a minimum, he would certainly have wanted to check that his Earlier and Later Limits would have given advance warning of eclipses that had been seen and entered into the official record. We may stress that advance warning was the main requirement for success, even if that warning was not always followed by an observed and recorded eclipse. Given a system that gave such warning, Liu Hong is unlikely to have seen it as a major problem if a few eclipses were predicted but not recorded as having been observed. In the first place, an eclipse that should in theory have been observable at the capital may have been missed because of poor weather conditions.[77] In the second place, to be included in the list it is not actually necessary for an eclipse to have been visible at Luoyang: as noted above, a considerable number certainly were reported from observations made elsewhere. A predicted eclipse unsupported by a record of observation may therefore still have been visible *somewhere*, but simply not have seen by anybody likely to send in a report. Further, an eclipse that was predicted to occur during the hours of darkness could never have been seen at all, even if it had occurred, and so the absence of a record is not significant. Similarly, given the difficulty of making a precise prediction of the moment of true conjunction (a difficulty which would have been obvious from

[76] See Pankenier (2012), which discusses the complete list of Han eclipses.

[77] I made systematic naked-eye observations in Paris at the time of a partial eclipse of the sun on 20 March 2015, at which time the sky was completely covered in cloud so that the sun was never visible. The sun's angular diameter was 32′, and no perceptible decrease in the level of illumination was seen before 10:15 Paris time, by when the moon's disc is calculated to have overlapped the sun by 22′, implying an eclipse magnitude of 0.69. So on a cloudy day a quite large solar eclipse in daytime might be completely missed.

a comparison of the predicted times of conjunction and of observed eclipses), the failure to see an eclipse predicted for just after dawn or before sunset might simply mean that it had occurred at a slightly different time from prediction, so that the sun was below the horizon at the time. Finally, as we shall see later, there is evidence that specialists in astronomical systems did not feel that the occasional failure to observe a predicted solar eclipse indicated a grave fault on their part: given the ominous nature of a solar eclipse on an important occasion, it was more important to be able to warn that an eclipse was likely than to give assurance that it definitely would take place.[78]

In two cases (9 February and 4 August 175 CE) Liu Hong's system warned of eclipses of which the *Hou Han shu* has no record, and which did not in fact occur, although both of these predictions involved the apparently strong feature of being so well within eclipse limits that the predicted latitude was less than 0.5 *du*. But since the predicted time of both eclipses, as defined by the instants of true conjunction according to the Uranic Manifestation, fell well within the hours of darkness, neither eclipse could have been observed even if it had occurred as predicted.

Two further eclipses were also apparently strongly predicted on 23 May 160 CE and 11 April 172 CE , but did not occur. But both were predicted close to the horizon, like a weaker prediction on 4 November 161 CE. Differences between predicted true conjunction times and the actual times of the eleven eclipses observed were, however, as large as four hours in four cases out of the eleven, so it was still quite possible that these eclipses had occurred at night, even if the sun was clearly visible near the horizon on that day. For comparison, according to the Uranic Manifestation, the eclipses observed on 18 February 166 CE and 23 April 171 CE should not have been visible at all, since they were predicted to occur before sunrise. They were, however, late enough in comparison with prediction to be recorded as visible. One somewhat bizarre case is the eclipse recorded as 'reported by the Chancellor of Liang' on 4 May 170 CE; an eclipse was predicted for this date—but the prediction was for 01:15, and it actually took place at 02:19, so it could not have been seen at all. Could it be that a prediction—perhaps even a prediction by Liu Hong—had somehow been written up as a report?

[78] See below, section 8.3.4. Sivin (1969), 25 speculates along similar lines: '... the judicial astrology of the time must have incorporated a rule of this sort: An eclipse seen but not predicted is an omen; that an eclipse is predicted but not seen has no astrological significance.' Perhaps unsurprisingly, there appears to be no explicit reference to such a rule in any ancient Chinese source.

On the basis of his success in predicting all recorded eclipses, it would have been reasonable for Liu Hong to conclude that his system was able to indicate when a solar eclipse was likely to be seen *somewhere*—even if the official sky-watchers in the capital were unable to see it. He was not of course aware how inadequate his system might seem to an astronomer like Ptolemy, accustomed to making predictions for an observer located at any given position on a spherical earth, since the concept of a spherical earth, and of the lunar parallax that followed from it, was not part of his world-picture. He could only go on the record of success of his system when tested against observations of solar eclipses known to him—and in those terms, his system must have seemed quite successful. From the point of view of a Han specialist in astronomical systems, he was entitled to conclude that he had 'found [his system] to correspond to the phenomena' as Yuan Shansong said he had.

Close inspection of the Uranic Manifestation predictions for the observed eclipses reveals a striking feature of the latitude predictions. Ten out of the eleven recorded eclipses in this period are associated with a 'Yin' latitude prediction, that is, a prediction of a northern latitude of the moon. The exception is eclipse number 5, on 23 June 168 CE, which at −0.02 *du* was predicted to be just south of the ecliptic; in fact this eclipse as observed would have been only just capable of being seen at Luoyang, since the moon only overlapped about 2′ into the sun's 15′ diameter disc. Predicted values of northern latitude, on the other hand, are in excess of 1 *du* in four cases of observed eclipses, and more than 0.9 *du* or more in two further cases. In modern terms, this northern bias is clearly attributable to the lunar parallax resulting from the northern latitude of Luoyang. These results strongly recall a well-known but somewhat cryptic statement attributed to Zhang Zixin around 560 CE—a statement that modern scholars have commonly taken as the first explicit reference to the effects of lunar parallax in China:[79]

合朔月在日道裏,則日食.若在日道外,雖交不虧.月望值交則虧,不問表裏.

If at conjunction the moon is within the path of the sun [i.e. to the north of the ecliptic], then there [may be] an eclipse, [but] if it is outside the path of the sun [i.e. to the south of the ecliptic], then [sun and moon] may meet with no obscuration occurring. But if the moon at opposition lines up opposite to the sun then there will be an obscuration, no matter whether it is within or outside. (*Sui shu* 20, 561)

[79] On Zhang Zixin, see chapter 1, note 39. For the interpretation of this passage as referring to lunar parallax, see Tang Quan 唐泉 (2011), 168–70.

In interpreting this passage, it is essential to distinguish between the moon's *predicted* position according to (for instance) the Uranic Manifestation system, and its *observed* (or 'apparent') position. The situation is as follows:

(a) A system such as the Uranic Manifestation system will *predict* that solar eclipses are possible whenever the true conjunction takes place close enough to a node, and that lunar eclipses are possible whenever the sun and moon in opposition are both close enough to nodes. Since the system predicts a symmetrical variation in lunar latitude to the north and to the south of the ecliptic, eclipses will be predicted in equal numbers of cases where the moon's *predicted* position is to the north of the ecliptic, and to the south.

(b) But in fact, because the moon's *observed* positions relative to the sun are in general displaced southwards relative to its predicted positions (assuming an observer in China), nearly all the eclipses actually observed will be associated with *predicted* positions of the moon to the north of the ecliptic, and few, if any, with predicted positions to the south of the ecliptic.

(c) If, however, the moon is predicted to be exactly opposite the sun at full moon (which will happen when both bodies are close to a node), then the shift of the moon's apparent position north or south due to parallax has no effect, since the moon will still be seen in the earth's shadow, and hence a lunar eclipse takes place.

This is precisely the situation that Zhang Zixin describes. But, as we have seen, this relation between observation and prediction would already have been evident to Liu Hong if he had simply compared the predictions of his system with observed solar eclipses, as it seems very likely he did.

Further, if Liu Hong had actually observed a few of the eclipses in question (and we know that he was in Luoyang during the relevant period), it would be evident that the appearance of the eclipses suggested in nearly every case that the actual latitude was considerably to the south of what had been predicted. Take for instance the eclipse of 17 December 168 CE. The predicted latitude of the moon was 0.5 *du* for a predicted true conjunction at 14:59 Luoyang time, which was in this case almost exactly the time of the actual conjunction. If the predicted latitude had been correct, the moon's disc should then have barely grazed the northern edge of the sun. But in fact, the appearance of the eclipse would have made it obvious to any observer that the moon was on the southern side of the sun, with a small value of latitude (in fact it was −0.11 *du*). Such experiences would certainly have been suggestive of the general fact that predicted latitudes were considerably too northerly—most strikingly in a case such as 4 July 167 CE, when the predicted value of latitude, 1.16 *du*, should have made any eclipse impossible since the moon would have been much too far to the north of the sun, but the actual latitude was only 0.2 *du*, resulting in a large partial eclipse.

We cannot assume that Liu Hong actually measured the latitude of the moon at eclipses. If, however, he had asked himself by how much his system was overestimating latitudes, he might have tried seeing how far his predicted latitudes were from the symmetrical variation about zero that was the basis of his yin-yang tabulation. It is not difficult to see that a shift of about 0.75 *du* (equivalent to $\frac{9}{12}$ *du*, and hence to a decrease of 9 in the Total Numbers) would spread the latitude predictions fairly evenly between north and south.

By 177 CE, Liu Hong could have felt some assurance that he had a system that should enable observers to avoid being taken by surprise by a solar eclipse, and moreover to give a fairly good idea of when one was likely to occur. He was, as we have seen, officially charged around that time with reviewing the work of people who claimed to have improved methods of predicting lunar eclipses—so why not attempt a solar eclipse prediction?

Let me be quite clear: we cannot simply argue that because Liu Hong had constructed a system with the elements needed to predict solar eclipses—that is, methods to predict the distance of the moon from a node at conjunction, and to find its latitude—he actually did so. An argument that runs:

1. Person X had the means to do Y
2. Y would have been a desirable thing for Person X to do
3. Therefore Person X must have done Y

is always dubious, and it is particularly dubious in the history of science, in part because we should be very cautious in assuming that we understand what X (rather than we ourselves) might have seen as desirable.

If, however, we find that in addition to the first condition above:

(a) Person X is reputed to have actually done Y by near contemporaries in his own culture, and
(b) Shortly after Person X's supposed activities, others in his culture begin to be reported to have done Y, and
(c) The means possessed by Person X are explicitly said by other near contemporaries to be suitable for doing Y

then the case is very different, and we may reasonably say that it looks rather as if Person X probably did do Y. We have already seen that condition (c) has been met, since we know that a few decades after Liu Hong's time Yang Wei stated that eclipse limits like those of Liu Hong can be used to predict solar eclipses. We shall now pass on to evidence that conditions (a) and (b) were likewise fulfilled.

The records of the preceding two decades suggested clearly that any successful prediction by Liu Hong's system would require a predicted northern 'Yin' latitude for the moon at conjunction, and perhaps a latitude value around $^{9}/_{12}$ *du* so that the actual observed latitude would be somewhere not far from zero. In fact, in 177 CE Liu Hong would not have to look very far ahead to find two eclipse predictions fulfilling these conditions. What is more, both eclipses were predicted to fall well into the hours of daylight, thus reducing the chances that a successful eclipse prediction might be rendered unobservable by taking place at night, when the sun is below the horizon. Moreover, a person who had had close contact with Liu Hong stated that he had made such a prediction, and that it was recognized widely as having succeeded, as we shall now see.

8.3.3 Liu Hong's solar eclipse prediction

We have already heard of Xu Yue as an important link in the transmission of the Uranic Manifestation system, no doubt based in part on local connections with Liu Hong's home region.[80] In the record of a discussion said to have taken place in the state of Wei 魏 (which included the region from which Liu Hong originated) in 223 CE,[81] he is recorded as saying:

> 今韓翊所造, 皆用洪法 … 至於日蝕, 有不盡效. 效曆之要, 要在日蝕. 熹平之際, 時洪為郎, 欲改四分, 先上驗日蝕: 日蝕在晏, 加時在辰, 蝕從下上, 三分侵二. 事御之後如洪言, 海內識真, 莫不聞見, 劉歆以來, 未有洪比.
>
> Now [the system] that Han Yi has devised uses [Liu] Hong's methods throughout … when it comes to solar eclipses, in some cases [his system] is not completely verified. Of the important points in verifying a system, the most important relates to solar eclipses.[82] At the end of the Xiping reign period

[80] See above, section 8.1.

[81] See above, section 8.1.2.

[82] A similar point was made centuries later in the report on the *Shou shi* 授時 'Season Granting' astronomical system (1280) in *Yuan shi* 元史 *(History of Yuan dynasty)*. (Completed 1370, punctuated edition of 1976). Song Lian 宋濂 (1310–1381), Beijing, Zhonghua Press: 曆法疏密, 驗在交食 'It is through eclipse [predictions] that the accuracy of astronomical systems may be checked' (*Yuan shi* 53, 1153); compare Sivin (2009), 311. It is clear from the rest of this statement that it refers specifically to solar eclipses.

(172–177 CE), when [Liu] Hong was a court gentleman, he wanted to reform the Quarter Remainder system, so he first checked [his methods] by a solar eclipse:[83] the sun was eclipsed in clear daylight, and the hour of occurrence was *chen* [7 a.m. to 9 a.m.]; the eclipse was from below to above, and encroached upon two parts out of three. When the reports came in they were as [Liu] Hong had said. Everybody in the empire [literally 'within the seas'] recognized the truth of it, and there was nobody who did not hear of it or see it. Since the time of Liu Xin, there has been nobody like Hong. (*Jin shu* 17, 499–500)

Xu continues with a comparison of the accuracy of the Uranic Manifestation system, which he evidently knew well, with the results of Han Yi's system.[84] When Xu Yue tells the story of Liu Hong's eclipse prediction—for a prediction it clearly was—his close connection with Liu suggests strongly that he is not merely repeating a rumour, but recounting something that Liu Hong actually said he had done, and in which he was felt to have succeeded in an impressive manner.[85]

Now unfortunately Xu Yue does not tell us the date of the eclipse that Liu Hong predicted. In order to verify his account, we therefore need to find a suitable solar eclipse that fulfils two principal conditions:

1. It has to be predicted by Liu Hong's system, and to fall at or after the end of the Xiping period, hence in 178 CE or later.[86]

[83] It is not entirely clear what role the word *shang* 上 plays in this sentence. It cannot bear one of its known senses, that is 'going back into the past', since what follows clearly refers to a prediction of a future event. It seems possible that it is being used in the sense of 'previous' *qian* 前, a usage noted by the commentator Gao Xiu 高誘 (fl. C. 210 CE) with reference to a passage in *Lü shi chun qiu* 呂氏春秋 *(Mr. Lü's Spring and Autumn [Annals])*, 10, 6b, in which case it serves to strengthen *xian* 先 as indicating that this is how Liu Hong started his test. Or it may refer to the fact that the verification of the prediction would be seen 'high up' in the heavens, rather than being sought in a written record of past observations. I am grateful to Professor Shi Yunli 石云里 for discussing this passage with me (private communication, December 2016).

[84] See above, section 8.1.2. The wider context of the discussion in which this material occurs is examined in Morgan (2013) chapter 4.

[85] The fact that the prediction was made at the end of the Xiping 熹平 period (172–177 CE) may explain why there is no mention of it in the documentary collection compiled by Liu Hong and Cai Yong, which was said to have been edited in the first year of the Guanghe 光和 period (178–183 CE). See chapter 6, 158, and *Hou Han shu, zhi* 3, 3082; Cullen 2017, 234. More significantly, we must recall that all the documents in that collection have official status as memorials, edicts, or reports made in an official context. If Liu Hong's prediction was made as part of his private researches, it would have had no place in the archives.

[86] Professor Lü Lingfeng has pointed out to me (private communication, December 2016) that we cannot take it completely for granted that the system Liu Hong had at his disposal around 178 CE was identical to the one that he transmitted to Zheng Xuan in 196 CE, or to what he apparently put into final form as the Uranic Manifestation system in 206 CE.

2. It has to have been recorded as having been observed—since an event of this kind, discussed as widely as Xu Yue claims, was unlikely not have been included in the official record.

What would it mean for Liu Hong's system to predict a solar eclipse? As we have seen, one method of prediction is built into his system, but a second would have been suggested quite strongly if he had carried out the check of his predictions against observation that is ascribed to him:

(a) If the conjunction of sun and moon occurs when the moon is within either the Earlier or Later Limits in the yin-yang sequence, then an eclipse may happen.
(b) Based on a test of prediction against observation, the latitude should be predicted as Yin, that is, it places the moon north of the ecliptic. Further, as we have seen, a predicted northern latitude of around $^{9}/_{12}$ *du*, or 0.75 *du*, might seem a favourable value for the prediction.

Box 8.2 has already calculated one example that fulfils these conditions just after the end of the Xiping period, and it may easily be verified that it was closely followed by another. The data for these eclipses may be summarised as:

A: Guanghe, year 1, civil month 10, last day (27 November 178 CE): predicted latitude 0.72 *du*, Yin; true conjunction predicted at 13:51 Luoyang time.

B: Guanghe, year 2, civil month 4, first day (24 May 179 CE): predicted latitude 0.73 *du*. Yin; true conjunction predicted at 11:50 Luoyang time.

It would have been reasonable for Liu Hong to have predicted a solar eclipse in either or both of these cases. Eclipse A does, however, seem a more likely candidate than B. In the first place, it is the first eclipse Liu Hong could have predicted if he was at the end of the Xiping period and looking ahead into the future. Both of these eclipses were predicted to fall within the hours of daylight—and both of them were in fact observed and recorded:

A: [光和元年]十月丙子晦, 日有蝕之, 在箕四度
[In the first year of the Guanghe period,] tenth month, day *bingzi*.13, last day of the month [27 November 178 CE], the sun was eclipsed, at the fourth *du* of Winnower. (*Hou Han shu, zhi* 18, 3370, also mentioned in *Hou Han shu, ji* [annals] 8, 341–2)

B: [光和]二年四月甲戌朔, 日有蝕之
In the second year [of the Guanghe period], fourth month, day *jiaxu*.11, first day of the month [24 May 179 CE], the sun was eclipsed. (*Hou Han shu, zhi* 18, 3370)

Modern calculations confirm that both these eclipses did actually occur and would have been visible from Luoyang.

To decide which of them could have been the one to which Xu Yue's story refers, let us look at the details he gives.

1. The sun was eclipsed in clear daylight,
2. The hour of occurrence was the fifth double-hour, *chen* [7 a.m. to 9 a.m.];
3. The eclipse was 'from below to above'.
4. The moon covered two parts of the sun out of three.

The first problem that confronts us is that we do not really know what kind of details these are, or where they came from. It is possible, but perhaps not very likely, that Xu Yue could have obtained this information from direct personal observation of the eclipse as it occurred, since he is speaking in 223 CE, about 45 years after the events of 178/179 CE. Nor does it seem likely that he had privileged access to official records containing detailed information on times of observation etc. not listed in the summary copied into the *Hou Han shu* monograph by Sima Biao, assuming that any such detailed records survived the fall of the Han dynasty. A more likely source is Liu Hong himself, who would certainly have recalled what was, if this story is true, a crucial test of the results of ten or 20 years' work. Xu Yue's role as a link in the transmission of Liu Hong's work renders it highly probable that he had personal contact with him at some stage, and they certainly shared the same local connections.[87] This contact could have taken place as late as 206 CE, 17 years earlier, when Liu Hong was finalizing the text of the Uranic Manifestation. The details in Xu Yue's account may not be accurate in every respect, but they are unlikely to be wholly fictional.

The first part of the description does not seem to pose any problems of interpretation, and is fulfilled by both eclipses—otherwise they could not have been observed. This is a non-negligible point: as we have seen, a solar eclipse may take place when the sun is below the horizon, in which case it cannot be seen.

As for the second condition, modern calculations of the times of the eclipses (Luoyang local time) are:

A: first contact 08:55; maximum 09:44; last contact 11:05.
B: first contact 11:22; maximum 13:02; last contact 14:31.

[87] We may recall that the *Shu shu ji yi* represented Xu Yue as saying explicitly that he knew Liu Hong: see section 8.1.

The start of A certainly fits the time given, since it falls just before the end of the fifth double-hour, *chen* [7–9 a.m.];[88] B does not start until some way into the seventh double-hour, *wu* [11 a.m.–1 p.m.].

As for the third condition, we are dependent on the interpretation of the phrase *shi cong xia shang* 蝕從下上 literally 'eclipse from below upwards'. Unfortunately, we do not have a repertoire of similarly detailed contemporary eclipse descriptions to compare this with. An observer at Luoyang would have seen:

> A: The moon first touched the sun at a position on the upper west side of the sun, at a position corresponding to about 2 o' clock on a vertically oriented clock-face. The moon then moved downwards and a little eastwards relative to the sun, until its last contact took place close to the lower edge of the sun, at about the 6 o'clock position.
>
> B: The moon first touched the sun on its western edge, at about the 3 o'clock position. It then passed across the sun, and left it while moving upwards, so that it last touched the sun at about the 12 o'clock position.

If the description simply means that the maximum 'bite' taken out of the sun was in its lower half, that is, the moon's coverage of the sun reached upwards from below then A is the better fit; an interpretation in terms of motion across the disc requires us to assume more detailed observation during the eclipse, but might favour B.

As for the fourth condition, we have:

> A: At maximum contact, the inner edge of the moon's disc is about halfway along a solar radius, and thus a quarter of the way across the sun.
>
> B: At maximum contact, only a thin crescent of the sun remained visible.

The two situations are compared in Figure 8.7.

Neither eclipse fits the stated condition very well; the coverage of A is less than stated, while that of B is much more, nearly a total eclipse, a circumstance that would surely have been noted by Liu Hong and remarked on in Xu Yue's description had this been the eclipse he predicted. It may be that somewhere between Liu Hong and the *Jin shu* as we have it today a character *yi* 一 'one' in the original

[88] Of course we are dependent here on the means used to time the eclipse, for which a clepsydra is the most likely candidate. Xu Yue's record places the start of the eclipse just within the double-hour *chen*, but others may have timed it differently. A memorial by Lu Zhi 盧植 (d. 192 CE) discussing the significance of this eclipse as an omen that indicated problems with the imperial government says 而閒者日食自巳過午, 既食之後, 雲霧晻曖. 'Now this recent eclipse was from [the double-hour] *si* [9–11 a.m.] and went on into [the double-hour] *wu* [11 a.m.—1 p.m.], and after it had finished [the weather] was cloudy and foggy' *Hou Han shu* 64, 2117. That timing is just consistent with the end of the modern timing estimate, while Xu Yue's timing fits the modern estimate for its start.

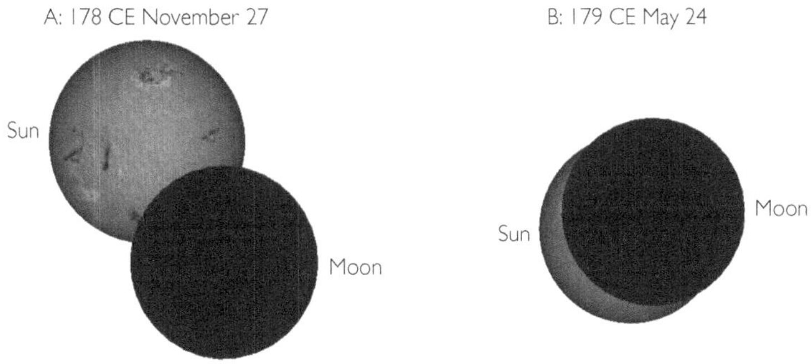

Figure 8.7 178 CE and 179 CE eclipse maxima compared. The view is that of a terrestrial observer, looking approximately south-east from Luoyang at an elevation of 29° for A, and approximately south-west at an elevation of 68° for B.

statement was garbled into *er* 二 'two' so that a one-third overlap (which is plausible for A given the means of observation available) became two-thirds.

On the basis of this discussion, the points of correspondence between eclipse A and Xu Yue's description make it considerably more probable that Xu Yue was referring to eclipse A than eclipse B.

Finally, the observation report for eclipse A on 27 November 178 ce places the sun in the 4th *du* of Winnower. This is in fact identical to the position of the sun on the date of the eclipse, calculated according to the Han Quarter Remainder system, which was the official system at the time Liu Hong made his prediction, and whose constants were re-used with only minor modifications in the Uranic Manifestation system as we have it today. This position could not have been a result of direct observation, since the stars would have been blotted out by the sun's light. Had observation been possible, it would have been seen that the sun was in fact only just approaching the start of the lodge Winnower at the time of the eclipse; it would have not have reached the 4th *du* until about 1 December. But this undetectable discrepancy could not have reduced the impression made by an otherwise successful prediction.

So on the last day of the tenth month of the first year of the Guanghe reign period, 27 November 178 CE, there was observed and recorded an eclipse of the sun that Liu Hong could perfectly well have predicted with a reasonable degree of confidence at the end of the preceding Xiping period, given the evidence of the success of his method in predicting all recorded eclipses from 158 CE onwards. The details of the eclipse in the written record as we now have it, such as its precise timing, were not all, as Xu Yue claims 'as Hong had said' *ru Hong yan* 如洪言. But even in the simpler case of lunar eclipses, as discussions above show,[89] the essential point was that the eclipse was predicted on the right day. No-one at the period discussed in this book seems to have criticized a lunar eclipse prediction on the basis of a discrepancy between the observed and predicted timing, or between the observed and predicted eclipse magnitude.

8.3.4 After Liu Hong

In section 8.3.2 we reviewed the kind of evidence that would be required to make it seem likely that Liu Hong might have predicted a solar eclipse. So far, two of the conditions then set appear to have been fulfilled: Liu Hong had certainly constructed methods of a type that were explicitly said by others not long after his time to have been capable of predicting solar eclipses (section 8.3.1). Further, we have just seen that Xu Yue, who is known to have had contacts with Liu Hong, claimed that he had made such a prediction. But what about the third condition? Are there any signs that what Liu Hong is claimed to have done was also done by others?

The evidence we require is not hard to find. A few years after Liu Hong had finalized his system in 206 CE, someone else made a solar eclipse prediction, and the prediction they made was clearly indicated by the methods of the Uranic Manifestation system. The predicted eclipse was timed for the first day of the first civil month—New Year's Day. If a solar eclipse were to take place on that day, during the great court ceremonies with the emperor at their centre, it would have been an extremely bad omen—so what was to be done about it? Centuries later the Kangxi emperor cancelled the usual New Year's banquets when he received such a prediction,[90] and it is clear that similar measures were considered on this occasion:

> . 建安中 [...] 太史上言: 「正旦當日蝕. 」劭時在尚書令荀彧所, 坐者數十人, 或云當廢朝, 或云宜卻會. 劭曰: 「梓慎, 裨竈, 古之良史, 猶占水火, 錯失天時. 禮記曰諸侯旅見天子, 及門不得終禮者四, 日蝕在一. 然則聖人垂制, 不為變〔異〕豫廢朝禮者, 或災消異伏, 或推術謬誤也. 」彧善其言. 敕朝會如舊, 日亦不蝕.

> In the Jian'an period [196–219 CE] ... the Grand Clerk sent up a memorial saying 'There should be a solar eclipse on New Year's Day'. At the time, [Liu] Shao[91] was at the office of Sun Yu, Director of the Imperial Secretariat [who had

[89] See for instance chapter 6, section 6.1.
[90] See the account in Jami (2012), 222–9.
[91] This was 劉劭 (c. 170–c. 245 CE).

> received the memorial], and several tens of people were also present. Some said that the [New Year's] court sitting should be suspended, and some said that it should still be held. [Liu] Shao said 'Zi Shen and Pi Zao were excellent Clerks of antiquity [in the Spring and Autumn Period]. Yet when divining on water and fire, [even] they mistook the seasons of heaven. The *Li ji* says that when the feudal lords have journeyed to have audience with the Son of Heaven, there are four things that may prevent the ceremonies being carried through after they have reached the gate. A solar eclipse is amongst them.[92] But there is nothing saying that when a sage [emperor] has sent down his commands, the court ceremonies may be cancelled because of a prediction of some portent—it may be that the disaster will fade away and the portent be obscured, or it may be that the methods of prediction will prove to be in error.' Sun Yu approved of this speech, and gave orders that the session of the court should proceed in accordance with the old practice. And in fact the sun was not eclipsed.

Since Sun Yu 荀彧 died in 212 (*San guo zhi* 10, 317–18), this incident cannot have happened later than that year. In fact in none of the preceding years of the Jian'an 建安 period had the Uranic Manifestation eclipse limits test warned of the possibility of an eclipse on New Year's Day; the first instance of this occurred on 20 February 212 CE, the first day of the first month of the 17th year, the year in the course of which Sun Yu met his end. Up to that date, there had been four recorded solar eclipses in the Jian'an period, in the fifth, sixth, 13th and 15th years. All had been predicted by the Uranic Manifestation's eclipse limits.

We do not know whether the Grand Clerk had calculated every possible eclipse during this period, as opposed to just verifying those that had been observed—had he done so, he would have seen that several of those predictions had been predicted to fall in the night, and hence were uncheckable. But one eclipse, with a predicted Yin latitude of 0.44 *du*, had been predicted for 26 August 211, and although the time of the predicted conjunction (12:22) was well within daylight hours, no eclipse had occurred. However, even assuming that the sky was clear and that the lack of an eclipse had been evident during that preceding August the Grand Clerk might reasonably have felt that it was his duty to warn that the New Year's ceremony was at risk of the appearance of a baleful omen that might have led to it being abandoned in confusion.

[92] On this anthology of material related to ritual, of uncertain date, see Loewe (1993), 293–307. The statement (by Confucius) referred to is found in *Li ji* 禮記 18, 24a in *Shi san jing zhu shu*, vol. 5, 369b. The other three causes are fire in the ancestral temple, death of the empress, and the participants' robes being soaked by rain.

There is no record of what was said to the Grand Clerk after his failed prediction. But when the *Song shu* tells the same story, it continues the narrative with an account of a further incident of a similar kind a few decades later:

> 魏高貴鄉公正元二年三月朔, 太史奏日蝕而不蝕. 晉文王時為大將軍, 大推史官不驗之負. 史官答曰: 「合朔之時, 或有日掩月, 或有月掩日. 月掩日, 則蔽障日體, 使光景有虧, 故謂之日蝕. 日掩月, 則日於月上過, 謂之陰不侵陽, 雖交無變. 日月相掩必食之理, 無術以知 [...] 自漢故事, 以為日蝕必當於交. 每至其時, 申警百官, 以備日變. 故甲寅詔有備蝕之制, 無考負之法. 古來黃帝, 顓頊, 夏, 殷, 周, 魯六歷, 皆無推日蝕法, 但有考課疏密而已. 負坐之條, 由本無術可課, 非司事之罪.」乃止.

> In the second year of the Zhengyuan reign period of the Duke of Gaogui of Wei, on the first day of the third month [24 April 255 CE], the Grand Clerk memorialized that the sun would be eclipsed, [but] it was not eclipsed. At that time, [the future] King Wen of Jin[93] was Generalissimo, and made much of the responsibility of the Clerk's officials for this failure [of the prediction] to be verified. The Clerk's officials replied: 'When there is a conjunction, sometimes the sun covers the moon, and sometimes the moon covers the sun. When the moon covers the sun, then it hides the body of the sun, so that its brilliance is diminished, and we call that a solar eclipse. When the sun covers the moon, then the sun goes across the moon, and we say that the yin is not encroaching on the yang, so there is no change even though [their paths] cross. As for the principles [determining whether] the sun or the moon is covered, or the certainty of an eclipse, there is no procedure to obtain that information. [...] In the former practice of the Han, [it was known that] solar eclipses must always fall on the conjunction. So whenever that time arrived, a warning was issued to all officials, in order to prepare them for [a possible] solar portent. Thus the *jiayin*.51 edict[94] laid down regulations for preparing for an eclipse [at a conjunction], but no rules were prescribed for investigating responsibility [for failed predictions]. The six systems of the Yellow Emperor, Zhuan Xu, Xia, Yin, Zhou and Lu which have come down from antiquity all lacked methods for predicting solar eclipses; there was only the [use of] recorded results [of eclipses] to check their accuracy. To enter into judgment about responsibility, when there is basically no method that can be checked implies no fault on the part of the officials in charge.' Thereupon the matter was dropped. (*Song shu* 14, 351)

[93] This was Sima Zhao 司馬昭 (211–265 CE)

[94] This is in all probability the edict issue on a *jiayin*.51 day in the second month of the second year of the Yuanhe reign period (18 March 85 CE), in which the use of the Han Quarter Remainder system was commanded. Part of this edict is quoted in *Hou Han shu, zhi* 2, 3026–4027; see also Cullen 2017, 377–9. The same edict appears to be referred to elsewhere in similar terms: see *Hou Han shu, zhi* 2, 3037, Cullen 2017, 403.

Although the Grand Clerk of Wei was by this time certainly using the Jing chu system, that system was, as Xu Yue had remarked, based on Liu Hong's system so far as solar eclipse prediction was concerned, and Liu Hong's system did indeed warn of a solar eclipse on the date in question, with entry into the yin-yang sequence at day 14, 4,342 parts, and Yin lunar latitude 0.15 *du*. This was certainly a reasonable possibility for an eclipse. However, when the conjunction actually occurred about ten minutes after sunrise, the latitude of the moon was about −0.64° and no eclipse took place.[95]

The view expressed by the Clerks in their reply to the attack on their competence seems to have been adapted from what Wang Chong had written one and a half centuries earlier (chapter 6, section 6.3.1). But whereas in Wang's time the idea that the sun might pass in front of the moon at a conjunction may have simply been a way of explaining why there was not a solar eclipse at every conjunction, here the Clerks are applying this explanation to the more limited problem of why there is not an eclipse on every occasion when the conjunction occurs at or very near a node. For them, the question at issue was whether it was reasonable to demand that they should be able to give a reliable prediction of a forthcoming solar eclipse, rather than simply warning that one was likely.

For the historian, however, the point is a different one. Although the Clerks wanted to deny that they could make such predictions reliably, their denial was not based on any claim that the motions of the sun and moon in terms of their positions on the celestial sphere were too complex for them to predict. On the contrary, it seems that by the middle of the third century CE, some specialists in *li* were convinced they could warn when conjunction of sun and moon would occur near a node, making a solar eclipse possible. It was the fact that they believed this, but were as yet unaware of the effects of lunar parallax, that forced them to find a reason why solar eclipses did not always occur as predicted. Liu Hong's work had clearly given the Clerk's officials a level of confidence in their ability to calculate the motions of the celestial bodies that none of their predecessors had shown signs of possessing.

Looking back to the conditions set out earlier (in section 8.3.2) in relation to the plausibility of stories of solar eclipse prediction, we may sum up as follows. We know that by c. 206 CE, Liu Hong had constructed a series of procedures that, if successful, would have enabled him to attempt to predict solar eclipses.

95 Interestingly, this puts the moon almost exactly where it would be if we applied the rough correction of $\frac{9}{12}$ *du* for shift southwards from predicted latitude suggested on the basis of the eclipse records from 158 to 177 CE discussed in section 8.3.2.

A few decades later, procedures very similar to his were explicitly claimed, by Yang Wei, to be capable of predicting solar eclipses. We have seen that someone who had contacts with Liu Hong stated that he had predicted a solar eclipse, and we have now seen that the practice of official astronomers attempting to predict solar eclipses (though not always successfully) began not long after his time. Taking all this together, while we cannot claim with absolute certainty that Liu Hong made a successful solar eclipse prediction in 178 or 179 CE, it seems considerably more likely than not that he actually did so.[96]

[96] One thing we can say with certainty is that the story of Liu Hong's solar eclipse prediction is vastly more plausible than the frequently discussed story that Thales of Miletus made such a prediction early in the sixth century BC. On this, see for instance Neugebauer (1975), 604 and F. Richard Stephenson and Louay J. Fatoohi (1997) 'Thales's Prediction of a Solar Eclipse.' *Journal for the History of Astronomy* 28 (4): 279–82.

CHAPTER 9

Epilogue

By the end of the Han, specialists in the analysis and prediction of the motions of the heavenly bodies had established ways of working which were, in their essentials, to be continued and developed without radical discontinuity for the rest of the imperial age. The stability of these methods was underpinned by their incorporation into the ideology of imperial legitimacy, which claimed that the emperor had not simply the right, but the duty, to 'grant the seasons' to his people—a duty which was imagined to have been fulfilled by the model rulers of high antiquity.[1]

An essential element in the working practices of these specialists was the notion of a *li* 曆 '[astronomical] system'. We have discussed the technical structures of the systems that were created in the period we have studied; those technical features broadly persisted in the centuries that followed, and provided a strong framework for the developments that took place after the Han. But an astronomical system did not write itself, and the technical structures of a system are insufficient to explain the way those structures changed and developed over the centuries that followed. And change and develop they certainly did: just as the empires of Qin and Han, were, after a period of division, succeeded by the empires of Sui 隋 (581–618) and Tang 唐 (618–907), followed in turn by Song 宋 (960–1279), Yuan 元 (1271–1368), Ming 明 (1368–1644), and finally Qing 清 (1644–1911), so were the astronomical systems we have studied replaced by others,[2] and the specialists who created them found successors who modified and developed what they had created.

[1] See chapter 1.

[2] Yabuuti Kiyosi 藪内清 (1969), 388–391 lists 48 named astronomical systems, of which only the first three have been discussed in this book. The last of these was officially adopted in 1645. By including other systems known to have been constructed but not necessarily adopted for official use, Sivin (2009), 43–53 brings the number up to 98. In some cases a system might amount to no more than a slight modification of one of its predecessors.

It was human beings who created and modified astronomical systems, and like all human beings they functioned within particular social institutions, institutions that shaped and conditioned the ways in which they acted and were able to act, and which were in turned shaped and conditioned by those who worked within them. One of the objects of this book has been to show how those institutions functioned in a well-defined and well-documented period that set the pattern for much of what followed—the early imperial age. The continuing story of astronomical systems and their creators from the end of Han to the end of imperial China would require more than one book to tell it at the same level of detail.[3] The broad social institutions within which the techniques of astronomical systems were developed were always wider and more complex than the official structures within which the Grand Clerks and their staff of observers, record-keepers, calculators and omen-interpreters worked. There were always others in the government service who had no official responsibilities in relation to the heavens, but who still contributed their insights and expertise. In addition there were private persons who, when they do appear in the official record, reveal a level of knowledge and skill that makes one wonder how many others there were of whom we have never heard. In the period studied in this book, we have seen that it was normal for open and critical discussion of technical matters to take place, in a context where the ultimate test of an existing or newly proposed system was its ability to match observation. And all this was embedded within a culture of record-keeping and the archiving of documentary resources that not only gave specialists in astronomical systems an extended span of time over which their ideas could be tested, but also enables modern scholars to study the work of those specialists in both breadth and depth.

The social institutions within which astronomical systems were used and refined were strong, versatile and resilient. They continued to function despite the great cultural, intellectual and social changes that accompanied the widespread adoption of Buddhism during the period of division that followed

[3] The account given by Joseph Needham in the 1950s remains the only attempt in English at a comprehensive summary of the history of Chinese astronomy. Needham's wide-ranging concern with social, historical and cultural context does, however, give his account considerable enduring value. As we have seen, however, Needham gives the topic of astronomical systems no more than a minor place in his story. See Needham and Wang Ling (1959), 171–461. Of course Needham's idea of a summary amounted to what most would consider a fairly substantial book. Even then, the detailed portion of his account ends with the refitting of the imperial observatory by the Jesuit Ferdinand Verbiest in the early Qing dynasty. For a discussion of Needham's approach to astronomy, see Christopher Cullen (1980) 'Joseph Needham on Chinese Astronomy.' *Past & Present*, (87): 39–53.

the fall of Han. In the eighth century we find that the office of Grand Clerk was filled by Qutan Xida 瞿曇悉達 whose name (a rendering of Gautama Siddharta) shows his membership of an Indian clan that had served at the Tang court for several generations. It is to him that we owe the great divinatory and astronomical compendium known as the *Kai yuan zhan jing* 開元占經 'Kaiyuan reign period manual of divination' cited several times in this book, without which our knowledge of pre-Tang astronomy would be considerably poorer.[4] The astronomical expert who conducted a great meridian survey of the empire in 725 CE was a Buddhist monk, Yixing 一行 (683–727); he was also responsible for the creation of the Great Expansion system *Da yan li* 大衍曆.[5] Under the Yuan, Guo Shoujing 郭守敬 (1231–1316) produced his 'Season-granting' system *Shou shi li* 授時曆, of which a comprehensive account has been given by Nathan Sivin.[6] Guo was an official of a cosmopolitan empire whose rulers also made use of the talent of Muslim astronomers such as Jamāl al-Dīn, known to us under his Chinese name of Zhama Luding 札馬魯丁.[7] While the Mongol rulers of the Yuan may have not encouraged contact between the various groups who served them as advisers, the Ming dynasty, which took power after the expulsion of the Mongols in 1368, felt confident enough to launch a programme to ensure that books on astronomy from the Islamic world were translated and read in China. The Ming government set up a special department for Islamic astronomy (the 'Muslim section' *Hui hui ke* 回回科) within the astronomical bureau. There was no suggestion that this bureau should work in isolation: quite the reverse. In fact, it appears that the possibility of using two different approaches to the heavens in parallel was seen as a useful precautionary move.[8]

[4] *Kai yuan zhan jing*. Chapter 104 of this work includes a translation of the Indian Navagraha 'Nine Upholders' system into Chinese; see also *Xin Tang shu* 28B, 691–692.

[5] A. Beer, Ho Ping-Yü [Ho Peng Yoke], Lu Gwei-djen, J. Needham, E.G. Pulleyblank and G.I. Thompson (1961) 'An 8th-century meridian line: I-Hsing's chain of gnomons and the pre-history of the metric system.' *Vistas in Astronomy* iv: 3–28 and Cullen (1982b).

[6] Sivin (2009).

[7] W. Hartner (1950) 'The astronomical instruments of Cha-ma-lu-ting, their identification, and their relations to the instruments of the observatory of Marāgha.' *Isis* 41, (124): 184–194; see the summary of Hartner's work in Needham and Wang Ling (1959), 372–375.

[8] Sivin (2009), 220–225. The Muslim section was one of four set up in the third year of the Hongwu 洪武 emperor (1370). The other three were the *Tian wen* 天文 ('Heavenly writings/patterns/signs') section, on which see Chapter 1, 16, the Clepsydra section, and the *Da tong li* 大統曆 (Great Concordance system) section, which carried out calculations in accordance with the official astronomical system, on which see below. The translation effort was launched in the 15th year, 1382. See *Ming shi* 明史 (History of the Ming dynasty, 1368–1644). (1739; punctuated edition of 1974). Zhang Tingyu 張廷玉 (1672–1755) et al., Beijing, Zhonghua Press, 31, 516–517.

The Muslim section was still in place near the end of the dynasty, when in 1601 Matteo Ricci (Li Madou 利瑪竇, 1552–1610), the first Jesuit missionary in China, arrived in Beijing. Astronomy, like mathematics, was one of the most successful arenas in which Ricci, and the Jesuits who followed him to China, marketed themselves as possessors of different, and (they claimed) superior, kinds of knowledge. One of the Jesuits' early converts, Xu Guangqi 徐光啟 (1562–1633), was a key agent in this complex process.[9] As a result of the success of the Jesuit effort in the late Ming and early Qing, they not only succeeded in gaining official posts in the imperial structures that managed astronomy, but also contrived to have western astronomical methods entirely substituted for the systems, both indigenous and Muslim, that had been in place on their arrival. The former was—as we have seen—not unprecedented; for foreigners the second certainly was.

Did the Jesuit take-over mark a definitive rupture with the complex of technical, social and intellectual institutions, centred round astronomical systems that had operated in preceding centuries? I would suggest that in significant respects it did not. In the first place, Jesuit success in winning office was gained through processes that we have already seen in operation in the period covered by the present book: competition to show that one system made better predictions than the other. Given the last substantive topic discussed in the previous chapter, it is interesting to note that it was the prediction of solar eclipses that defined one of the main arenas in which the Jesuits and their Chinese supporters competed with those opposing them. In this, however, they were simply joining in a dispute that had already begun in the decades before any qualified Jesuit astronomers arrived in China.[10] To their audience of supporters at home, the Jesuits presented their success in being allowed to apply 'European astronomy' in their work as a hopeful precursor of the ultimate success of the adoption of their religion in China.[11] For the Qing emperors who employed them, they were, however, simply another

[9] See for instance Keizo Hashimoto (1988) *Hsü Kuang-ch'i and astronomical reform: The process of the Chinese acceptance of Western astronomy, 1629–1635*. Osaka, Kansai University Press, and Catherine Jami, Peter M. Engelfriet and Gregory Blue, Eds. (2001), *Statecraft and intellectual renewal in late Ming China: the cross-cultural synthesis of Xu Guangqi (1562–1633)*. Sinica Leidensia, v. 50. Leiden; Boston, Brill.

[10] Lü Lingfeng 吕凌峰 (2007) 'Eclipses and the Victory of European Astronomy in China.' *East Asian Science, Technology, and Medicine*, (27): 127–145, particularly 128–130.

[11] See for instance Verbiest's conclusion in Ferdinand Verbiest and Noël Golvers (1993) *The Astronomia Europaea of Ferdinand Verbiest, S.J. (Dillingen, 1687): text, translation, notes and commentaries*. Nettetal, Steyler Verlag, chapter 28, 132, where he represents 'the Christian Religion in China . . . as a most August Queen who appears publicly with her arms leaning upon Astronomy . . . because she was first introduced into China through Astronomy . . .'.

group of useful foreign technical experts who, like all experts, would have to adapt themselves to the requirements of their official roles. And this they did—preparing annual calendars as had all their predecessors, complete with all the hemerological apparatus required for the divination of 'lucky days', and, with occasional misgivings, providing interpretations of all celestial omens exactly as their predecessors had done.[12] In the end, despite all their hopes and ambitions, the institution reconfigured (or perhaps we might even say 'digested') those who had entered it so triumphantly, and the hopeful propagators of a new world-view found themselves limited to roles which would have been in many ways familiar to their ancient colleagues who had filled the office of Grand Clerk one and a half millennia earlier. Xu Guangqi had anticipated this process to some degree, when he said that it should be possible to:

> 鎔彼方之材質, 入大統之型模
>
> Melt down the substance of their [western] materials, and [cast it] into the mould of the [Chinese] Da tong [system]. (Xu Guangqi 徐光啓 (1562–1633) 1984: 8, 374–375)

Thus the world's most enduring institution concerned with the quantitative analysis, prediction and observation of natural phenomena continued into the early modern age.

After 1836, the last Catholic missionaries left the *Qin tian jian* 欽天監 'The bureau for reverencing heaven' (a term which the Jesuits had preferred to translate by less 'superstitious' titles such as 'the Tribunal of Mathematics'), and asked permission to return to Europe.[13] And with the end of the Qing empire in the early 20th century, the official structures within and around which Sima Qian, Liu Xin, Bian Xin, Jia Kui, Liu Hong and so many others had worked were gone for ever. The modern scientific institutions that replaced them were of a different kind, and had different aims. But even though there is now no emperor to fill the role of 'granting the seasons to the people' as did Yao in remote antiquity, the old

[12] Back in Rome, there was considerable suspicion of possible 'superstitious' activities by Jesuits fulfilling official roles in China, fuelled in one case by a denunciation of this aspect of Adam Schall's work as an imperial astronomer by another Jesuit in China, Gabriel de Maghalães: Jami (2012), 38–40.

[13] Ping-Ying Chang (2015). 'Chinese hereditary mathematician families of the Astronomical Bureau (1620–1850).' City University of New York, PhD, 214–220. As Chang points out, the normal operation of the bureau was maintained by the continued service of the hereditary Chinese post-holders whose families had provided most of its personnel during the Qing, and did so up to its end.

luni-solar seasons are now 'granted' by the flourishing industry of traditional almanac makers and publishers in all countries of East Asia. Earlier in this book I voiced the suspicion that the traditional top-down view of 'season-granting' might not be the whole of the story. It seems to me that people's persistent demand for the 'seasons' in their traditional form when there is no longer any authority to grant them lends some weight to that view.

BIBLIOGRAPHIES

Premodern Bibliography

Bei tang shu chao 北堂書鈔 (Extracts from books copied in the Northern Hall). (1966 repr. of 1877 edn. based on traced Song edition.) 虞世南, Yu Shinan, (c. 639 CE). Taibei, Wen hai chu ban she 文海出版社.

Chu xue ji 初學記 (Records for initial study). (1962 repr.). Xu Jian 徐堅 (c.700 CE), Beijing, Zhonghua Press.

Chun qiu fan lu 春秋繁露 (Luxuriant dew of the Spring and Autumn annals). (1592 woodblock edition, photographic reprint 1978). Dong Zhongshu 董仲舒 (c. 179–c. 104 BCE), Kyōto, Chūbun shuppansha.

Claudii Ptolemaei opera quae exstant omnia. Vol.1, Pts I and II, *Syntaxis mathematica* (Greek text of Ptolemy's Almagest) edidit J.L. Heiberg. (1898). Ptolemy, Claudius, Lipsiae, In aedibus B.G. Teubneri, 1898–1903.

Clouds. (1998). Aristophanes (c. 446–c. 386 BCE) and Jeffrey Henderson (tr. and ed.), Cambridge, MA; London, Harvard University Press.

Diodorus Siculus (fl. 65–30 BCE) and R. M. Geer (tr.) (1947). *Diodorus of Sicily: with an English translation* (Loeb). London.

Dong Han hui yao 東1漢會要 (Collected essentials of Eastern Han [institutions]). (1978). Xu Tianlin 徐天麟 (*jin shi* 1205 CE), Shanghai, Shanghai gu ji chu ban she 上海古籍出版社.

Geminos's introduction to the phenomena: a translation and study of a Hellenistic survey of astronomy. (2006). Geminos (c. 70 BCE), tr. James Evans and J. L. Berggren, Princeton, NJ and Oxford, Princeton University Press.

Guo yu 國語 *(Discourses from the states)*. (1922 onwards). Uncertain authorship (compiled before late fourth century BCE). Shanghai, Commercial Press, Si bu cong kan 四部叢刊 series.

Han shu 漢書 *(History of Western Han dynasty)*. (probably completed in present form c.110 CE, punctuated edition of 1962). Ban Gu 班固 (32–92 CE), Beijing, Zhonghua Shuju.

Herodotus (c. 484–c. 425 BCE) (1920). *Histories, Vol 1:Books I and II* (Loeb). New York.

Hipparchus (1894). *Ipparchou tōn Aratou kai Eudoxou phainomenōn exēgēseōs biblia tria = Hipparchi in Arati et Eudoxi phaenomena commentariorum libri tres/ad codicum fidem recensuit, germanica interpretatione et commentariis instruxit Carolus Manitius.* (Hipparchus and Eudoxus on the Phenomena). Lipsiae: in aedibus B.G. Teubneri, 1894.

Hippocrates (of Kos, born c. 460 BCE), tr. and int. G. E. R. Lloyd (1983). *Hippocratic writings.* Harmondsworth, Penguin.

Hou Han ji 後漢紀 (Annals of the Eastern Han dynasty). (1922 onwards). Yuan Hong 袁宏 (328–376 CE), Shanghai, Commercial Press, Si bu cong kan 四部叢刊 series.

Hou Han shu 後漢書 (History of Eastern Han dynasty). (Main text completed c. 450 CE, monographs by Sima Biao 司馬彪 (c. 240–c. 306 CE) added later, punctuated edition of 1963). Fan Ye 范曄 (398–445 CE), Beijing, Zhonghua Shuju.

Huai nan hong lie ji jie 淮南鴻烈集解 (Collected commentaries on the great work of [the prince of] Huai nan). (1989). Liu An 劉安 (c. 179–122 BCE), Beijing, Zhong Hua press.

Jin shu 晉書 (History of the Jin dynasty, 265–419 CE). (c. 648 CE. Punctuated edition of 1974). Fang Xuanling 房玄齡 (579–648 CE), Beijing, Zhonghua Press.

Kai yuan zhan jing 開元占經 (Divinatory manual of the Kaiyuan reign period). (c. 725 CE). Qutan Xida 瞿曇悉達, Taipei, Si ku quan shu MS edn. c. 1782, photorepr. Commercial Press 1983–1986.

Lü shi chun qiu 呂氏春秋 (Mr. Lü's Spring and Autumn [Annals]). (completed 239 BCE, 1922 reprint of Ming woodblock edition). Lü Buwei 呂不韋 (?–235 BCE), Shanghai, Commercial Press, Si bu cong kan 四部叢刊 collection, 420–4.

Lun heng 論衡 (Discourses weighed in the balance). (1922 onwards, reprint of Ming woodblock edition). Wang Chong 王充 (27–c. 100 CE), Commercial Press, Si bu cong kan 四部叢刊 series.

Ming shi 明史 (History of the Ming dynasty, 1368–1644). (1739; punctuated edition of 1974). Zhang Tingyu 張廷玉 (1672–1755) et al., Beijing, Zhonghua Press.

Natural history (vol 1). (1938). Pliny (Gaius Plinius Secundus 23–79 CE), H. Rackham tr, et al., (Loeb) London.

Qian fu lun 潛夫論 (Comments of a recluse) (after 147 CE). Wang Fu 王符 (c.85–163 CE), Taipei, Si ku quan shu MS edn. c. 1782, photorepr. Commercial Press 1983–1986.

Republic. (2013). Plato (428/427 or 424/423–348/347 BCE), Cambridge, MA; London, Harvard University Press.

Shi ji 史記 (Records of the Historian). (completed c. 90 BCE, punctuated edition 1959). Sima Qian 司馬遷 (c. 145–86 BCE), Beijing, Zhonghua Shuju. Tr. Chavannes et al. (2015).

Shi san jing zhu shu 十三經註疏 (The thirteen classics with commentaries and subcommentaries). (1973 reprint of original of 1815). Ruan Yuan 阮元 (1764–1849), Taipei, Yiwen Press.

Tai ping yu lan 太平御覽 (Imperial readings of the Taiping reign period). (983 CE, 1960 reprint of Song edition). Li Fang 李昉 et al., Beijing, Zhong Hua.

Timaeus; Critias; Cleitophon; Menexenus; Epistles, with an English translation by R.G. Bury. (1989). Plato (428/427 or 424/423–348/347 BCE), Cambridge, MA, Harvard University Press.

Wen xuan 文選 (Selected literature). (1965 reprint of 1936 edition). Xiao Tong 蕭統 (501–531 CE), Hong Kong, Commercial Press.

Yuan shi 元史 (History of Yuan dynasty). (completed 1370, punctuated edition of 1976). Song Lian 宋濂 (1310–1381), Beijing, Zhonghua Press.

Modern Bibliography

Aaboe, Asger (1972). 'Remarks on the theoretical treatment of eclipses in antiquity.' *Journal for the History of Astronomy* **3**(2): 105–18.

Allan, Sarah (2003). 'The Great One, water, and the Laozi: new light from Guodian.' *T'oung Pao* **89**(4–5): 237–85.

Anhui cultural relics working group 安徽省文物工作队 (1978). '阜阳双古堆西汉汝阴侯墓发掘简报 Fu yang shuang gu dui Xi Han Ru Yin hou mu fa jue jian bao (An outline report on excavations at the Western Han tomb of the marquis of Ru Yin at Shuang Gu Dui, Fu Yang).' *Wen wu* (8): 12–30 & 98–9.

Anon. (1978). 'Han Wei Luo Yang cheng nan jiao de ling tai yi zhi 汉魏洛阳城南郊的灵台遗址 (The remains of the observatory to the south of the Han-Wei city wall of Luoyang).' *Wen wu* (1): 54–7 & 73–5.

Anon. (1989). *Zhong guo gu dai tian wen wen wu lun ji* 中国古代天文文无论集 (Collected articles on cultural relics relating to ancient Chinese astronomy). Beijing, Wenwu publishing house.

Beer, A., Ho Ping-Yü [Ho Peng Yoke], et al. (1961). 'An 8th-century meridian line: I-Hsing's chain of gnomons and the pre-history of the metric system.' *Vistas in Astronomy* **iv**: 3–28.

Bickerman, E. J. (1980). *Chronology of the ancient world*. London, Thames and Hudson.

Bielenstein, Hans (1980). *The bureaucracy of Han times*. Cambridge [etc.], Cambridge University Press.

Bodnár, István M. (2007). *Oenopides of Chius: A survey of the modern literature with a collection of the ancient testimonia (preprint 227)*. Berlin, Max-Planck-Institut Für Wissenschaftsgeschichte.

Britton, J.P. (2007). Calendars, intercalations and year-lengths in Mesopotamian astronomy. *Calendars and Years: Astronomy and Time in the Ancient Near East*. J.M. Steele. Oxford, Oxbow Books: 115–32.

Brown, David (2000). *Mesopotamian planetary astronomy-astrology*. Groningen, Styx.

Bullock, Jeffrey S. (2011). *Yang Xiong, philosophy of the Fa yan: a Confucian hermit in the Han imperial court*. Highlands, NC, Mountain Mind Press.

Chang, Hasok (2011). 'How historical experiments can improve scientific knowledge and science education: the cases of boiling water and electrochemistry.' *Science & Education* **20**(3–4): 317–41.

Chang, Leo S. and Yu Feng (1998). *The four political treatises of the Yellow Emperor: original Mawangdui texts with complete English translations and an introduction*. Honolulu, University of Hawai'i Press.

Chang, Ping-Ying (2015). 'Chinese hereditary mathematician families of the astronomical bureau (1620–1850).' City University of New York, PhD.

Chavannes, Édouard et al. tr. (2015). *Les mémoires historiques de Se-ma Ts'ien* (translation of *Shi ji* 史記 'Records of the Historian', completed c. 90 BCE). Paris, You Feng.

Chemla, Karine and Shuchun Guo (2004). *Les neuf chapitres: le classique mathématique de la Chine ancienne et ses commentaires*. Paris, Dunod.

Chen Jiujin 陳久金 and Chen Meidong 陈美东 (1989). Cong Yuan Guang li pu ji Ma Wang Dui bo shu tian wen zi liao shi tan Zhuan Xu li wen ti 从元光历谱及马王堆帛书试探颛顼历问题 (An investigation of the Zhuan Xu system on the basis of the Yuanguang almanac and the Mawangdui silk manuscript). *Zhong guo gu dai tian wen wen wu lun ji* 中国古代天文文无论集 *(Collected articles on cultural relics relating to ancient Chinese astronomy)*. Anon. Beijing, Wenwu Publishing House: 83–103.

Chen, Meidong 陈美东 (1995). *Gu li xin tan* 古历新探 *(New investigations of old astronomical systems)*. Shenyang, Liaoning educational press.

Chen, Meidong 陈美东 (1989). Shi lun Xi Han lou hu de ruo gan wen ti 试论西汉漏壶的若干问题 (On some questions relating to Western Han clepsydras). *Zhong guo gu dai tian wen wen wu lun ji* 中国古代天文文物论集 *(Collected articles on astronomical artefacts from ancient China)*. Chinese Academy of Social Sciences Institute of Archaeology. Beijing, Wenwu Press: 137–44.

Chen Wei (2016). Guan yu Qin Han 'zhi ri' de xin kao cha 关于秦汉'质日'的新考察 (A new examination of the expression *zhi ri* in Qin and Han times). International Conference on Mantic Arts in China, Erlangen, Germany.

Chen Wei 陳偉, Ed. (2014). *Qin jian du heji* 秦簡牘合集 *(Compendium of Qin documents on bamboo strips and planches)*. Wuhan, Wuhan University.

Chinese Academy of Social Sciences Institute of Archaeology (1980). *Zhong guo gu dai tian wen wen wu tu ji* 中国古代天文文物图集 *(Collected illustrations of objects connected wth astronomy in ancient China)*. Beijing, Wenwu press.

Couplet, Philippe, Prospero Intorcetta, et al. (1687). *Confucius Sinarum philosophus, sive scientia sinensis latine exposita: Studio & opera Prosperi Intorcetta (...) Adjecta est tabula chronologica sinicæ monarchiæ ab hujus exordio ad hæc usque tempora*. Parisiis, Apud D. Horthemels.

Cullen, Christopher (1976). 'A Chinese Eratosthenes of the flat earth: a study of a fragment of cosmology in *Huai nan tzu*.' *Bulletin of the School of Oriental and African Studies* **39**(1): 106–27.

Cullen, Christopher (1977). 'Cosmographical discussions in China from early times up to the T'ang Dynasty.' School of Oriental and African Studies, PhD. Text available online through Academia.edu.

Cullen, Christopher (1980). 'Joseph Needham on Chinese astronomy.' *Past & Present* 87: 39–53.

Cullen, Christopher (1981). 'Some further points on the shih.' *Early China* **6**: 31–46.

Cullen, Christopher (1982a), September 1982. Early Chinese measurements of right ascension before the armillary sphere. First International Conference on the History of Chinese Science, Louvain, Belgium.

Cullen, Christopher (1982b). 'An eighth century Chinese table of tangents.' *Chinese Science* **5**: 1–33.

Cullen, Christopher (1993). 'Motivations for scientific change in Ancient China: Emperor Wu and the Grand Inception astronomical reforms of 104 BC.' *Journal for the History of Astronomy* **24**(3): 185–203.

Cullen, Christopher (1996). *Astronomy and mathematics in ancient China: the Zhou bi suan jing*. Cambridge/New York, Cambridge University Press.

Cullen, Christopher (2000). 'Seeing the appearances: ecliptic and equator in the Eastern Han.' *Zi ran ke xue shi yan jiu* 自然科學史研究 *(Studies in the History of Natural Sciences)* **xix**(4): 352–82.

Cullen, Christopher (2002). 'The first complete Chinese theory of the moon: the innovations of Liu Hong c. AD 200.' *Journal for the History of Astronomy* **33**: 1–24.

Cullen, Christopher (2004). 'The birthday of the Old Man of Jiang County and other puzzles: work in progress on Liu Xin's *Canon of the Ages*.' *Asia Major* **xiv**(2): 27–70.

Cullen, Christopher (2007a). 'Huo Rong's observation programme of AD 102 and the *Han li* solar table.' *Journal for the History of Astronomy* **38**(1): 75–98.

Cullen, Christopher (2007b). 'Actors, networks and "disturbing spectacles" in institutional science: 2nd century Chinese debates on astronomy.' *Antiquorum Philosophia* **1**: 237–68.

Cullen, Christopher (2009). People and numbers in early imperial China. *Oxford Handbook of the History of Mathematics*. Eleanor Robson and Jackie Stedall. Oxford, Oxford University Press: 591–618.

Cullen, Christopher (2011a). 'Wu xing zhan 五星占 "Prognostics of the Five Planets".' *SCIAMVS* **12**: 193–249.

Cullen, Christopher (2011b). 'Understanding the planets in Ancient China: prediction and divination in the Wu xing zhan.' *Early Science and Medicine* **16**: 218–51.

Cullen, Christopher (2011c), 11 October. Daily life and cosmic time: excavated calendrical documents and their significance. Seminar presentation, The Fitzwilliam Museum conference on Life and Afterlife of Han China, Cambridge.

Cullen, Christopher (2014), 14 January. Étoiles et saisons: peut-on reconstituer l'observation du ciel dans la Chine ancienne? Seminaire: Histoire des sciences, des techniques et de la médecine en Asie orientale, EHESS, Paris.

Cullen, Christopher (2015). 'Lu Jiuci yu Chen Qi de dui hua he zao qi de Zhong Guo shu xue shi 鲁久次与陈起的对话和早期的中国数学史 "The dialogue of Lu Jiuci and Chen Qi, and the early history of mathematics in China".' *Zi ran ke xue shi yan jiu 自然科学史研究 (Studies in the History of Natural Science)* **34**(2): 254–7.

Cullen, Christopher (2017). *The foundations of celestial reckoning: three Ancient Chinese astronomical systems*. London, Routledge.

Cullen, Christopher and Anne S. L. Farrer (1983). 'On the term "Hsüan Chi" and the flanged trilobate jade discs.' *Bulletin of the School of Oriental and African Studies, University of London* **46**(1): 52–76.

De Crespigny, Rafe (1976). *Portents of protest in the later Han dynasty: the memorials of Hsiang K'ai to Emperor Huan in 166 A.D.* Canberra, Australian National University Press in association with the Faculty of Asian Studies.

De Crespigny, Rafe (2007). *A biographical dictionary of Later Han to the Three Kingdoms (23–220 AD)*. Leiden, Brill.

Diels, Hermann (1903). *Die Fragmente der Vorsokratiker: Griechisch und Deutsch*. Berlin, Weidmann.

Dull, Jack L. (1966). 'A historical introduction to the apocryphal, *ch'an-wei*, texts of the Han Dynasty.' University of Washington, PhD.

Eberhard, Wolfram (1970). *Sternkunde und Weltbild im alten China: gesammelte Aufsätze*. Taipei, Chinese Materials and Research Aids Service Center.

Eberhard, Wolfram and Rolf Mueller (1936). 'Contributions to the astronomy of the Han Period III: Astronomy of the Later Han Period.' *Harvard Journal of Asiatic Studies* **1**(2): 194–241.

Evans, James (1998). *The history and practice of ancient astronomy*. New York/Oxford, Oxford University Press.

Feng Yunpeng 馮雲鵬 and Feng Yunyuan 馮雲鵷 (1929 (first edn. 1823)). *Jin shi suo 金石索 'Repertory of [inscriptions on] metal and stone'*. Shanghai, Commercial Press.

Fermor, J. and J. M. Steele (2000). 'The design of Babylonian waterclocks: Astronomical and experimental evidence.' *Centaurus* **42**(3): 210–22.

Gingerich, Owen (2005). *The book nobody read: chasing the revolutions of Nicolaus Copernicus*. London, Arrow.

Goldin, Paul R. (2013). 'Heng Xian and the problem of studying looted artifacts.' *Dao* **12**(2): 153–60.

Golvers, Noël (1993). *The Astronomia Europaea of Ferdinand Verbiest, S.J. (Dillingen, 1687): Text, translation, notes and commentaries*. Nettetal, Steyler Verlag.

Goodrich, Chauncey S. (1957). 'The reign of Wang Mang: Hsin or New?' *Oriens*: 114–18.

Graham, A. C. (1989). *Disputers of the Tao: philosophical argument in ancient China*. La Salle, Ill., Open Court.

Guo Shuchun 郭書春 (2004). *Hui jiao jiu zhang suan shu* 匯校九章算術 *(Comprehensively collated [edition of the text] Mathematical procedures in nine sections]*. Shenyang 瀋陽, Liao ning jiao yu chu ban she 遼寧教育出版社 (Liaoning educational publishers).

Guo Shuchun 郭書春 and Liu Dun 劉鈍, Eds. (2001). *Suan jing shi shu* 算經十書 *(The ten mathematical classics)*. Taibei, Jiu zhang chu ban she 九章出版社.

Guthrie, W. K. C. (1978). *A history of Greek philosophy. Volume 1, The earlier Presocratics and the Pythagoreans*. Cambridge, Cambridge University Press.

Han Wei 韩巍 (2015). 'Bei Da cang Qin jian "Lu Jiuci wen shu yu Chen Qi" 北大藏秦简《鲁久次问数于陈起》初读 (A brief study of the text entitled "Lu Jiuci asks Chen Qi about numbers" on the Qin bamboo slips collected by Peking University).' *Bei jing da xue xue bao: zhe xue she hui ke xue ban* 北京大学学报:哲学社会科学版》 *(Journal of Peking University: Humanities and Social Sciences)* (2): 29–36.

Hannah, Robert (2005). *Greek and Roman calendars: constructions of time in the classical world*. London, Duckworth.

Harper, Donald and Marc Kalinowski, Eds. (2017, forthcoming). *Books of fate and popular culture in Early China. The daybook manuscripts of the warring states, Qin, and Han*. Leiden, Brill.

Harper, Donald John (1999). Warring States natural philosophy and occult thought. *The Cambridge history of ancient China: from the origins of civilization to 221 B.C.* Michael Loewe and Edward L. Shaughnessy. Cambridge, UK/New York, Cambridge University Press: 813–84.

Hartner, W. (1950). 'The astronomical instruments of Cha-ma-lu-ting, their identification, and their relations to the instruments of the observatory of Marāgha.' *Isis* **41**(124): 184–94.

Hashimoto, Keizo (1988). *Hsü Kuang-ch'i and astronomical reform: The process of the Chinese acceptance of Western astronomy, 1629–1635*. Osaka, Kansai University Press.

Heath, T. L. (1925, second edition of original 1908 publication, revised with additions; repr. Dover, 1956, New York.). *The thirteen books of Euclid's elements, translated from the text of Heiberg*. Cambridge.

Holford-Strevens, Leofranc (2005). *The history of time: a very short introduction*. Oxford, Oxford University Press.

Hucker, Charles O. (1985). *A dictionary of official titles in imperial China*. Stanford, Stanford University Press.

Hunger, Hermann (2002). *Astrological reports to Assyrian kings, state archives of Assyria, Vol VIII* Helsinki, Helsinki University Press.

Hunger, Hermann and David Edwin Pingree (1989). *MUL.APIN: an astronomical compendium in cuneiform*. Horn, Austria, F. Berger.

Jami, Catherine (2012). *The Emperor's new mathematics: Western learning and imperial authority in China during the Kangxi reign (1662–1722)*. Oxford, Oxford University Press.

Jami, Catherine, Peter M. Engelfriet, et al., Eds. (2001). *Statecraft and intellectual renewal in late Ming China: the cross-cultural synthesis of Xu Guangqi (1562–1633)*. Sinica Leidensia, v. 50. Leiden; Boston, Brill.

Jones, Alexander (1999). *Astronomical papyri from Oxyrhynchus*: (P. Oxy. 4133-4300a) / edited with translations and commentaries by Alexander Jones. Philadelphia, Philadelphia: American Philosophical Society, 1999.

Jones, Alexander (2002). 'Eratosthenes, Hipparchus, and the obliquity of the ecliptic.' *Journal for the History of Astronomy* **33**(1): 15–19.

Kalinowski, Marc (2004). 'Fonctionalité calendaire dans les cosmogonies anciennes de la Chine.' *Etudes chinoises* **23**: 87–122.

Kalinowski, Marc and Phyllis Brooks (1999). 'The Xingde 刑德 texts from Mawangdui.' *Early China* **23/24**: 125–202.

Kawahara Hideki 川原秀城 (1989). San tong li yu Liu Xin de shi jie guan 三統曆與劉歆的世界觀 'The Triple Concordance system and the world-view of Liu Xin'. *Chūgoku kodai kagakushi ron* 中国古代科學史論 *'Articles on ancient Chinese science'*. Yamada Keiji 山田慶二. Kyoto, Kyoto University Research Institute for Humanistic Studies: 121–38.

Kepler, Johannes (1621). *Prodromus dissertationum cosmographicarum, continens mysterium cosmographicum . . .* Francofurti : Recusus typis Erasmi Kempferi, sumptibus Godefridi Tampachii.

Kepler, Johannes and Tycho Brahe (1627). *Tabulae Rudolphinae, quibus astronomicae scientiae . . . restauratio continentur; a . . . Tychone . . . primum animo concepta et destinata . . . Tabulas ipsas . . . continuavit . . . perfecit, absolvit; adque causarum et calculi perennis formulam traduxit Joanne Keplerus*. Ulmae, Typis Jonae Saurii.

Knoblock, John and Jeffrey K. Riegel (2000). *The annals of Lü Buwei = [Lü shi chun qiu]: a complete translation and study*. Stanford, Stanford University Press.

Koyré, Alexandre (1973, first published 1961 in French). *The astronomical revolution: Copernicus, Kepler, Borelli*. London, Methuen.

Kühnert, Franz (1888). 'Das Kalendarwesen bei den Chinesen.' *Österreichische Monatsschrift für den Orient* (8): 111–16.

Lagerwey, John and Marc Kalinowski (2009). *Early Chinese religion*. Leiden/Boston, Brill.

Latour, Bruno (1987). *Science in action: how to follow scientists and engineers through society*. Cambridge, Mass., Harvard University Press.

Legge, James (1872a). *The Chinese Classics, Volume V: The Ch'un Ts'ew, with the Tso Chuen. Part 1: Dukes Yin, Hwan, Chwang, Min, He, Wan, Suen and Ch'ing, and the Prolegomena*. Hong Kong and London.

Legge, James (1872b). *The Chinese Classics, Volume V: The Ch'un Ts'ew, with the Tso Chuen. Part 2: Dukes Seang, Ch'aou, Ting, and Gae, with Tso's Appendix; and the Indexes*. Hong Kong and London.

Li Chongzhi 李崇智 (2001). *Zhong guo li dai nian hao kao* 中國曆代年號考 *(An investigation of reign names in China through the ages)*. Beijing, Zhonghua Shuju.

Li Rui 李銳 (1768–1817) (1993). Li shi yi shu 李氏遺書 (Transmitted works of Mr Li). *Zhong guo ke xue ji shu dian ji tong hui, tian wen juan* 中國科學技術典籍通彙天文卷 *(Compendium of classic texts of Chinese science and technology, astronomical section)*. Bo Shuren 薄樹人. Beijing, Henan educational press. **2**: 701–818.

Li Zhonglin 李忠林 (2012). 'Qin zhi Han chu (qian 246 zhi qian 104) li fa yan jiu—yi chu tu li jian wei zhong xin 秦至汉初（前 246 至前104）历法研究 － 以出土历简为中心 (Researches on astronomical systems from Qin to early Han (246 BCE to 104 BCE)—centering on excavated calendrical bamboo slips).' *Zhong guo shi yan jiu* 中國史研究 *'Studies in Chinese history'* **2**(134): 17–70.

Li Ling (1995). 'An archaeological study of Taiyi (Grand One) worship.' *Early medieval China* **2**: 1.

Li Yan and Du Shiran (1987). *Chinese mathematics: a concise history*. Oxford, Clarendon.

Libbrecht, Ulrich (1973). *Chinese mathematics in the thirteenth century: the Shu-shu chiu-chang of Ch'in Chiu-shao*. Cambridge, MIT Press.

Lien, Y. Edmund (2012). 'Zhang Heng's *Huntian yi zhu* revisited.' *T'oung Pao* **98**(1–3): 31–64.

Liu Lexian 刘乐贤 (2011). 'Qin Han li ri de nei rong ji gong yong 秦汉曆日的内容及功用 (On the contents and uses of day-sequences of the Qin and Han).' *Faguo Hanxue* 法國漢學 *'French sinology'* **14**: 351–86.

Liu Tan 劉坦 (1955). *Lun xing sui ji nian* 論星歲紀年 *(On dating years by the Jupiter cycle)* Beijing, Science Press.

Lloyd, G. E. R. and Nathan Sivin (2002). *The way and the word: science and medicine in early China and Greece*. New Haven; London, Yale University Press.

Loewe, M. (1960). 'The orders of aristocratic rank of Han China.' *T'oung Pao* **48**(1/3): 97–174.

Loewe, M.A.N. (1961). 'The measurement of grain during the Han period.' *T'oung Pao* **49**: 64–95.

Loewe, M.A.N. (1993). *Early Chinese texts: a bibliographical guide*. Cambridge, Cambridge University Press.

Loewe, Michael (1979). *Ways to paradise: the Chinese quest for immortality*. London, Allen & Unwin.

Loewe, Michael (1994). *Divination, mythology and monarchy in Han China*. Cambridge, Cambridge University Press.

Loewe, Michael (2000). *A biographical dictionary of the Qin, former Han and Xin periods (221 BC–AD 24)*. Leiden, Brill.

Loewe, Michael (2006). *The government of the Qin and Han Empires, 221 BCE–220 CE*. Indianapolis and Cambridge, Hackett.

Loewe, Michael (2015). Liu Xiang and Liu Xin. *Chang'an 26 BCE: an Augustan Age in China*. M. Nylan et al. Seattle, University of Washington: 369–89.

Loewe, Michael and Edward L. Shaughnessy (1999). *The Cambridge history of ancient China: from the origins of civilization to 221 B.C.* Cambridge, UK/New York, Cambridge University Press.

Lü Lingfeng 吕凌峰 (2007). 'Eclipses and the victory of European astronomy in China.' *East Asian Science, Technology, and Medicine* (27): 127–45.

Major, John S., with an appendix by Christopher Cullen (1993). *Heaven and earth in early Han thought: chapters three, four and five of the Huainanzi* Albany, State University of New York Press.

Mansvelt Beck, B. J. (1990). *The treatises of later Han: their author, sources, contents, and place in Chinese historiography*. Leiden, Brill.

Martzloff, Jean-Claude (2009). *Le calendrier chinois: structure et calculs, 104 av. J.-C.-1644. Indétermination céleste et réforme permanente. La construction chinoise officielle du temps quotidien discret à partir d'un temps mathématique caché, linéaire et continu*. Paris, H. Champion.

Maspero, Henri (1950, but written c. 1932–1939). 'L'astronomie dans la Chine ancienne: histoire des instruments et des découvertes.' *Mélanges posthumes, volume III Études historiques*. P. Demiéville. Paris, Musée Guimet: 13–34.

Meeus, Jean (1991). *Astronomical algorithms*. Richmond, Virginia.

Meeus, Jean and Denis Savoie (1992). 'The history of the tropical year.' *Journal of the British Astronomical Association* **102**(1): 40–2.

Morgan, Daniel P. (2013). 'Knowing heaven: astronomy, the calendar, and the sagecraft of science in Early Imperial China.' University of Chicago, PhD.

Nylan, Michael, Vankeerberghen, Griet, et al., Eds., (2015). *Chang'an 26 BCE: an Augustan age in China*. Seattle, University of Washington Press.

Nakayama, Shigeru (1969). *A history of Japanese astronomy. Chinese background and Western impact*. Cambridge, MA, Harvard University Press.

Needham, Joseph and Wang Ling (1956). *Science and civilisation in China, volume 2: History of scientific thought*. Cambridge, Cambridge University Press.

Needham, Joseph and Wang Ling (1959). *Science and civilisation in China, volume 3: Mathematics and the sciences of the heavens and the earth*. Cambridge, Cambridge University Press.

Needham, Joseph, Ling Wang, and K. G. Robinson (1962). *Science and civilisation in China, volume 4: Physics and physical technology. Part 1: Physics*. Cambridge, Cambridge University Press.

Needham, Joseph, Ling Wang, and Derek J. de Solla Price (1986). *Heavenly clockwork: the great astronomical clocks of medieval China*. Cambridge, UK/New York, Cambridge University Press.

Needham, Joseph, Lu Gwei-djen, John H. Combridge, and John Major (1986). *The hall of heavenly records: Korean astronomical instruments and clocks, 1380–1780*. Cambridge, Cambridge University Press.

Neugebauer, O. (1962). 'Thâbit ben Qurra "on the solar year" and "on the motion of the eighth sphere".' *Proceedings of the American Philosophical Society* **106**(3): 264–99.

Neugebauer, Otto (1969). *The exact sciences in antiquity*. New York/London, Dover Publications; Constable.

Neugebauer, Otto Eduard (1975). *A history of ancient mathematical astronomy*. Berlin/New York, Springer-Verlag.

Nienhauser, William H. (1994–). *The grand scribe's records*. Bloomington, Indiana University Press.

Nylan, Michael (2009). Classics without canonization, learning and authority in Qin (221–210 BC) and Han (206 BC–AD 220). *Early Chinese Religion, Part 1*. John Lagerwey and Marc Kalinowski. Leiden/Boston, Brill: 721–776.

Pankenier, David W. (2013). *Astrology and cosmology in early China: conforming earth to heaven*. New York, Cambridge University Press.

Pankenier, David W. (2012). 'On the reliability of Han dynasty solar eclipse records.' *Journal of Astronomical History and Heritage* **15**(3): 200–12.

Pokora, Timoteus (1975). *Hsin-lun (New treatise), and other writings by Huan T'an (43 BC–28 AD): an annotated translation with index*. Ann Arbor, Center for Chinese Studies, University of Michigan.

Qin ding shu jing tu shuo 欽定書經圖說 *(The Book of Documents illustrated and explained: imperially commissioned)*. (1905). Sun Jianai 孫家鼐 et al. (edited) and Zhan Xiullin 詹秀林 et al. (illustrations), Beijing, Da xue tang bian shu ju 大學堂編書局.

Qu Anjing 曲安京 (2005). *Zhong guo li fa yu shu xue* 中国历法与数学 *(Chinese astronomical systems and mathematics)*. Beijing, Science Press.

Qu Anjing 曲安京 (2008). *Zhong guo shu li tian wen xue* 中国数理天文学 *(Chinese mathematical astronomy)*. Beijing, Science Press.

Robinson, Andrew (2016). 'Use a toad to catch a quake.' *New Scientist* **232** (3102 December 3): 42–43.

Robinson, M. (2009). 'Ardua et astra: on the calculation of the dates of the rising and setting of stars.' *Classical Philology* **104**: 354–75.

Rochberg, Francesca (2004). *The heavenly writing: divination, horoscopy, and astronomy in Mesopotamian culture*. Cambridge, Cambridge University Press.

Samuel, Alan Edouard (1972). *Greek and Roman chronology; calendars and years in classical antiquity*. München, Beck.

Sanft, Charles (2011). 'Edict of monthly ordinances for the four seasons in fifty articles from 5 CE: introduction to the wall inscription discovered at Xuanquanzhi, with annotated translation.' *Early China* (32, dated 2008–2009, actually appeared 2011): 125–208.

Schuessler, Axel (2007). *ABC etymological dictionary of Old Chinese*. Honolulu, University of Hawai'i Press.

Schuessler, Axel and Bernhard Karlgren (2009). *Minimal Old Chinese and later Han Chinese: a companion to Grammata serica recensa*. Honolulu, University of Hawai'i Press.

Shapin, Steven and Simon Schaffer (1985). *Leviathan and the air-pump: Hobbes, Boyle, and the experimental life*. Princeton/Oxford, Princeton University Press.

Shaughnessy, Edward L. (1991). *Sources of Western Zhou history: inscribed bronze vessels*. Berkeley/Oxford, University of California Press.

Shen, Kangshen, J. N. Crossley, and Anthony W. C. Lun (1999). *The nine chapters on the mathematical art: companion and commentary*. Oxford, Oxford University Press.

Shi Yunli 石云里 and Xing Gang 邢钢 (2006). 'Zhong guo Han dai ri yue shi ji suan ji qi dui xing zhan guan de ying xiang 中国汉代的日月食计算及其对星占观的影响 (The calculation and divination of luni-solar eclipses of the Han Dynasty and its impact on the astrological thoughts).' *Zi ran bian zheng fa tong xun* 自然辩证法通讯 *(Journal of the Dialectics of Nature)* **28**(2): 79–85.

Sivin, Nathan (1969). *Cosmos and computation in early Chinese mathematical astronomy*. Leiden, E. J. Brill.

Sivin, Nathan (2009). *Granting the seasons: the Chinese astronomical reform of 1280, with a study of its many dimensions and a translation of its records*. New York, Springer.

Smart, W. M. and Robin M. Green (1979 (reprint of 6th edition 1977)). *Textbook on spherical astronomy*. Cambridge/New York, Cambridge University Press.

Standaerdt, Nicolas (2012). 'Jesuit accounts of Chinese history and chronology and their Chinese sources.' *East Asian Science, Technology and Medicine* (35): 11–87.

Steele, J. M. (1997). 'Solar eclipse times predicted by the Babylonians.' *Journal for the History of Astronomy* **28**(2): 133–9.

Steele, John M. (2000b). *Observations and predictions of eclipse times by early astronomers*. Dordrecht/Boston, Kluwer Academic Publishers.

Steele, John M. (2000a) 'Eclipse prediction in Mesopotamia.' *Archive for History of Exact Sciences* **54**(5): 421–54.

Stephenson, F. R. and K. K. C. Yau (1992). 'Astronomical records in the Chun-Chiu chronicle.' *Journal for the History of Astronomy* **23**: 31.

Stephenson, F. R. and L. V. Morrison (1995). 'Long-term fluctuations in the Earth's rotation: 700 BC to AD 1990.' *Philosophical Transactions of the Royal Society of London. Series A: Physical and Engineering Sciences* **351**(1695): 165–202.

Stephenson, F. R. and L. J. Fatoohi (1997). 'Thales's prediction of a solar eclipse.' *Journal for the History of Astronomy* **28**(4): 279–82.

Su Junlin 苏俊林 (2010). 'Guan yu "zhi ri" jian de ming cheng yu xing zhi 关于'质日"简的名称与性质 (The name and nature of the '*zhi ri*' bamboo strips).' *Journal of Hunan University (Social Science Edition)* **4**: 17–22.

Sun Zhanyu 孙占宇 (2010). 'Zhan guo Qin Han shi qi jian chu shu tao lun 战国秦汉时期建除术讨论 (Discussion on Jian-Chu divination of the Warring States and Qin-Han Dynasty).' 西安财经学院学报 *Xi'an Caijing xueyuan xuebao (Journal of Xi'an University of Finance and Economics)* **23**(5): 88–93.

Sun, Xiaochun and Jacob Kistemaker (1997). *The Chinese sky during the Han: constellating stars and society*. Leiden/New York, Brill.

Tang Quan 唐泉 (2011). *Ri shi yu shi cha* 日食与视差 (Solar eclipses and parallax). Beijing, Science Press.

Tchang, Mathias (1905 (repr. Taipei 1967)). *Synchronismes chinois: chronologie complète et concordance avec l'ère chrétienne de toutes les dates concernant l'histoire de l'Extrême Orient*. Shanghai, Imprimerie de la mission Catholique.

Teboul, Michel (1983). *Les premières théories planétaires chinoises*. Paris, Collège de France Institut des hautes études chinoises.

Toomer, G. J. (1974a). Meton. *Dictionary of scientific biography*. Frederic Lawrence Holmes and Charles Coulston Gillispie. New York, Scribner's: 337–9.

Toomer, G. J. (1974b). 'Hipparchus on the distances of the Sun and Moon.' *Archive for History of Exact Sciences* 14: 126.

Toomer, G. J. (1998). *Ptolemy's Almagest*. Princeton, N.J., Princeton University Press.

Tsien, Tsuen-hsuin (2004). *Written on bamboo & silk: the beginnings of Chinese books & inscriptions*. Chicago; London, University of Chicago Press.

Tuckerman, Bryant (1962). *Planetary, lunar and solar positions 601 BC to AD 1, at five-day and ten-day intervals*. Philadelphia, American Philosophical Society.

Twitchett, Denis (1979). *The Cambridge history of China. Vol 3, Sui and T'ang China, 589–906, Part I*. Cambridge, Cambridge University Press.

Twitchett, Denis Crispin, Michael Loewe, and John King Fairbank (1986). *The Cambridge history of China. Vol.1, The Ch'in and Han Empires, 221 BC–AD 220*. Cambridge, Cambridge University Press.

van der Loon, P. (1952). 'On the transmission of Kuan-tzŭ.' *T'oung Pao* **41**(4/5): 357–93.

Verbiest, Ferdinand (1674). *Xin zhi yi xiang tu*. 新製儀象圖 (Illustrations of newly manufactured astronomical instruments), Beijing.

Verbiest, Ferdinand (1687). *Astronomia Europaea sub imperatore Tartaro Sinico Cam Hy appellato ex umbra in lucem revocata a P.R. Ferdinando Verbiest Flandro-Belga e Societate Jesu*. Dillingen, Joannis Caspari Bencard, per Joannem Federle.

Verbiest, Ferdinand and Noël Golvers (1993). *The Astronomia Europaea of Ferdinand Verbiest, S.J. (Dillingen, 1687): text, translation, notes and commentaries*. Nettetal, Steyler Verlag.

Vissière, Isabelle and Vissière, Jean-Louis (1979). *Lettres édifiantes et curieuses de Chine: 1702–1776*. Paris, Garnier-Flammarion.

Wang Jianmin 王健民 and Liu Jinyi 刘金沂 (1989). Xi Han Ruyin hou mu chu tu yuan pan shang er shi ba xiu gu ju du de yan jiu 西汉汝阴侯墓出土圆盘二十八宿古距度的研究 (Researches on the old *du* widths of the 28 lodges on the basis of the disc excavated from the Western Han tomb of the Marquis of Ruyin). *Zhongguo gudai tianwen wenwu lunji*. Beijing, Wenwu publishing house: 59–68.

Wang Jianmin 王健民, Liang Zhu 梁柱, and Wang Shengli 王胜利 (1979). 'Zeng hou Yi mu chu tu de er shi ba xiu qing long bai hu tu xiang 曾侯乙墓出土的二十八宿青龙白虎图像.' *Wenwu* (7): 40–5.

Wang Xianqian 王先謙 (1900, reprinted 1959). *Han shu bu zhu* 漢書補注 *(The Han shu, with addenda and comments)*. Beijing, (reprint) Commercial Press.

Wen Shaofeng 温少峰 and Yuan Tingdong 袁庭栋 (1983). *Yin xu bu ci yan jiu: ke xue ji shu pian* 殷墟卜辞研究—科学技术篇 *(Researches on Shang oracle inscriptions: science and technology section)*. Chengdu, Sichuan social science press.

Wilhelm, Richard and Cary F. Baynes (1967). *The I ching: or, Book of changes*. Princeton, N.J., Princeton University Press.

Wilkinson, Endymion (2000). *Chinese history: a manual. Revised and enlarged.* Cambridge, Mass., Harvard University Press.

Wilkinson, Endymion and Asia Center Harvard University (1998). *Chinese history: a manual.* Cambridge, Mass./London, Harvard University Asia Center; Harvard University Press [distributor].

Wittmann, A. (1979). 'The obliquity of the ecliptic.' *Astronomy and Astrophysics*, **73**(1–2): 129–31.

Wright, Mary C. (1958). 'What's in a reign name: the uses of history and philology.' *The Journal of Asian Studies* **18**(1): 103–6.

Wu, Hung (1989). *The Wu Liang Shrine: the ideology of early Chinese pictorial art.* Stanford, Calif., Stanford University Press.

Xiao Congli 肖從禮 (2011). 'Qin Han jian du "zhi ri" kao 秦漢簡牘"質日"考' (An investigation of the 'zhi ri' texts on Qin and Han bamboo strips and wooden planches.) from http://www.gwz.fudan.edu.cn/SrcShow.asp?Src_ID=1427.

Xu Guangqi 徐光啓 (1562–1633) (1984). *Xu Guangqi ji* 徐光啟集 *(Collected works of Xu Guangqi)*. Shanghai, Shanghai gu ji chu ban she 上海古籍出版社.

Xu Zhentao 徐振韬 (2009). *Zhong guo gu dai tian wen xue ci dian* 中国古代天文学词典 *(A dictionary of pre-modern Chinese astronomy)*. Beijing, China science and technology press.

Yabuuti Kiyosi 薮内清 (1963). Astronomical tables in China from the Han to the T'ang dynasties. *Chūgoku chūsei kagaku gijutsushi no kenyū* 中国中世科学技術史の研究 *(Research on the history of science and technology in medieval China)*. Yabuuti Kiyosi 薮内清. Tokyo: 445–92.

Yabuuti Kiyosi 藪内清 (1969). *Chūgoku no tenmon rekihō* 中国の天文暦法 *(Chinese mathematical astronomy)*. Tokyo, Heibonsha.

Yabuuti Kiyosi (1974). 'The calendar reforms in the Han dynasties and ideas in their backgrounds.' *Archives internationales d'histoire des sciences* **24**: 51–65.

Yan Dunjie 嚴敦傑 (1958). 'Zhong guo gu dai de huang chi dao cha ji suan fa 中國古代的黃赤道差計算法 (Method for reduction of the ecliptic as used by the Ancient Chinese astronomer).' *Kexueshi jikan* 科學史集刊 **1**: 47–58.

Yan Kejun 嚴可均 (1762–1843) (1965, repr. of 1958 edn. plus indices). *Quan shang gu san dai Qin Han San guo Liu chao wen* 全上古三代秦漢三國六朝文 *(Complete texts of High Antiquity, Three Dynasties, Qin, Han, Three Kingdoms and Six Dynasties)*. Beijing, Zhonghua Shuju.

Yasui Kōzan 安居香山 and Nakamura Shōhachi 中村璋八 (1959– (8 vols)). *Isho shūsei* 緯書集成 *(Comprehensive collection of [fragments of] weft books)*. Tokyo, Kangi Bunka Kenyūkai 漢魏文化研究会.

Yin Difei 殷滌非 (1978). 'Xi Han Ru Yin hou mu chu tu de zhan pan he tian wen yi qi 西汉汝阴侯墓出土的占盘和天文仪器 (The divining board and the astronomical instrument excavated from the Western Han tomb of the Marquis of Ru Yin).' *Kao gu* **5**: 338–43.

Zhang Peiyu 张培育 (2007). 'Gen ju xin chu tu li ri jian du shi lun Qin he Han chu de li fa 根据新出历日简牍试论秦和汉初的历法 (An investigation of the calendrical systems of the Qin and early Han, based on recently discovered strip and planche calendars).' 中原文物 (5): 62–77.

Zhang Peiyu 張培瑜 (1990). *San qian wu bai nian li ri tian xiang* 三千五百年历日天象 *(An astronomical calendar for 3500 years)*. Henan, Henan jiaoyu chubanshe.

Zhu Hanmin 朱汉民 and Chen Songchang 陈松张 (2010). *Yue lu shu yuan zang Qin jian* 岳麓书院藏秦简 *(Qin dynasty bamboo strips preserved at the Yuelu Academy)*. Shanghai, Shanghai cishu chubanshe 上海辞书出版社.

INDEX

Notes
Tables, figures, and boxes are indicated by an italic *t*, *f*, and *b* following the page number.

M

N

O

P

T

U

V

W

The manufacturer's authorised representative in the EU for product safety is Oxford University Press España S.A. of El Parque Empresarial San Fernando de Henares, Avenida de Castilla, 2 - 28830 Madrid (www.oup.es/en or product.safety@oup.com). OUP España S.A. also acts as importer into Spain of products made by the manufacturer.
Printed and bound by CPI Group (UK) Ltd, Croydon, CR0 4YY
14/09/2026
02216305-0016